"Clearly our present approaches to achieving 'sustainability' are failing. And we are rapidly running out of time. This book offers a realistic, optimistic way forward based on complex systems thinking – multi-level approaches, reducing barriers, empowering actors through societies – all aimed at achieving the global societal tipping point that is required."

–Will Steffen
Emeritus Professor, The Australian National University
Senior Fellow, Stockholm Resilience Centre

"After several years of passionate talking, here comes a book by Christian Berg grounded in ethical reasoning and critical concern for the common good, fair trade and justice for all. He convincingly gives practical and measurable steps which remove man-made barriers, lack of political will, citizens' inaction for sustainable development with the aim that no one will be left behind."

–Monsignor Prof Obiora Ike
Executive Director, Globethics.net; President,
Club of Rome, Nigeria Chapter

"An important contribution at a critical time. People stand up for the future everywhere. A transition towards sustainability needs, however, an integrated approach towards its barriers and concrete guidance. This book is made so valuable due to its discussion of both factors."

–Professor Dr Uwe Schneidewind
President and Chief Research Executive, Wuppertal Institut

T0186275

SUSTAINABLE ACTION

In this timely exploration of sustainable actions, Christian Berg unpacks the complexity in understanding the barriers we face in moving towards a sustainable future, providing solution perspectives for every level, from individuals to governments and supra-national organizations offering a lucid vision of a long-term and achievable goal for sustainability.

While the 2030 Agenda has already set ambitious targets for humanity, it offers little guidance for concrete actions. Although much is already being done, progress seems slow and some actions aiming at sustainability may be counterproductive. Different disciplines, societal actors, governmental departments and NGOs attribute the slow progress to a number of different causes, from the corruption of politicians to the wrong incentive structures.

Sustainable Action surveys all the fields involved in sustainability to provide action principles which speak to actors of different kinds, not just those professionally mandated with such changes. It offers a road map to all those who might not constantly think about systems change but who are concerned and want to contribute to a sustainable future in a meaningful way.

This book will be of great interest to students and scholars of sustainability issues, as well as those looking for a framework for how to change their systems at work to impact the quadruple bottom line: environment, economy, society, and future generations.

Christian Berg lectures on sustainability at different German universities (TU Clausthal, Saarland University, Kiel University). He has worked in business for more than a decade, among others as Chief Sustainability Architect at SAP. He has published several books on sustainability-related topics and has led the task force on Sustainable Economic Activity and Growth within German Chancellor Angela Merkel's Future Dialogue. He holds degrees in physics (Dipl.-Phys.), philosophy (MA), theology (Mag. Theol. and Dr Theol.), and engineering (Dr-Ing.). For further information, visit his website at www.christianberg.net.

Routledge Studies in Sustainability

For more information on this series, please visit www.routledge.com/Routledge-Studies-in-Sustainability/book-series/RSSTY

SUSTAINABLE ACTION

Overcoming the Barriers

Christian Berg

Routledge
Taylor & Francis Group

LONDON AND NEW YORK

earthscan
from Routledge

THE CLUB OF ROME

First published 2020
by Routledge
2 Park Square, Milton Park, Abingdon, Oxon OX14 4RN

and by Routledge
52 Vanderbilt Avenue, New York, NY 10017

Routledge is an imprint of the Taylor & Francis Group, an informa business

British Library Cataloguing-in-Publication Data
A catalogue record for this book is available from the British Library

Library of Congress Cataloging-in-Publication Data
Names: Berg, Christian, 1967- author.
Title: Sustainable action : overcoming the barriers / Christian Berg.
Description: Milton Park, Abingdon, Oxon ; New York, NY : Routledge,
2020. |
Series: Routledge studies in sustainability | Includes bibliographical references
and index.
Identifiers: LCCN 2019031431 (print) | LCCN 2019031432 (ebook) |
ISBN 9780367183202 (hbk) | ISBN 9780367183219 (pbk) | ISBN
9780429060786 (ebk)
Subjects: LCSH: Sustainable development. | Sustainability. | Social policy. |
Globalization–Environmental aspects.
Classification: LCC HC79.E5 B4555 2020 (print) | LCC HC79.E5
(ebook) | DDC 338.9/27–dc23
LC record available at https://lccn.loc.gov/2019031431
LC ebook record available at https://lccn.loc.gov/2019031432

ISBN: 978-0-367-18320-2 (hbk)
ISBN: 978-0-367-18321-9 (pbk)
ISBN: 978-0-429-06078-6 (ebk)

Typeset in Bembo
by Swales & Willis, Exeter, Devon, UK

MIX
Paper from
responsible sources
FSC
www.fsc.org FSC® C013985

Printed in the United Kingdom
by Henry Ling Limited

To Daniela –
Partner
Counsellor
Best friend

CONTENTS

ILLUSTRATIONS

ACKNOWLEDGEMENTS

I am grateful to several friends and colleagues who provided valuable comments on certain parts of the manuscript: Ino Augsberg, Christian Calliess, Michael F. Jischa, Rolf J. Langhammer, Christa Liedtke, Nele Matz-Lück, Uli Mayer-Johanssen, Konrad Ott, Maike Sippel, Thomas Turek and Jutta Zimmermann.

I am thankful for the opportunity to discuss parts of the concept in brownbag sessions within the Institute for Social Sciences as well as in the philosophical research colloquium of Ludger Heidbrink and Konrad Ott, both at the Kiel University.

I want to thank the anonymous reviewers' at different stages of the manuscript: the publisher's reviewers of the book outline as well as the Club of Rome reviewers who recommended this book as a report to the Club of Rome. I appreciate their work and careful consideration of my manuscript. I deliberated on each suggestion they made for improvement, although I could not conform with all of them.

Although the current book was not written as a textbook in the usual sense of the word, I often had my students in mind while writing. They come from a great variety of disciplines since I teach in different programs at different universities: mainly students of engineering at Clausthal University of Technology, students from literally any possible background in the MBA programme at Saarland University, and students mostly from the social sciences or one of the sustainability programs (e.g. Sustainability, Society, and Environment) at Kiel University. I appreciate the dedication and commitment to sustainability which I have often seen and the critical questions I have often been asked.

Despite this great number of reviewers and sparring partners, the full responsibility for the text in its present form as well as for any mistakes or inconsistencies it might still contain does solely lie with the author.

I thank K16, the communication agency which supported me with the illustrations.

I want to thank Robert Raabe for his great editorial support, the publisher Rebecca Brennon and the editorial assistant Julia Pollacco of Routledge for their continuous guidance and advice during the writing process.

A final gratitude shall be expressed to my family. As I grow older and meanwhile have teenage kids myself, I appreciate even more the love, the energy and the values which I have experienced from my parents, Gisela and Eberhard Berg, throughout my life.

The utmost appreciation is dedicated to my wife Daniela for her continuous patience, enduring support and never-ending positive outlook during the troublesome process of writing this book, and to our sons Melvin and Gabriel. The future is yours!

FOREWORD "SUSTAINABLE ACTION – OVERCOMING THE BARRIERS"

When the Club of Rome was founded in 1968, the term "sustainability" referred exclusively to the forestry sector, in particular to the silvicultural principle of replanting as many trees as one harvested. The concept of sustainable development was otherwise of scarce significance in the public discourse. Only half as many people as today were living on our planet, CO_2 concentration in our atmosphere was still below 330 parts per million and continuous economic growth seemed to be accompanied by unprecedented prosperity, at least in Western industrialized countries. Wages were increasing, and by extension, disposable income for private consumption grew year on year. Education, health, social security and general living standards appeared to be constantly improving.

The first report to the Club of Rome, *The Limits to Growth* (1972), was the first international report to consider the impact of growth on our society, by questioning how long this positive development trend could continue. It concluded by warning that if the growth rates in world population, industrialization, pollution, food production, and resource depletion continued unchanged, the limits to growth on this planet would be reached sometime within the next one hundred years.

This contradiction of unlimited and uninhibited material growth on a planet with limited resources came as a real bombshell: more than 12 million copies of the book, in more than 30 languages, were sold. In the course of this controversial debate, including attacks and dis-creditation from those who saw their interests threatened, the idea of sustainability of human activities emerged.

In the years that followed, the international environmental movement emerged, states established Ministries of the Environment and there was a growing acknowledgement of the interdependence of nature and the economy. It was in this context that the concept of sustainability further developed.

The authors of *The Limits to Growth* were particularly insistent on the fact that it is possible to change growth trends and determine new conditions for a just and desirable world, which allow for stability and global balance. Today, it is no longer a question of *if* we can achieve a global standard of living that does not exceed the limits of our planet, but rather *how* we can achieve this goal. While conditions for achieving this goal have arguably become more favourable, new and complex challenges have also emerged.

The adoption of the Paris Agreement and the SDGs in 2015, coupled with increasingly ambitious commitments to climate neutrality by states, regions and major cities around the world, are signs that the international community takes the complexity of social and environmental sustainability more seriously. However, the current speed and extent of action to match these commitments – and indeed the utter *inaction* of many others – is simply not enough to ensure a stable, liveable and just future.

This report to the Club of Rome, *Sustainable Action – Overcoming the Barriers*, provides a conceptual, analytical, moral, philosophical and even historical assessment of the term "sustainability" in a remarkably holistic way. The author, Christian Berg, identifies the obstacles to the urgently needed ecological transition, addresses the questions of principles and responsibility, and suggests concrete and abstract solutions by developing his own concept of "Futeranity", which defines sustainability as a utopian ideal and an overarching, common goal.

Throughout, Berg refers to the current debates about the challenges of the future of our planet and humanity. The report identifies the points of tension of our time: hope, cynicism, radicalism and despair, individual and collective responsibility, "Fridays for Future" and "fake news", central power structures and local initiatives, moral obligations and global inaction and inertia.

A decisive reason behind the Executive Committee of the Club of Rome's endorsement of *Sustainable Action – Overcoming the Barriers* as a report to the Club of Rome, was the author's analytical thought processes and the systemic lens through which he considers complex issues, concepts, interrelations and solutions. Acknowledging the complexity of sustainability is the author's starting point and premise, enabling him to share a wide range of thoughts, findings, insights and facts with his reader, including the concept of "Futeranity".

We find ourselves at a critical turning point in history, a reality which comes out very clearly in the report. In our collective consciousness, understanding increases of the fundamental question of what the world we want to live in looks like, as does the existential risk from not heeding the warning bells around climate and species extinction. That is, a global society which is able to live sustainably with the resources of our finite planet, and those which our human ingenuity provides; and a just society based on genuine prosperity and happiness. This is the vision of the Club of Rome in an enlightened modern world, based on the values of collective action, equity and the sustainable distribution of resources for all eco systems and species.

The world of the future can be a safer and more resilient place than it is today. Humanity has the tools, science and technology as well as the insights at its disposal to overcome the current systemic crisis and transition to a better world. Achieving this will depend on each and every one of us and on the actions we collectively take as a society. *Sustainable Action – Overcoming the Barriers* is an excellent starting point to reframe our understanding of sustainability today and in the future.

Winterthur, Switzerland, 11 July 2019

Dr Mamphela Rampehele & Sandrine Dixson-Declève

Co-Presidents of the Club of Rome

PREFACE

Yet another book about sustainability? There is already so much literature on sustainable development and its manifold burning issues that one may doubt the need for any more. On a generic level, we know what to do. Combating the climate crisis, for instance, would imply decarbonizing our civilization, i.e. our energy systems, transportation systems, our systems of production and consumption, etc. There are countless suggestions for solutions and numerous guidebooks for sustainable consumption. For most single issues, technologies are at hand, sometimes already at competitive prices. It seems as if all this would "simply" need to be applied. It seems as if we do not have a cognition problem but an execution problem, as is often heard. In a way this is true, but it is only part of the problem. All this is not happening, as I will argue in this book, because the transition towards a more sustainable society is apparently much more complex than often assumed. It is not just an execution problem. Since we do not *know* how this execution can be realized, we still have a problem of *cognition*. This is the starting point of this book: any transition towards more sustainable societies will require a comprehensive look at the barriers to sustainability.

Despite the impressive literature on sustainability, there is an astounding lack of analysis of the barriers to sustainability. With very few exceptions, Michael Hulme's commendable book *Why We Disagree On Climate Change* being one of them, I am not aware of comprehensive accounts of why sustainability is so difficult to achieve. This is presumably due to the fact that such a book would have to cover a broad variety of disciplines, a difficult undertaking when the main incentive systems in our academic culture still reward specialization. Furthermore, the reader might doubt the added value of such an endeavour – since scholars dedicated to sustainability in their own field of study are well aware of the issues from within their discipline (overutilization of global commons, for instance, is closely studied by environmental economists), so why bother about sustainability barriers in other fields?

The reason is, in my view, that we need an integrated, systemic view of the barriers. We need to see the interrelations of the issues, the connection between the barriers and the synergies between different solutions. Being involved in issues of sustainability in a variety of roles for twenty years now – in business, as an academic teacher, as political advisor, and within civil society – I have quite often experienced great ideas, smart brains, excellent projects and compassionate people – and yet I had to realize that change is so difficult to achieve. The more I thought about the reasons why we are on a non-sustainable path, the more convinced I have become that we need a systemic, integrated view of the barriers in order to facilitate change. Everything is interconnected. This is the reason why I dared to take up the challenge of such a highly interdisciplinary book, and that's why I hope people from different backgrounds, contexts and disciplines will find this book useful.

The systemic view of the barriers towards sustainability needs to be complemented, however, by an actor's view. What can actors on different levels – be they individuals, corporations, government agencies, or others – contribute to the systemic change needed? What can they contribute to more sustainability? The 2030 Agenda with its 17 Sustainable Development Goals (SDGs) sets ambitious sustainability targets. However, due to the complexity of the issues, their interrelations and the trade-offs among them, it is by no means a trivial matter to figure out how *individual actions* can contribute to the achievement of the SDGs. Fighting hunger, poverty, or climate change are essential and noble goals, but how can individual actors contribute to that? Quite often the best-intended actions produce unwanted (side) effects because the complexity of the issue was underestimated.

There are libraries full of books which address specific aspects of sustainability, different contexts, or actor groups. Advice is given, for instance, to consumers, to corporations, to investors, or procurement offices. These are all valuable and important since they give guidance for concrete situations. However, what might be valid in one instance could be proven wrong in another. "Purchase locally" is mostly good advice, but there are contexts in which other options might be more sustainable, as will be discussed later. Moreover, quite often "sustainable" is equated with "green", which is then further reduced to "carbon free". Yet the sole focus on one issue might impede the achievement of others and ultimately be more harmful than beneficial. There is notably little guidance which is generally applicable but still practical. There is a need for something in between the general validity of the categorical imperative and the advice to purchase locally. This is why I suggest the development of principles for sustainable action.

It is these two aspects by which this book hopes to contribute to the sustainability discourse: a comprehensive view of the barriers to sustainability and a set of sustainable action principles which can be applied by a variety of different actors on multiple levels.

I am fully aware that this book cannot claim to be complete or perfect, neither in its account of barriers nor regarding sustainable action principles. Future research needs to discuss, correct and complement the suggestions made here. Nevertheless, it is my hope that the general approach will facilitate a more comprehensive view of barriers and their solution perspectives and the development of concrete guidance by action principles for sustainability.

The book does not need to be read sequentially. Since a key message of the book is that sustainable development will only be possible by an integrated view of the barriers, there is no preferred or logical starting point – each chapter should be understandable on its own. The general idea of the book is summarized at the end of the first chapter, its main results are outlined at the beginning of the final chapter and this summary, in 17.1, might also be read as an executive summary.

ABBREVIATIONS

Abbreviations/Acronyms

BOP	Base of the Pyramid
CDR	Carbon Dioxide Removal
CPI	Corruption Perception Index
EU	European Union
EROI /ERoEI	Energy Return on Energy Invested
ESD	Education for Sustainable Development
GDP	Gross Domestic Product
GHG	Greenhouse-gas
HDI	Human Development Index
ICC	International Chamber of Commerce
IGO	International Governmental Organization
IMF	International Monetary Fund
ILO	International Labor Organization
IPCC	Intergovernmental Panel on Climate Change
LIC	Low-income country
MDGs	Millennium Development Goals
MNE(s)	Multi-national Enterprise(s)
MIC	Middle-income country
NDCs	Nationally Determined Contributions
ODA	Official Development Assistance
POPs	Persistent Organic Pollutants
SDGs	Sustainable Development Goals
SDSN	Sustainable Development Solutions Network
UN	United Nations
UNCED	United nations Conference on Environment and Development (="Rio Conference" 1992)

UNEP	United Nations Environment Programme
UNFCCC	UN Framework Convention on Climate Change
WCED	World Commission on Environment and Development
WSSD	World Summit on Sustainable Development ("Rio+10" in Johannesburg)
WHO	World Health Organization
WTO	World Trade Organization

1

INTRODUCTION

Sustainability – a utopian ideal?

Contents

1.1 Sustainability – an "exhausted" concept?

The concept of sustainable development has had a remarkable career. In 1987 the World Commission on Environment and Development (WCED) published their final report with a definition of the concept, which has since then been referred to as the "Brundtland definition" (WCED 1987). It has shaped not only politics but civil society and business throughout the world. In 1992 at the Rio Summit, basically every nation on earth reached agreement on making sustainability their common goal for the future path of humanity (UNCED 1992). In 2015, just 23 years after Rio, the world community agreed on comprehensive, concrete and measurable targets for the period up until 2030: the Sustainable Development Goals (SDGs) (UN 2015). Furthermore, in the same year, 2015, the Conference of the Parties (COP) within the United Nations Framework Convention on Climate Change (UNFCCC) was able to come to a global agreement on the reduction of climate change (UNFCCC 2015).

There are a multitude of sustainability activities all over the world, among governments, civil society and business. The first countries have decided to phase out coal, the first car manufacturers have declared they would become carbon neutral, the first companies have announced they would become entirely waste free. This is all great and important – and should not be neglected.

At the same time, however, one may justifiably ask about the results of all these discussions, agreements and commitments. Humans have so dominantly shaped the face of the earth that they have become the dominant influence even in geological terms – we have entered the Anthropocene (Crutzen 2002). Is there any progress in protecting our natural livelihood? Have we managed to distribute resources more equitably? Some progress is measurable – the Millennium Development Goals (MDGs), for instance, have at least in certain respects and in part been accomplished. But what about our ecosystems, what about climate change, what about biodiversity?

Looking at the concentration of CO_2 within the atmosphere, which the Keeling Curve has measured since 1958 (see Figure 1.1), one can only sadly say: failed. Despite half a century of discussion on environmental issues (Rachel Carson 1962) and more than twenty summits on climate change, emissions continue to rise as if we had not done anything. Where on this curve can we see any effects of the Rio Conference, the Kyoto Protocol (1997), or the Paris Agreement (COP 21, 2015)? The Keeling Curve shows a continuous increase in CO_2 concentration with only one or two minimal qualifications: the gradient decreases slightly in the early 1970s and slightly more in the early 1990s, but these are due to the economic downturns after the oil crisis and the collapse of the Soviet Union. Does the Keeling Curve not document the striking failure of our sustainability policies – or does it even call into question the concept of sustainability as such? What is the value of all our commitments, agreements and best-intended actions if they show no result? Are we fooling ourselves?

Moreover, climate change is, of course, only one of several environmental issues, potentially not even the most threatening one. Biodiversity loss, potentially

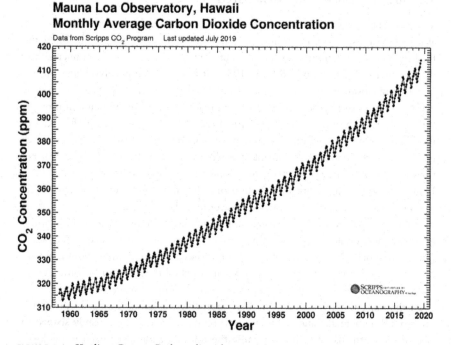

FIGURE 1.1 Keeling Curve: Carbon dioxide concentration at Mauna Loa

Source: Scripps Institution of Oceanography (2019), https://scripps.ucsd.edu/.

even more threatening than climate change (Rockström et al. 2009; Steffen et al. 2015), has dramatically increased in recent decades. In several aspects, humanity is exceeding the planetary boundaries, which implies increased likelihood of irreversible harmful developments (see Figure 1.2).

The 2018 Living Planet Index of WWF documents a 60% fall in just over 40 years (WWF 2018). The state of the oceans is alarming due to acidification, temperature rise and pollution (World Ocean Review 2017); the tropical rain forests and their indigenous people are severely threatened by deforestation and land use change (Martin 2015), and we are plundering the planet by exploiting its natural resources (Bardi 2014).

The status of human development has improved globally between 1990 and 2017 – but strong differences remain with regard to regions (sub-Saharan Africa lags behind with an Human Development Index (HDI) of 34.9% compared to the global average of 72.8%) and gender (the worldwide HDI average for women is almost 6% lower than that for men); and inequalities reduce HDI improvements significantly (inequality adjusted HDI is 58.2% instead of 72.8 in global average) (UNDP 2018). Roughly 60% of the global population do not have access to safely managed sanitation services, and 30% have no access to safe drinking water (UNESCO 2019, 1).

The overall account of the Sustainable Development Report 2019 is that "Four years after the adoption of the SDGs and the Paris Agreement no country is on track to meeting all the goals. We are losing ground in many areas" (Sachs et al. 2019, viii).

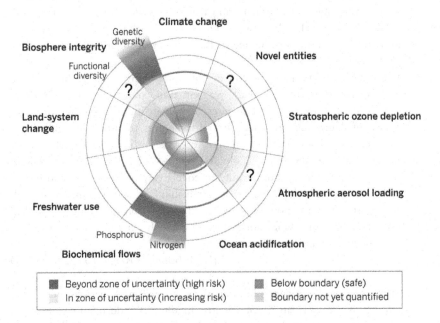

FIGURE 1.2   Planetary boundaries

Source: Steffen et al., Planetary boundaries: Guiding human development on a changing planet, Science Vol. 347. 13 February 2015, Issue 6223, doi: 10.1126/science.1259855.

And while the scientific accounts of climate change, climate migration, bio-diversity loss, pollution and deforestation call for increased urgency of action (see Steffen et al. 2018), the sustainability discourse is challenged from a mostly unexpected angle: populism. Populists and their agitation absorb so much attention, cause setbacks in international treaties, provoke distrust in scientific evidence, denounce the media, heat up societal polarization to such an extent that they might become a bigger threat to sustainability than, for instance, the "ambition gap" of the Paris Agreement.[1] This rise of right-wing populism in many regions of the globe and our prevailing unsustainability might even have overlapping root causes, as the feeling of insecurity which makes people turn to populism is partly rooted in rapid changes, growing inequalities, and the perception that an "elitist" political class seems incapable of addressing the real problems (J.-W. Müller 2016; Dibley 2018; Lockwood 2018). We will return to this in a later section (see 4.3).

What are the consequences? Let us look at three different responses to this situation.

1.1.1 Abandon the concept of sustainability?

Dennis Meadows, co-author of the first report to the Club of Rome[2] *The Limits to Growth* (Meadows et al. 1972) and a pioneer of sustainability long before this concept had entered the international discourse, noted in 2000 that it would be too late for sustainable development, and that we should rather strive for "survival development" (D. L. Meadows 2000, 147f.).[3]

Benson and Craig proclaim the end of the concept of sustainability:

> It is time to move past the concept of sustainability. The realities of the Anthropocene (Crutzen 2002) warrant this conclusion. They include unprecedented and irreversible rates of human induced biodiversity loss, exponential increases in per-capita resource consumption, and global climate change. These factors combine to create an increasing likelihood of rapid, nonlinear, social and ecological regime changes. The recent failure of Rio 20 provides an opportunity to collectively re-examine – and ultimately move past – the concept of sustainability as an environmental goal. We must face the impossibility of defining – let alone pursuing – a goal of "sustainability" in a world characterized by such extreme complexity, radical uncertainty and lack of stationarity.
>
> *(Harm Benson & Kundis Craig 2014, 777)*

They rather propose "resilience thinking as one possible new orientation".

According to the German social scientist Ingolfur Blühdorn, sustainability as a "road map for a structural transformation of socially and ecologically self-destructive consumer societies" would be "exhausted" (Blühdorn 2017).

Of course, these authors have good reasons for their arguments – there is far too little progress, measured against both the necessary and the possible. However, should we abandon the ideal that the peoples of the world can prosper in harmony with each other and with nature? Have we underestimated the complexity of the issues? Have we relied too much on the idea that insight would lead to action – even though every one of us can experience the opposite in his or her own life?[4] Have we perhaps not found the right governance model for sustainability? Have we underestimated the inertia of systems, the resistance to change, the lack of institutional support for cross-sectoral, cross-disciplinary initiatives? The answers to all these questions is unfortunately in the affirmative OK: yes! But does this disqualify the goal as such? Abandoning the goal of sustainability because it would be too late or because it would be unrealistic sounds to me like committing suicide in fear of death. It is a sad irony that the frustration with the concept of sustainability seems to increase around (or shortly after) the time when the global consensus on the need for progress has reached an all-time high. Both the Paris Agreement and the 2030 Agenda are significant milestones in the history of global collaboration – however piecemeal, unambitious, or unrealistic they might be considered.

This book will argue that we need to keep sustainability as an ideal; an ideal which we might never reach, which might be utopian, but still a necessary one.

1.1.2 Abandon liberal democracy?

Humankind has apparently not found the right governance model for sustainability. Some scholars argue that the Western liberal democracies have obviously not been able to tackle the problems, and that we will see an increasing "environmental authoritarianism" (Beeson 2010;[5] Chen und Lees 2018; Blühdorn 2019). Blühdorn and Deflorian argue that the prevailing policy approaches to sustainability have failed and should rather be seen as "the collaborative management of sustained unsustainability". The authors "firmly hold on to the belief that a radical socio-ecological transformation is urgently required" and cannot really offer an alternative (yet) but want to facilitate the urgently needed transformation by deconstructing prevailing narratives (Blühdorn & Deflorian 2019, 13). In his 2018 book, *The Sustainable State*, Chandran Nair, founder of the Global Institute for Tomorrow, a Hong-Kong-based think-tank, is likewise critical of the "laissez-faire, industrialized Western model" which is not sustainable. Instead, he is advocating strong interventionist governments to shape the future (Nair 2018). Nair emphasizes that he is not justifying China's policies – but he does acknowledge its great potential in tackling issues which the Western countries have so far not come to grips with:

> [O]ne can disagree with China's nondemocratic nature, but it remains true that the country has had a much better track record in improving the lives of its ordinary people – and in a much shorter period – than many other

> developing countries. It is also not clear how China could have improved these standards of living so quickly without strong state intervention.
>
> *(Nair 2018, 177)*

I agree with many of these authors' observations – although I do not follow all of their conclusions. However, as a privileged citizen of a Western country which has achieved some limited success in protecting its own environment at the cost of externalizing much of its ecological and social footprint to other parts of the world (see 5.1; Peters, Davis & Andrew 2012, 3273), I feel obliged to take up this discussion of the relationship between liberal values and the role of the state. I do share the abovementioned authors' concern about the question whether liberal democracies will be able to tackle our global challenges effectively. But we have to be reminded that our freedom needs to end where we deprive others of their freedom, as John Stuart Mill nicely put it: "The only freedom which deserves the name is that of pursuing our own good in our own way, so long as we do not attempt to deprive others of theirs, or impede their efforts to obtain it" (Mill 1869, 27).

I submit that we will need to challenge luxury consumption patterns at the expense of the common good. Calling for or hankering after an eco-dictatorship would, however, not resolve the issue – and would again be simplistic. First, one can question that authoritarian concepts which might function at state level would also do so on a global level. Furthermore, even a global dictator would need to resolve the challenge of feeding 8 billion people, and securing the stability of our societies and ecosystems. Apart from the many good arguments why such a regime would most likely be unsuccessful in the long run (see H. Müller 2008), the fate of actual national dictatorships give little hope that such governance would be any better for the people or the environment. More important, however, is the second point. We cannot pave the way for humanity (which in my view will always have to embrace sustainability) with inhumane measures – thus disqualifying any dictatorship. We cannot work for peaceful and liberal societies for tomorrow if we sacrifice them today. We need to protect and defend our basic personal liberties – not just for us, but also for those already deprived of them. It is the complex interrelation of the state's role and personal liberties which warrants further investigation. Since only the state can provide the framework in which personal liberties can be realized, one of the most critical questions of sustainability is to what extent the state may (need to) limit personal liberties for the sake of the common good – and how the latter is to be determined. This question will be taken up later (see Chapter 7). In any case, abandoning liberal democracy is not suggested as an answer.

That's why we need to seek another solution.

1.1.3 Fierce claims versus frank denial

A third kind of response to the increasing frustration about the lack of progress in sustainability are the fiercer claims and more drastic measures of those concerned that sustainability gain at least some attention. Children boycott school (#Fridaysforfuture) or call out to adults: "Stop talking, start planting", as the campaign of a children's NGO, Plant-for-the-Planet, demanded. The tone is also growing more fierce in the public discourse. In a panel contribution to the 2019 World Economic Forum (WEF) in Davos, the Dutch historian Rutger Bregman vigorously demanded to talk about taxes, in particular the marginal tax rates for the rich. In the rather distinguished context of the Davos meeting he stated: "Taxes, taxes, taxes – all the rest is bullshit" (The Guardian News 2019).

The measures suggested to slow our destructive development are becoming more extreme, too. In their 2016 Report to the Club of Rome, Jorgen Randers and Graeme Maexton, for instance, suggest women could be paid 80,000 USD if they had just one child at age of fifty (Randers & Maxton 2016, ch. 9). Inspired by this idea, Verena Brunschweiger, a German school teacher, pointed out that "Children are the worst thing you can do to the environment" (DER WESTEN 2019), an observation which stimulated heated debate.

Of course, human existence arguably *is* the worst thing that has happened to the other species on this planet, and each and every human being contributes to consumption and contributes to the deterioration of our environment. For the environment it would probably be best if humans ceased to exist – but nobody can really want that. We cannot discuss here the call to stop having children, but in my view it does, as do the previous two arguments, throw out the baby with the bath water.[6]

I sympathize with the concern expressed by these suggestions, but apart from objections in substance I doubt that more strident, moralistic demands will help get the message across – maybe they are even counterproductive because populists' objection to political correctness reveals that such claims might be perceived as patronizing the moralism of an elitist class. Everybody knows from personal daily experience that additional information does not necessarily lead people to change their behaviour. This is true in particular for people with extremist views, like the "climate sceptics". Their position cannot be countered by empirical evidence but needs to be understood in sociological or psychological, if not religious terms (e.g. Jaspal, Nerlich & van Vuure 2016; Hobson & Niemeyer 2012).[7]

Despite the sympathy I have for the seriousness with which sustainability is taken in such extreme positions, and apart from the fact that they are contributing to the polarization of society, I think they mostly make the mistake of neglecting the complexity of both issues and solution approaches by absolutizing *one* measure to *address the one issue*, which is considered greatest.

The focus on a single issue (be it climate change, migration, plastics in the ocean, or something else) absorbs our attention and prevents us from seeing the different kinds of barriers which block the path to sustainability (see Figure 1.3).

FIGURE 1.3 Single issues absorb our attention and prevent us from seeing the manifold barriers to sustainability

Source: Own illustration.

As I will argue in the remainder of the book, a more promising (and more realistic) approach is to reach a perspective on the barriers to sustainability which is as comprehensive as possible and to develop principles which can guide several kinds of actors on several different levels, working independently of each other but following the same set of principles. The hope is that if multiple actors in multiple contexts follow multiple action principles on multiple levels, change can be facilitated much more effectively (see Figure 1.4).

What is the way forward? Sustainability, and climate change in particular, was for too long seen as a concern reserved to natural scientists (partly due to the fact that the most prominent public voices on climate change *are* natural scientists). Hardly anybody had anticipated the recent challenges to climate change politics due to the upswing of populism. I submit we are much better at deliberating the social *consequences* of global change than understanding the *social conditions* for addressing them.[8] We have apparently not yet come to grips with the complexity of the challenges we face. If a complex system were determined by ten parameters and we change just one or two but neglect the others, we will not succeed. Only an integrated, systemic approach which addresses a variety of different barriers working on different levels and involving different actors can effectuate change. If this is given, change can occur quite rapidly and with amazing results. This process is similar to a phase transition in physics, which we will now explore.

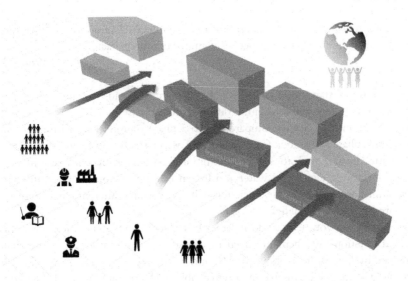

FIGURE 1.4 When multiple actors address multiple barriers, the transition towards sustainability is facilitated

Source: Own illustration.

1.2 Phase transition towards sustainability

The changes required for a transition towards sustainability are substantial. Incremental improvements here and there will not suffice: "What is needed is to challenge the fundamental features of industrial modernity: we need a new theme, not another variation on the existing one" (Kanger & Schot 2018). We will need fundamentally new forms of production and consumption, a refined market framework, a better global governance, a more equitable distribution of goods – to name just a few. What is needed are, in short, socio-technical transitions towards sustainability.

According to Geels (2011), our contemporary

environmental problems, such as climate change, loss of biodiversity, and resource depletion (clean water, oil, forests, fish stocks, etc.) present formidable societal challenges. Addressing these problems requires factor 10 or more improvements in environmental performance which can only be realized by deep-structural changes in transport, energy, agri-food and other systems … These systemic changes are often called "socio-technical transitions", because they involve alterations in the overall configuration of transport, energy, and agri-food systems, which entail technology, policy, markets, consumer practices, infrastructure, cultural meaning and scientific knowledge … These elements are reproduced, maintained and transformed by actors such as firms and industries, policy makers and politicians,

> consumers, civil society, engineers and researchers. Transitions are therefore complex and long-term processes comprising multiple actors.
>
> *(Geels 2011, 24)*

None of these transitions will be easy to achieve, and the desired result, a sustainable society, will look quite different from today.

By use of analogy, let us compare this to a phase transition in nature. We all know such phase transitions from school – if water freezes, it changes its aggregate status from liquid to solid. The physical characteristics of the different aggregate statuses, e.g. liquid water and frozen ice, are obviously quite different. In a similar way, sustainable societies will have quite different characteristics compared to today's societies.

Now what do you need to do in order to facilitate a phase transition? Isn't this a stupid question? We certainly need to change the temperature, i.e. water will evaporate by heating it up above 100°C, and it will freeze – and its molecules "all of a sudden" arrange themselves in a remarkably well-structured grid – if you cool it below the freezing point. That is true. But it's not the full truth. You can also reach the phase transition at *constant temperature* by changing the *surrounding pressure*. At high absolute altitudes (e.g. mountains), water vaporizes below 100°C; at high pressures (e.g. in a pressure cooker) it boils above that. Water even remains liquid below 0°C if the pressure is high enough. The simple reason is that this phase transition is dependent on *two independent variables*. This means that a phase transition can be facilitated and comes about even more quickly if the two parameters are changed synergetically, i.e. *by changing the two variables at the same time*. On the other hand, such a transition can be prevented if somebody else is increasing the pressure while you increase the temperature. It is therefore critically important to know all the key parameters if you want to trigger a phase transition. This is not only valid in physics – I also hold it to be true with regard to any sustainability transition we wish to achieve. In order to facilitate such a transition, it will be decisive to understand which *barriers* stand between our current situation and the desired status. Triggering a phase transition towards sustainability will therefore crucially depend on the knowledge of the most critical factors impeding change; it will require a solid understanding of the barriers to sustainability.

This matches well with insights from transition research.

For instance, the analogy of a phase transition which is dependent on the isochronic condition of several parameters independent of each other nicely parallels the view of historians on societal transformations. Historical transformations in societies can neither be understood as attributed to *one* actor, or *one* event, or *one* driver, but are rather triggered by a "concurrence of multiple change" (Osterhammel 2009; WBGU 2011, 5).

Similarly, talking about unsustainable resource use, Hirschnitz-Garbers et al. see strong evidence that "driver effects play out in multi-directional, dynamic processes with complex interactions between different drivers in many directions" (Hirschnitz-Garbers et al. 2016, 25).

Geels and Schot (2007), and Geels (2011) argue for a "multi-level perspective" in understanding sustainability transitions, which "does away with simple causality in transitions. There is no single 'cause' or driver. Instead, there are processes in multiple dimensions and at different levels which link up with, and reinforce, each other ('circular causality')" (Geels 2011, 29).[9]

Again, looking at historical transitions, Raskin et al. also see multi-causality:

> Historical transitions are complex junctures, in which the entire cultural matrix and the relationship of humanity to nature are transformed. At critical thresholds, gradual processes of change working across multiple dimensions – technology, consciousness and institutions – reinforce and amplify. The structure of the socioecological system stabilizes in a revised state where new dynamics drive the continuing process of change. But not for all. Change radiates from centres of novelty only gradually through the mechanisms of conquest, emulation and assimilation.
>
> *(Raskin et al. 2002, 3)*

In her 2019 Report to the Club of Rome, Künkel (2019) stresses the need for a systemic approach to the issues; we cannot expect any single solution to fix the issues. Künkel calls for a new understanding of leadership, which complex adaptive systems (such as our socio-political-ecological system) require: "There is no 'one right way' to bring about the change envisaged. Given the complexity of the systems, multiple efforts, from multiple sources, at multiple levels, with multiple different approaches will be needed" (Künkel 2019, 14).

In line with Grin, Rotmans and Schot (2010) we will view transitions as "co-evolution processes that require multiple changes in socio-technical systems or configurations". They are "multi-actor processes, which entail interactions between social groups, scientific communities, policymakers, social movements, and special interest groups", they are "radical shifts" in terms of scope (not regarding speed of change) and they are "long-term processes (40–50 years)" ((Grin, Rotmans & Schot 2010, 11; see also (Köhler et al. 2019)).

In sum, we need to acknowledge that a transition to sustainability will be a *multi-dimensional challenge which requires an approach which is*

- *multi-dimensional, i.e. addresses social, environmental and economic concerns;*
- *multi-level, i.e. includes all levels from local to global;*
- *multi-sectoral, i.e. involves governments, civil society, science as well as business;*
- *multi-actor (incentivizes actors of different kinds and levels (see Künkel 2019, e.g. 8, 263).[10]*

Differing from any other developments in human history, this is the first time that such transitions need to take place as deliberate action, on a global scale, in the (relative) absence of external pressure but anticipating coming future pressure.

1.3 Understanding the barriers to sustainability

Despite several decades of sustainability science, there are relatively few contributions systematically exploring sustainability barriers. Mike Hulme's book *Why We Disagree About Climate Change* (Hulme 2009) is an important exception. Hulme argues that there are several different answers to this question. His starting point is the view of the issue from different societal sectors: science, religion, media, etc. Different perceptions of science, different value and belief systems, different priorities and interests as well as different concepts of responsibility for future generations would illuminate the disagreement and lack of progress on climate change issues. This is an interesting and reasonable approach. However, while this current proposal follows a similar objective (i.e. gaining an overview of the barriers to sustainability), it follows a different structuring logic reflecting a systematic, conceptual view and documented in a typology. Furthermore, Hulme focuses on climate change and does not consider sustainability in general.

Howes et al. (2017) investigate the *policy implementation failures* of environmental sustainability policies (Howes et al. 2017).[11] As with all policy implementation, there are a number of conditions which have to be met so that policies can be effectively implemented: the goals need to be clearly defined, they need to be measurable, they need to be monitored, there must be sufficient budget, and what to do in case the target is not met must be defined. Such implementation failures are certainly relevant for a sustainability transition because all of the failures just mentioned apply for sustainability policies as well. Budget constraints, resistance among certain stakeholder groups, ignorance and lack of knowledge are often impediments to progress. However, in the following we need to suppress these implementation issues since we want to look into the *conceptual* challenges.

This is also the reason why not all factual barriers to sustainability will be discussed even though they might be critical in practice. Population growth, for instance, is one of them. To be sure, almost any issue of non-sustainability is exacerbated in a rapidly growing population – certainly resource demand, pollution, and development. However, I do not see this as a *conceptual* challenge and do not consider this a barrier in itself. It will certainly be involved indirectly since, for instance, the reduction of humanity's environmental impact is much more difficult in a growing population.[12]

In their 2011 report on *World in Transition. A Social Contract for Sustainability*, the German Advisory Council for Global Change (WBGU) discusses barriers or obstacles to sustainability (WBGU 2011). To achieve a "Great Transformation", "[v]arious multilevel path dependencies and obstacles must be overcome" (WBGU 2011, 1). The report demands that the first step of this Great Transformation needs to be *the overcoming of transformation barriers* (WBGU 2011, 6.21.77ff.). They suggest the following five barriers: path dependencies, tight time frame, barriers obstructing global cooperation, rapid urbanization, and easily available cheap coal supply (WBGU 2011, 6). Obviously, these barriers differ in kind. Some are systemic ones (path dependencies), some are social (urbanization), some are economic (cheap coal supply). While the authors do not systematically elaborate on the different kinds of barriers, they do broach the issue of sustainability barriers in the context of the value-action gap,[13] and they do talk about investment barriers and about barriers to change in societal transformations in recent history. But there is no systematic reflection and account of barriers to a sustainability transformation as such, although the authors acknowledge: "Once the decisive barriers have been overcome, the move towards low carbon can be expected to develop its own dynamics" (WBGU 2011, 6).

But what are these barriers? How can we get a structured and comprehensive understanding of them? This is the starting point of the current book. If we have a comprehensive and structured view of the barriers to sustainability this will be, as the WBGU says in Kantian terminology, a "condition of possibility" for the transition to sustainability (WBGU 2011, 2). In other words, if we are ignorant of the barriers, no such transition will be possible. Bosselmann also argues for first understanding the problem, only then considering solutions (Bosselmann 2017, 42f.).

1.4 Developing action principles for sustainability

What is the role of actors if you look into the systemic barriers to sustainability? If you consider the externalities of the market system or the lacking global governance?

This concern has repeatedly been raised against the systemic view of transitions which Geels and others have proposed with the their "multi-level perspective" (MLP). Geels and Schot (2007, 414): "The MLP typically is a global model that maps the entire transition process. Such a global model tends to give less attention to actors." Geels (2011) admits that the multi-level perspective "has been criticized for underplaying the role of agency in transitions" (Geels 2011, 29). Writing in the same tradition, Kanger and Schot (2018) concede that "some readers may feel that we have offered a structural explanation here that leaves far too little room for agency." Wittmayer et al. (2017, 53) argue "that the transitions field to date lacks a suitable vocabulary to analyse the (changing) interactions and relations of *actors* as part of a sustainability transition" (emphasis added).

This also relates to the very practical question of how actors can get involved. *It is certainly essential to see the transition to sustainability from a systemic perspective because the issues are systemic.* But at the same time, what does this imply for each and every actor in concrete decisions? Very few people work at a policy level, as decision makers in business or government and have the chance to deliberate about systemic changes. But it is not enough to rely on those in power – because those in power are amazingly powerless if they have to rely on their decision alone, if they do not manage to get the people behind them.

One lesson from complex adaptive systems is that you cannot direct them or steer them as you might steer or control a machine. But this does not mean that you cannot influence these systems. You can certainly change their behaviour if you change the relationships among the parts, if you change the components. As David Stroh in his commendable book *Systems Thinking for Social Change* explains, you need to improve the relationships among the parts if you want to optimize a complex system (Stroh 2015, 15).

It is therefore critical to supplement the systems view with an actor-oriented view – because it is the agents and their interdependence which shape the system's behaviour. There is no single steersman – it is the hugely complex inter-action of the multitude of actors on different levels that determines the system's behaviour. Furthermore, the decision makers and policy makers are also rarely "free-floating" and independent in their decisions. They depend on the agents supporting their view, and they need support from the bottom up, from other policy and decision makers.

Apart from this, there is also a more practical, pragmatic reason for involving the actors. Each one of us committed to sustainability cannot just rely on or wait for systemic change but needs to know what concrete action is needed in the different roles we play, as consumers in our personal lives, but also in our different positions in professional life. There needs to be concrete advice for the increasing number of concerned people – not just for the so-called "woman in the street" in her consumption pattern, but also for people in procurement, in research, in product development, in design, in the media – everywhere.

But that's the challenge – how to find advice that is both generic enough to be applicable for a wide range of people and roles but concrete enough to sup-port daily decisions at the same time?

To illustrate this, let us consider two extreme examples. On one end of the spectrum there are generic principles that claim universal validity. Kant's cat-egorical imperative is a good example: "Act only according to that maxim through which you can at the same time will that it become a universal law" (Kant 2002 (1785), 4:421). This is much more than just a golden rule and can truly guide ethical reflection. But Kant did not have future generations in mind, so one might wonder about whether it is applicable to sustainability. That's why Hans Jonas adjusted his principle and stated that any action should be compatible with enduring life on earth (Jonas 1984).

Again, Jonas' sustainability equivalent also claims universal validity, but it is adjusted to the situation today, in which humans have, for the first time, the power to destroy this life. However, if asked about daily choices such as: is the apple from Italy better than the one from New Zealand, this does not help the average con-sumer much. In other words, the generality of the principle is a challenge to its concrete application. To find more concrete, actionable advice, there are numerous guidebooks on the other end of the spectrum with very specific hints on how to become more sustainable – many of them giving very valuable advice which should be considered much more. However, any such advice faces two challenges. First, it needs to reduce the enormous complexity of sustainability to one or very few meas-ures and link it to concrete advice, such as: this is more carbon friendly. In the public discourse this is then often equated with sustainable. But "sustainable" is not just "green", "green" is not just "greenhouse gas (GHG) reduction", and GHG reduction is not equivalent to "carbon". The second challenge for very concrete advice might sound trivial: the advice must certainly be correct. This is much harder than one might think because quite often things that appear to be "sustainable" look somewhat different on closer inspection. For instance, as we will discuss later, "Purchase local!" is good advice in most instances. But if the environmental impact of logistics is way below the one from production, there might be cases in which global sourcing might also make ecological sense.[14]

Deliberating on sustainability in a complex world, Casey Brown nicely elaborates on the interplay between the systems level and the need to provide agents with principles:

> Thus to "redesign" the human-nature system so that it gravitates towards a configuration we deem desirable, one needs to influence the decisions and actions of the individual agents. In other words, one needs to provide the conditions or make interventions so that the emergent trait of the human-nature system is sustainable development. Ultimately, this requires interplay between the agents making up the system, and the structures that define how the system reacts to their actions. For example, for the people of the world to transform our trajectory toward sustainability, they must be motivated to take the correct actions and have the means to influence the trajectory with their actions. Motivation and the identification of correct actions are presumably achieved through government and science. Yet both government and science must be responsive to individual agents' responses.
>
> *(Brown 2008, 149)*

In sum, a systemic view is important, indeed vital, because systemic issues require systemic responses. But systems are influenced by their parts, their components and the interdependencies among them. This is where the actors come in. In the absence of any "steersman" everything depends upon the actors. But these actors are not just individuals, they are actors at various levels: individual actors, civil society actors, corporate actors, governmental actors, etc. They all have a role to play, they all have choices and they all need to orient themselves. Not all actors, to be sure, care about sustainability. But those who do should have a guide which meets the above-mentioned criterion: sufficiently concrete to be operational but sufficiently generic to be generally applicable.

Such action principles should be applicable to actors of different kinds and levels. Every one of us has several roles in which we act: as consumers, citizens, voters, professionals, friends, managers, etc., and I presume that the required phase transition towards sustainability can be much facilitated if people change their mind and start thinking and acting differently *in all their roles*. All of a sudden a business executive might change his mind because he realizes how difficult it is to justify his business decisions when he argues with his "rebellious" daughter about #fridaysforfuture at the dinner table.[15] All of a sudden people find free-riding uncool and disregard communal values. All of a sudden politicians feel that there is time for change – and initiate political changes that had been inconceivable before.

There are many such examples and there will be many more to come ...

This is the background for the action principles which we will discuss in the second part of this book.

1.5 Concept of sustainability

1.5.1 Heated debate

We briefly alluded to the concept of sustainability in the beginning – but what do we actually mean by talking about sustainability?

Most definitions relate in some way or another to the definition from the *Report of the World Commission on Environment and Development: Our Common Future* ("the Brundtland Report") according to which a development is sustainable if "it meets the needs of the present without compromising the ability of future generations to meet their own needs" (WCED 1987, section 27).

This definition has been the subject of heated debate for decades and we can only highlight some controversial points. Wolfgang Sachs, former researcher at the Wuppertal Institute for Climate, Environment and Energy, already stressed in an article from 1997 that the Brundtland definition would suggest that it's possible to have both nature and justice. Sachs concisely described this as the "catch-22 of sustainability": every attempt to mitigate the "natural crisis exacerbates the justice crisis" and vice versa (W. Sachs 1997, 98). The Brundtland definition would avoid facing the "justice challenge" (W. Sachs 1997, 100) because it would leave open whose needs and which needs. "Shall development address the request for water, land, and secure income or the appetite for air travel and shares?" (W. Sachs 1997, 99f.)

Karl-Werner Brandt, a German sociologist, criticizes the "determinedly anthropocentric perspective" of the Brundtland definition, which would transform the question of natural *protection* into a question of natural *utilization* (K.-W. Brandt 1997, 13). Along similar lines, James Rosenau detects a "significant change of meaning" which "the very idea of sustainability" has undergone: "Now it connotes 'sustainable development', with the emphasis on sustaining economies rather than nature ..." (Rosenau 2003, 13).

According to Dennis Meadows, the majority of people using the Brundtland definition to justify their work would be deluded twofold. First, the needs of the present generation would certainly not be met today, and secondly, the economic activities we are performing to meet present needs would definitely diminish the number of options which future generations will have in substantial ways (D. L. Meadows 2000, 126). Meadows continues that in his country, the United

States, a "developer" would be somebody who buys a piece of land, cuts down all the trees and constructs building and streets on it. Given this background, the phrase "sustainable development" would actually be an oxymoron (Meadows 2000, 127).

Sustainability is often understood in terms of capital, which should not be diminished but maintained. Only the regenerating surplus should be consumed. Several different forms of capital can be distinguished, e.g. financial capital, natural capital, social capital, human capital, etc. Different definitions of sustainability can now be distinguished "based on whether or not they allow that different forms of capital may be substituted for each other" (Figge 2005, 185). Whereas those advocating substitutability are called to follow a concept of "weak sustainability", adherents of "strong sustainability" do not consider that such substitution would be justified (see Daly 1996; Neumayer 2003; Ott 2014).

Figge argues that even the weak concept of sustainability would allow only for limited substitution of different forms of capital because too much substitution would increase the risks related to this. In a risk-averse society, even the weak concept is rather limited in allowing substitution because diversification has a risk-reducing effect (Figge 2005).

The economist (and physicist) Robert Ayres also argues against substitutability:

> One key insight that has emerged is that there are a number of services of nature that cannot, even in principle, be replaced by man-made capital or human labor. This is the essence of what is meant by "strong sustainability" … Yet strong sustainability is a controversial position, inasmuch as it has been explicitly contradicted by a number of reputable mainstream economists, who argue that human ingenuity and man-made capital can indeed replace virtually all such services … The key question, then is: what are the possibilities and limits of substitutability, not only between capital, labor and energy but also between different economic activities ranging from shelter to food and communication, and between human labor and man-made capital and other services of nature, from fresh water and clean air to topsoil and bio-diversity.
>
> *(Ayres 2008, 291)*

In his 2017 book, *The Principle of Sustainability*, Klaus Bosselmann also criticizes the vagueness of the Brundtland definition, which has

> opened up the possibility of downplaying sustainability. Hence, governments spread the message that we can have it all at the same time, i.e. economic growth, prospering societies and a healthy environment. No new ethic required. This so called weak version of sustainability is popular among governments, and business, but *profoundly wrong and not even weak*, as there is no alternative to preserving the earth's ecological integrity.
>
> *(Bosselmann 2017, 2; italics added)*

According to Bosselmann, it would be crucial to realize the

> ecological core of the concept ... There is only ecological sustainable development or no sustainable development at all. To perceive environmental, economic and social as equally important components of sustainable development is arguably the greatest misconception of sustainable development and the greatest obstacle for achieving social and economic justice.
>
> *(Bosselmann 2017, 21)*

The fundamental issue which several authors see in the Brundtland definition, that the conflicts between justice and nature, between development and natural preservation are not sufficiently addressed and that this definition would have a certain bias towards development, cannot be settled here. In my view, it is not the Brundtland definition itself (or any definition of sustainability) which makes the issues so contentious. Rather, it is *the underlying issues* for which we do not have answers, the *trade-offs* between the social and the ecological dimensions, the conflicting interests of the global South and the global North, the huge inequalities and matters of justice and fair distribution of resources, etc.

Much of the discussion around the right concept, the right "definition" of sustainability stems, in my opinion, from the inherent problems which every concept faces that wants to do justice to both present and future "needs" (what kind of needs?) of all people in harmony with nature. The 2030 Agenda for Sustainable Development faces the same problem in its attempt to provide "a plan of action for people, planet and prosperity", its preamble complemented by peace and partnership, to make up the 5 Ps (UN 2015). It calls to "balance the three dimensions of sustainable development: the economic, social and environmental" – the 17 SDGs and 169 targets should be "integrated and indivisible".

Moreover, since sustainability is an integrative concept and the SDGs are "integrated and indivisible", none of the 17 goals or 169 targets can claim to be sustainable in itself. To posit an extreme case: taking the 2030 Agenda literally, sustainability will only come together in the joint achievement of the 169 targets. However, since there is no single coordinating panel, nobody directing the development accordingly, and no actor able to address 169 targets simultaneously, actors will naturally focus on *subsets* of the SDGs – and while doing so with best intentions (hopefully) will claim to facilitate sustainability. But it is by no means guaranteed that this will ever lead to sustainability because the best-intended measures in focussing on a sub-goal might actually thwart the achievement of other goals. To return to the very simple analogy of the phase transition: in the case of boiling water we know that this depends on two parameters: pressure (p) and temperature (T). At a surrounding pressure of 1 bar, water boils at 100°C. The phase diagram of water tells us exactly at which values water will experience a phase transition from fluid to gaseous state and it would be stupid to expect anything else. How can we then be sure that 17 goals and their 169 targets can be effectively achieved together? We cannot. Nobody can. Of course, I am well aware that social systems cannot be described by physical laws. However, social systems also follow natural laws as limiting conditions[16] – and the critical question is whether the achievement of the social goals documented in the SDGs will even be possible at all.

To be very clear, it is one thing *to claim* the importance of goals and their simultaneous achievement – but quite another to *actually achieve them*. To do so requires, first of all, this to be feasible, which in the case of the 169 targets nobody can tell with any certainty. On the contrary, there are serious doubts that these targets can all be met (Scherer et al. 2018), one study even concludes that the SDGs "cannot be met sustainably" (IASS 2015, 4), basically because the aspired development would ruin the environment.

1.5.2 Working definition

The undoubted benefit of the Brundtland definition is the general global consensus it enjoys – at least outside academic circles. In the present book, we will therefore refer to this definition occasionally. The 2030 Agenda for Sustainable Development is also written in a similar spirit. However, one needs to acknowledge the severe shortcomings of weak concepts of sustainability, which do not consider the planetary boundaries. I agree with Bosselmann, who stated that the "notion of sustainable development, if words and their history have any meaning, is quite clear. It calls for development based on ecological sustainability in order to meet the needs of people living today and in the future" (Bosselmann 2017, 10). This definition favours (at least implicitly) a strong concept of sustainability, acknowledging the need to preserve ecological integrity.

However, despite my personal preference for a strong concept of sustainability, the argument of the current book does to a large extent not require such a strong concept – it would also hold for a weak one since we are not even on a path to weak sustainability, and the book will investigate the reasons for this, the barriers to sustainability.

As we will see in the next section, we have a much better understanding of what is *not* sustainable than indicating what it really *is*. The remainder of this book will argue that this has to do with the multi-dimensional nature of both the concept of sustainability as well as the barriers which prevent us from getting there.

1.5.3 We do not know how to reach sustainability – three reasons

The Brundtland definition silently presupposes some understanding of *the demands of future generations* as well as some understanding of *how not to compromise their ability* to meet their needs. With regard to the latter, as a minimum requirement for their ability to meet their needs we can assume that we should not exceed the planetary boundaries (Rockström et al. 2009), something we already know today because it would imply irreversible changes in our ecosystems which would very likely negatively impact the lives of future generations. There might be many more such planetary boundaries and we might miss important others but respecting the known ones should be a minimum principle. Therefore, taking deliberate precautions when interfering with our ecosystems and with changing the landscape of our earth would be appropriate.

But how do we assess future needs?

> A significant obstacle to deciding how sustainable development be achieved is our lack of understanding of the implications of actions taken now for the future. That is, we face the challenge of decision-making under great uncertainty. This makes the implementation of many seemingly wise and straightforward concepts of sustainability difficult and impossible.
>
> *(Brown 2008, 143)*

There have been some remarkable misconceptions of projections of future demand. Famous among them is the statement which the former CEO of IBM reportedly made in 1944, that he did not see any demand for more than five computers on this globe. Within just a few decades, hundreds of millions of people had a computer. Brown (2008) refers to the famous case of the 1894 horse manure crisis in New York, which was resolved by the invention of the automobile: "The world is complex and a great deal of understanding is needed to anticipate all the consequences of major interventions. However, it is likely that we will never have enough understanding to make accurate predictions" (Brown 2008, 143).

We need to acknowledge that it was the "experts" who were ignorant about the future – which is, of course, not the experts' fault but simply due to the inherent impossibility of future predictions. *How can we then dare to project what the needs of future generations will be?* This may even be more uncertain than the question of how to ensure we do not compromise their ability to meet their needs.

As a first rough approximation we might let ourselves be guided by our own requirements, but the speed of technological developments and their respective global roll-out is so enormous that future demand can be hardly projected (this becomes even more complicated because our innovations and technological developments will influence the demand for them).

Moreover, even *if we knew the needs* of future generations and even if we assumed that – as a minimum requirement – *no further destruction of our ecosystems must take place*, we would still be largely ignorant of how to get there. To illustrate this with reference to the SDGs: the 17 goals and 169 targets are all important and valuable. But nobody knows how to reach them all – because you cannot target them all at once and any targeting of a subset risks failing to meet the others.

Therefore, it is much easier to say how sustainability *cannot be reached* than how it can be reached. Dennis Meadows even frankly admits that "we do not know what sustainable development is. Yet we know much about what it is not. It does *not* mean destruction of natural resources, it does *not* mean the extinction of biological species, *not* the inefficient wastage of energy" (Meadows 2000, 128; transl. CB). He continues to argue that inasmuch as removing all symptoms does not guarantee health, we cannot be sure that a complete removal of the symptoms of non-sustainability would guarantee that a society has entered the path to sustainability. In a similar fashion, Fritz Reusswig, researcher at the Potsdam Institute for Climate Impact Research, wonders whether it might be more reasonable to investigate unsustainable developments rather than acting on sustainability objectives (Reusswig 1997, 89f.).

1.5.4 Do not charge policies with the pretension of ultimacy

Because our knowledge about what can really prove to be right in the long term is limited, we should be very cautious in deliberating the policies with which we want to change the course of our action for the better. Nobody is faultless and the more impactful our policies become, the more carefully they need to be considered. There are some quite impressive examples which evidence that the best intentions can yield devastating effects (see 2.2), mostly because the complexity of the related issues was underestimated and "side-effects" ignored. In an impressive TED Talk, Zimbabwean ecologist Allan Savory reports from his own experience. In the 1960s he studied desertification in his homeland Rhodesia (Zimbabwe today) and was seeking measures to stop land degradation. He had studied desertification of sub-Saharan Africa for years, had tried several different mitigation measures but nothing helped. Eventually he concluded that the true cause of land degradation must have been the elephants. He consulted international colleagues and was eventually convinced that only culling the elephants would help. He convinced his government to cull 40,000 elephants in order to stop desertification. But the problem got worse. It was not the elephants. Later he realized that he was terribly wrong and confessed that this was the greatest tragedy of his life, the burden of which he will carry to his grave (Savory 2013; Vos 2014). Meanwhile, the discussion about whether the elephant population in Africa's national parks need to be controlled has arisen again (EPMP 2010). The important point here is that a measure or policy which people had considered to be without alternative is condemned by the same group of people only a couple of decades later. In subsequent chapters we will discuss more examples of devastating effects of best-intended interference with ecosystems. I submit that the lesson from this is to be extremely careful if and when we interfere with complex systems, and that we should not charge our measures and policies with the ultimacy of the call for sustainability.

This does certainly not disqualify the targets, the goals – but it calls for humility and caution with regard to related policies and measures. For example, there is no doubt that humanity has to reduce its environmental footprint substantially and urgently, and there is all good reason to assume that diversified, decentralized and locally adapted renewable energy solutions would be a great leap forward in this regard. But the higher the impact on the respective (socio-ecological) systems, the greater the precaution required.

1.5.5 Sustainability as a utopian ideal which can marginally be reached

We can never be absolutely sure about the long-term consequences of our actions and policies. Expert judgements, initiatives or policies, even when backed by the respective scientific evidence of the respective time, might later need to be corrected. If even the smartest, best-intentioned people can be terribly wrong about what needs to be done, how can we ever dare to claim that an action or policy does not threaten the ability of future generations to meet their needs? Therefore, I dare to say that it is not possible to endorse any action of policy with the certificate "fully sustainable", simply because the long-term behaviour of complex systems is unpredictable.[17] Insofar as this will never be fully possible, the quest for sustainability is utopian. According to the German historian Thomas Nipperdey, a utopia is "the outline of a potential world which consciously exceeds the limits and options of the particular reality, targeting a substantially different world which is characterized by a high degree of perfection" (Nipperdey 1962, 359). Understanding sustainability in such a way as a utopian ideal has two consequences. On the one hand, it will shift the focus from chasing the "truly sustainable solution" (which is too often frustrated and thereby exhausts the concept) towards a more humble and modest but nevertheless energetic and realistic approach, i.e. addressing the non-sustainable and seeking to overcome it. The causes of non-sustainability are barriers to sustainability. Therefore, these barriers to sustainability move into the centre of attention. We have a much better understanding of what is *not sustainable* than what it is – so let us make sure we stop the unsustainable.

On the other hand, sustainability does not thereby lose its guiding role for our action. On the contrary – it is more needed than ever. The precaution regarding the measures can make us even more certain about the goal. We need the ideal of sustainability as a guiding vision for a world in which everybody can prosper without living at the expense of others, of the environment, and the future. As such, sustainability is a necessity for a humane world in harmony with nature. It is a utopian ideal.

In his book *Utopia for Realists*, the Dutch historian Rutger Bregman explicates why utopian thinking is actually much needed today. Bregman analyses abundance in our (Western) societies and wonders what the real problem for his generation would be (he was born in 1988). It is not that they are not doing well. It is not that they realize their children will be worse off than they themselves. No, the true problem is that we cannot imagine anything better (Bregman 2017). With his book he does not try to predict the future – but to push open the door to it. While the classical utopias have been destroyed by Hannah Arendt, Karl Popper and the postmodernists, Bregman pleads for another understanding of utopian thinking. The utopian thinking he advocates does not offer solutions, it offers signposts. Instead of forcing us into a straitjacket, it encourages us to change.

I see this kind of thinking also present in Martin Buber's *Paths in Utopia*, in which Buber argues that the realistic character of Utopia would mean "the unfolding of possibilities, latent in humanity's communal life, of a 'right' order." Utopia would not be a "mere cloud castle", it seeks "to stimulate or intensify in the reader or listener his critical relationship to the present". Utopia (and eschatology) seek to show the reader "perfection – in the light of the Absolute, but at the same time as something towards which an active path leads from the present" (Buber 1950, 8).

1.5.6 Planetary boundaries and social injustice

Before we look into the structure and procedure of the book, a few words are needed on the criteria for the unsustainable. By what measure will we call a condition or a development unsustainable? Generally speaking, a development can be called unsustainable if it does not contribute to meeting the needs of the present and/or if it compromises the ability of future generations to meet their needs. The latter will be measured in terms of planetary boundaries. A minimum criterion in the ecological dimension is that the planetary boundaries are maintained since the likelihood for irreversible developments and runaway effects increases sharply if these boundaries are exceeded. In the social dimension, it seems clear to me that the current distribution of resources, goods, and capital can by no means be called just – neither on the international level nor on most national levels. As justice is, in my view, a precondition for sustainability, any policy which does not address these social issues or even exacerbates them would need to be called unsustainable. Policies which neglect the needs of the present generation also qualify as barriers.

In addition, the violation of fundamental, well-established and commonly agreed laws, norms and standards would be another indication for a sustainability barrier, as much as any obvious damage of the common good.

In most cases, it will be enough to indicate minimum requirements because, as we will see in subsequent chapters, several conditions could be identified as barriers even under these minimum assumptions.

1.6 Structure of the book

So far we have said that

- the concept of sustainability is still widely accepted and more needed than ever;
- any transition towards sustainability will require a comprehensive understanding of the barriers;
- the systemic view of the barriers needs to be complemented by an actor perspective.

Along these lines the first part of the book will look into sustainability barriers from an inter- or multi-disciplinary perspective. The aim is to offer a comprehensive overview of the different kinds of barriers. Although it was my goal to illustrate the barriers as comprehensively as possible, I am fully aware that this account is certainly subjective and limited. In each chapter I have had to select what I consider most relevant – at the price of neglecting aspects which others might find crucial.

Nevertheless, I hope that such an attempt will help us to notice recurring patterns, hotspots, synergies and trade-offs among the barriers and action principles. It goes without saying that such an overview cannot go into much detail. This would not only inflate the volume beyond the practical – thereby scaring off many readers, which would run counter to its prime intent – it would certainly also exceed my capacities as an author because a more detailed look into the disciplines would require a much more intimate familiarity with the respective fields.

For each barrier, solution perspectives will be suggested. Although some barriers are inevitably tied to the concept of sustainability ("intrinsic barriers"), as will be explained, and cannot just be overcome, there are nevertheless ways to tackle them.

In the second part of the book, the perspective changes from a systems view to an actors view. I will suggest action principles for sustainability which can guide agents of different kind and level – from individual consumers to corporations, government agencies and policy makers. The multi-level perspective of the first part will be reflected by a multi-actor perspective on the action principles.

Such action principles help addressing the barriers. For instance, the "polluter pays" principle justifies a carbon tax, which is one possible realization of the internalization of external costs, which is a solution perspective for the barrier of market failures. The relation of barriers, solution perspectives, implementation measures and action principles is illustrated in Figure 1.5.

However, in most cases there is not just a bijection between barriers and principles. Rather, a given barrier might be addressed by multiple action principles as much as a given action principle will often help to address several barriers.

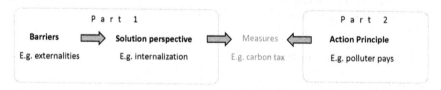

FIGURE 1.5 Relation of the two parts of the book: The first part takes a systems view and suggests solution perspectives for each of the barriers. These solution perspectives point to concrete, context-dependent implementation measures. The second part suggests action principles for sustainability, which support implementation measures from another side.

Source: Own illustration.

Relatively little will be said about concrete *implementation measures* such as taxes, subsidies, cap-and-trade mechanisms, etc. Why? First, it would be challenging to suggest such measures in a one-size-fits-all manner. The regional and cultural differences among our global societies are so substantial that locally adapted solutions are needed. Secondly, the details of the measures would require much more background on the subject which we cannot provide here. However, the measures are, as it were, the creative room where both the solution perspectives of the barriers and the action principles meet, coming from opposite directions.

This finally also implies that the structuring logic does not follow certain technologies. Obviously technologies will play a key role in any future development – be it sustainable or not. Technologies for energy production, storage and transportation, or recycling technologies are evidently of critical importance. In other cases it is certain that they will impact human life in the long run, but their effect is not clear. Digitization and related technologies like robotics, learning algorithms, artificial intelligence, 3D-printing, virtualization and miniaturization will literally change every aspect of life. There are huge threats for human life and certainly for sustainability related to them – up to the possibility that the entire existence of humankind might be at stake after the "technological singularity" (Kurzweil 2005; Bostrom 2014) – but there are great opportunities as well: for value creation with little resource consumption, for dematerialization, for efficiency gains, but also for more sufficient lifestyles, free information sharing, civil society and social values. We will relate to these developments where related barriers are already visible (e.g. acceleration and short-termism) although this will not materialize in the structure of the chapters.

1.7 Methodological approach

The current book attempts to give a highly interdisciplinary overview of barriers to and action principles for sustainability. I have long struggled with the question of whether even the attempt to do so would be megalomaniac because I can obviously not back up the arguments from the many fields with my own academic background.

However, as the reader will hopefully realize in the following, it is my conviction and the core of my approach that we need to reach a comprehensive understanding of the issues, and that it is not enough to delegate the critical issues to experts in the respective fields.

In the last twenty years, during which I have been involved in sustainability issues in several different roles in business, academia, political advisory or NGO work, I have experienced so many different reasons why the obviously needed and apparently nearby solution was not realized. In almost every case there were rational arguments available why this measure or that policy could not be implemented. Furthermore, quite often the failure was not caused by any faulty internal logic of the suggested measure. Rather, what prevented change came from underestimated (or even neglected) factors that had *not been sufficiently* addressed or that could not be addressed because the system's set-up did not allow it.

This made me realize that change will only be possible if we take a *systemic* view, if we realize the complexity of the issues and seek to address the multitude of barriers. It is not enough if sufficient knowledge is available *in principle* – it also has to be available in the same minds, at least to a certain extent (ignoring, for the time being, that knowledge alone is not sufficient either).

Since I am not aware that such an attempt already exists (with the exceptions discussed above), I felt I had to try to give such an account from my own perspective – as preliminary, tentative and erroneous as it might be.

This has a number of methodological consequences:

1. Although I can certainly not claim expert status on all the topics addressed in the following, I have tried my best to avoid any conflict with what appears as a well-established body of knowledge in each of the fields presented. To this end, I have tried to do justice to what Spangenberg calls the "basic law of interdisciplinarity", i.e. that "no discipline must build upon assumptions that are in flagrant contradiction to the established and undisputed body of knowledge of another discipline in charge of the issue at stake" (Spangenberg 2011, 278).
2. Due to the highly interdisciplinary character of this book, I have tried to keep the language as simple and straightforward as possible. We will seek to follow the advice of Brandt et al. that transdisciplinary research "should try to use as simple language as possible, shared by many disciplines and with results ultimately also understandable by civil society" (Brandt et al. 2013, 7). I want to emphasize the latter ambition, i.e. that the current book can be well understood by non-academic members of civil society. My whole approach assumes that change will require that a comprehensive understanding of barriers and action principles is known to as many and as different actors as possible.
3. Both previous points imply, however, that concepts used in the following should be taken as they are presented here – and not as any long reception history in certain fields might imply. This is due to a practical necessity – because I could simply not provide that, and even if I could it would bloat the book to beyond what average readers could digest – but it is also an exercise in interdisciplinary language because cross-disciplinary understanding requires some common ground in terms of language and concepts.
4. Since most readers will relate to certain fields of study more than to others, they should not expect original contributions in their own field. Nevertheless, the added value for them will hopefully result in discussing not only the other fields but also their potential interrelations.
5. Normativity: sustainability is, obviously, a normative concept. We can ignore the question whether science can, should, or must not deal with normative questions – since we can simply take the normativity of the concept as given due to the global agreements of the 2030 Agenda and their Sustainable Development Goals. I will try to indicate any of my own value judgements.

If not mentioned otherwise, all translations into English from German sources have been made by the author. Numbers in brackets refer to the previous quote.

1.8 Summary

1. *Despite its implementation difficulties, sustainability remains a critical concept, currently more than ever.*
2. *Lack of progress should not lead us to abandon the concept as such, call for an eco-dictatorship or lead us to ever more strident moralistic claims. Rather, we should sober-mindedly investigate the complexity of the issues, in particular the barriers, and seek solution perspectives for each.*
3. *In spite of its shortcomings, the Brundtland definition is still the most widely known concept of sustainability. I will therefore refer to it several times because the barriers to be discussed even impede the realization of a weak concept of sustainability. However, in addition to that one can argue that observing the planetary boundaries is certainly a minimum requirement even within the Brundtland frame, since this is a necessary condition that future generations will be able to meet their needs.*
4. *Any transition towards sustainability will require us*

   a *to embrace several dimensions, levels, sectors and actors,*
   b *to understand the diversity of barriers to sustainability,*
   c *to elaborate solution perspectives for each of the barriers, and*
   d *to provide guidance to actors (of different kind and level) how to orient their actions towards sustainability.*

5. *The action principles address actors of different kinds and levels and offer guidance towards sustainability. While the first part of the book argues from a systemic perspective, seeks barriers and how they can be addressed, the second part takes the actors' perspective. The solution perspectives for the barriers and the action principles both point to the realm of implementation measures, for which only few indications will be given since they vary depending on the context and require deep contextual understanding.*

Notes

1 Which means that the agreement postulates keeping the 2°-limit but the Nationally Determined Contributions are far too insufficient for that.
2 The Club of Rome, founded on initiative of Aurelio Peccei and Alexander King in Rome in 1968, is presumably the first global think tank dedicated to the great challenges of humankind. It has become renowned for the reports submitted to it by independent experts, above all the first one by Meadows et al. (Meadows et al. 1972; see The Club of Rome 2019).
3 According to Meadows, the

majority of people using the Brundtland definition for justifying their work participate in a double deception. Firstly, the needs of the present generation are certainly not met today. Secondly, the economic activities we are performing to meet present needs, definitely diminish in substantial ways the number of options which future generations will have.

(D. L. Meadows 2000, 126)

To my knowledge this text only appeared in German and was translated back to English by the author.

4 The 2014 IPCC report detects such an attitude regarding the climate adaptation planning, since "the framing of adaptation planning as a 'problem-free' process and underestimation of its social nature has contributed to the creation of unrealistic expectations in societies on the capacity for planning and mainstreaming climate adaptation" (Andersson & Keskitalo 2018, 76).

5 For Southeast Asia, Beeson sees "that the intensification of a range of environmental problems means that authoritarian rule is likely to become even more commonplace there in the future" (Beeson 2010, 276). Chen and Lees argue

that the shifting strategies of governance associated with green urbanization are evidence of the emergence of a distinct paradigm of authoritarian environmentalism, characterized by a re-centralization of state power and a reduction of local autonomy, in environmental policy making in China.

(Chen and Lees 2018, 212)

6 I have briefly explained my objection against this idea elsewhere (Berg 2017).

7 Hobson investigated the values and beliefs of climate sceptics and how entrenched these are in other beliefs. She concludes that rational arguments on the conceptual level will not have any effect. "In short, if 2 hours seeing (at times quite challenging) climate scenarios for your local region, and then 3 days spent deliberating cannot dispel the myriad of forms of climate scepticism, what will?" Climate scepticism is rather to be seen as a result of "political recalcitrance and a pervasive public discourse promoting inaction around mitigation" (Hobson and Niemeyer 2012, 409).

8 The 2014 IPCC report identified a similar imbalance in climate adaptation planning:

There is the risk of underestimating the complexity of adaptation planning as a social process, and it can lead to creating unrealistic expectations in societies, and overestimating the capacity of planning to deliver the intended outcome of preparing societies to adapt to the negative impacts of climate change.

(Mimura, Pulwarty and Duc, 874).

9 More precisely, Geels argues that transitions come about through an alignment of processes at three levels: niche-innovations, regime changes and pressure from the landscape level. Niche innovations bring novelty, but these innovations can only be successful if supported by pressure from the "landscape level" (which means changes in the broad political, social and economic landscape, e.g. by demographic shifts or rise of consumer culture) in combination with destabilization of the prevailing regime (Geels and Schot 2007, 400). The prevailing regime and its rules are both medium and outcome of action, as Geels describes according to Giddens.

On the one hand, actors enact, instantiate and draw upon rules in concrete actions in local practices; on the other hand, rules configure actors. Examples of regime rules are cognitive routines and shared beliefs, capabilities and competences, lifestyles and user practices, favourable institutional arrangements and regulations, and legally binding contracts.

(Geels 2011, 27)

10 "Successful sustainability transformations ultimately hinge on a broad range of actors to organize around stewarding transformative change" (Künkel 2019, 263).

11 The authors also stress the interdependence of issues: "Failure was rarely due to one isolated factor and was usually linked to a combination of interacting economic, legal and political factors (the three most cited factors)" (p. 165).

12 I owe this reference to the population topic to one of the anonymous reviewers of the book outline.

13 "In addition, the main barriers which make it more difficult for certain individuals or social environments to make sustainable decisions are considered to be the lack of long-term orientation, loss aversion, and path dependencies in general" (78).

14 In any case one can never really be sure that a certain action can truly claim to be sustainable because we simply cannot foresee the long-term consequences of actions in complex systems. The "butterfly effect" of chaotic systems does not shape our daily experience because most of our normal routines luckily exhibit a great deal of regularity (and are not chaotic). However, we can never really be sure (which is an argument against consequentialist ethics).

15 In this book I will try to use integrative language. Occasionally, when I think a specific form makes more sense because it captures reality better, I will choose just one form. In this case, I deliberately used the male form because I do see this, by trend, as a problem of men.

16 Every system has, of course, to do justice to the foundational laws of the level "below" – for instance, psychology functions in the limits of biology, which functions in the limits of chemistry, which functions in the limits of physics.

17 I argue that sustainability is thus a *comparative* concept. Although an absolute qualification as "sustainable" is impossible, it might well be possible to indicate which of two options is more sustainable than the other (although that might often be difficult enough as well).

Bibliography

Andersson, Elias, & E. Carina H. Keskitalo. "Adaptation to climate change? Why business-as-usual remains the logical choice in Swedish forestry." *Global Environmental Change* 48, 2018: 76–85.

Ayres, Robert U. "Sustainability economics: Where do we stand?" *Ecological Economics* 67 (2), 2008: 281–310.

Bardi, Ugo. *Extracted. How the Quest for Mineral Wealth Is Plundering the Planet.* White River Junction, VT: Chelsea Green Publishing, 2014.

Beeson, Mark. "The coming of environmental authoritarianism." *Environmental Politics* 19 (2), 2010: 276–294.

Berg, Christian. "Geld für weniger Kinder? Der falsche Anreiz." *Xing Klartext.* 30. 10. 2017. www.xing.com/news/klartext/geld-fur-weniger-kinder-der-falsche-ansatz-2179, (accessed July 15, 2019).

Blühdorn, Ingolfur. "Post-capitalism, post-growth, post-consumerism?" *Global Discourse* 2017: 42–61. DOI: 10.1080/23269995.2017.1300415

Blühdorn, Ingolfur & Michael Deflorian. "The collaborative management of sustained unsustainability: On the performance of participatory forms of environmental governance." *Sustainability 11*, 02 2019: 1189.

Bostrom, Nick. *Superintelligence. Paths, Dangers, Strategies*. Oxford: Oxford University Press 2014.

Bosselmann, Klaus. *The Principle of Sustainability. Transforming Law and Governance*. London, New York: Routledge, 2017.

Brandt, Karl-Werner. "Probleme und Potentiale einer Neubestimmuing des Projekts der Moderne unter dem Leitbild 'nachhaltige Entwicklung'. Zur Einführung." In *Nachhaltige Entwicklung*, Karl-Werner Brandt (ed.), 9–32. Opladen: Leske + Budrich, 1997.

Brandt, Patric, Anna Ernst, Fabienne Gralla, Christopher Luederitz, Daniel J. Lang, et al. "A review of transdisciplinary research in sustainability science." *Ecological Economics 92*, 2013: 1–15.

Bregman, Rutger. *Utopia for Realists*. London: Bloomsbury Publishing, 2017.

Brown, Casey. "Emergent sustainability: The concept of sustainable development in a complex world." In *Globalization and Environmental Challenges*, Hans Günter Brauch, Navnita Chadha Behera, Béchir Chourou, Pál Dunay, John Grin, et al. (eds.), 141–149. Berlin, Heidelberg, New York: Springer, 2008.

Buber, Martin. *Paths in Utopia*. New York: Macmillan, 1950.

Carson, Rachel, *Silent Spring*, Boston, MA: Houghton Mifflin Company, 1962.

Chen, Geoffrey C. & Charles Lees. "The new, green, urbanization in China: Between authoritarian environmentalism and decentralization." *Chinese Political Science Review*, 3 (2), 2018: 212–231.

Crutzen, Paul J. "Geology of mankind." *Nature 415*, 03. 01. 2002: 23.

Daly, Herman E. *Beyond Growth*. Boston, MA: Beacon Press, 1996.

DER WESTEN. *Lehrerin (38): "Kinder sind das Schlimmste" – diese krasse Forderung hat sie jetzt*. 08. 03. 2019. www.derwesten.de/panorama/lehrerin-verena-brunschweiger-38-kinder-sind-das-schlimmste-diese-krasse-forderung-hat-sie-jetzt-id216615983.html (accessed July 15, 2019).

Dibley, Arjuna. "Why the rise of populist nationalist leaders rewrites global climate talks." *Nature: Climate Change*, 5. 12 2018. https://phys.org/news/2018-12-populist-national ist-leaders-rewrites-global.html

EPMP. "Elephant population management project". 2010. www.elepmp.org/ (accessed July 07, 2019).

Figge, Frank. "Capital substitutability and weak sustainability revisited: The conditions for capital substitution in the presence of risk." *Environmental Values 14*, 2005: 185–201.

Geels, Frank W. "The multi-level perspective on sustainability transitions: Responses to seven criticisms." *Environmental Innovation and Societal Transitions 1*, 2011: 24–40.

Geels, Frank W., & Johan Schot. "Typology of sociotechnical transition pathways." *Research Policy 36*, 2007: 399–417.

Grin, John, Jan Rotmans, & Johan Schot. *Transitions to Sustainable Development: New Directions in the Study of Long Term Transformative Change* (Routledge Studies in Sustainability Transitions, Band 1). New York, London: Routledge, 2010.

Harm Benson, Melinda, & Robin Kundis Craig. "The end of sustainability." *Society and Natural Resources 1(6)* 2014:777–782 DOI: 10.1080/08941920.2014.901467

Hirschnitz-Garbers, Martin, Adrian R. Tan, Albrecht Gradmann, & Tanja Srebotnjak. "Key drivers for unsustainable resource use e categories, effects and policy pointers." *Journal of Cleaner Production 132*, 2016: 13–31.

Hobson, Kersty, & Simon Niemeyer. "'What sceptics believe': The effects of information and deliberation on climate change scepticism." *Public Understanding of Science 22 (4)*, 2012: 396–412.

Howes, Michael, Liana Wortley, Ruth Potts, Aysin Dedekorkut-Howes, Silvia Serrao-Neumann, Julie Davidson, Timothy Smith & Patrick Nunn, Environmental Sustainability: A Case of Policy Implementation Failure? in: *Sustainability*. 2017 9, 165, DOI: 10.3390/su9020165

Hulme, Michael. *Why We Disagree about Climate Change.* Cambridge: Cambridge University Press, 2009.

IASS. *The Role of Biomass in the Sustainable Development Goals: A Reality Check.* Potsdam: Institute for Advanced Sustainability Studies (IASS). 2015.

Jaspal, Rusi, Brigitte Nerlich, & Kitty van Vuure. "Embracing and resisting climate identities in the Australian press: Sceptics, scientists and politics." *Public Understanding of Science 10*, 2016: 807–824.

Jonas, Hans. *Das Prinzip Verantwortung.* Frankfurt a.M: Insel also: Suhrkamp, stw 1085, 1984.

Kanger, Laur, & Johan Schot. "Deep transitions: Theorizing the long-term patterns of sociotechnical change." *Environmental Innovation and Societal Transitions 32*, 2018, 7–21.

Kant, Immanuel. *Groundwork for the Metaphysics of Morals* (ed. by Allan W. Wood). New Haven, CT, London: Yale University Press, 2002 (1785).

Köhler, Jonathan, Frank W. Geels, Florian Kern, Jochen Markard, Elsie Onsogo, et al. "An agenda for sustainability transitions research: State of the art and future directions." *Environmental Innovation and Societal Transitions 31*, 1–32: 2019.

Künkel, Petra. *Stewarding Sustainability Transformation. An Emerging Theory and Practice of SDG Implementation.* Berlin: Springer Nature, 2019.

Kurzweil, Ray. *The Singularity Is Near. When Humans Transcend Biology.* New York, London: Penguin Books, 2005.

Lockwood, Matthew. "Right-wing populism and the climate change agenda: Exploring the linkages." *Environmental Politics 27* (4)2018: 721–732.

Martin, Claude. *On the Edge. The State and Fate of the World's Tropical Rain Forests.* Vancouver: Greystone Books, 2015.

Meadows, Dennis L. "Es ist zu spät für eine nachhaltige Entwicklung. Nun müssen wir für eine das Überleben sichernde Entwicklung kämpfen." In *Zukunftsstreit*, Wilhelm Krull (ed.), 125–149. Weilerswist: Velbrück Wissenschaft, 2000.

Meadows, Donella H., Dennis L. Meadows, Joergen Randers, & William W. Behrens, III. *The Limits to Growth. A Report for the Club of Rome's Project on the Predicament of Mankind.* London: Earth Island Ltd., 1972.

Mill, John Stuart. *On Liberty.* London: Longmans, Green, Reader and Dyer, 1869.

Mimura, Nobuo, Roger S. Pulwarty, D.M. Duc. Ibrahim Elshinnawy, Margaret Hiza Redsteer, He Qing Huang, Johnson Ndi Nkem, Roberto A. Sanchez Rodriguez, Richard Moss, Walter Vergara, Lisa S. Darby, and Sadahisa Kato "Adaptation planning and implementation." In *Climate Change 2014: Impacts, Adaptation and Vulnerability. Part A: Global and Sectoral Aspects. Working Group II Contribution to the Fifth Assessment Report of the Intergovernmental Panel on Climate Change*, IPCC (ed.), 869–898. Cambridge, New York: Cambridge University Press, 2014.

Müller, Harald. *Wie kann eine neue Weltordnung aussehen? Wege in eine nachhaltige Politik.* Frankfurt a.M: Fischer Taschenbuch Verlag, 2008.

Müller, Jan-Werner. *Was Ist Populismus?* Berlin: Suhrkamp, 2016.

Nair, Chandran. *The Sustainable State. The Future of Government, Economy, and Society.* Oakland, CA: Berrett-Koehler Publishers, 2018.

Neumayer, Eric. *Weak versus Strong Sustainability: Exploring the Limits of Two Opposing Paradigms.* Northampton, MA: Edward Elgar, 2003.

Nipperdey, Thomas. "Die Funktion der Utopie im politischen Denken der Neuzeit." *Archiv für Kulturgeschichte* 44 (3) 1962: 357–378.

Osterhammel, Jürgen, Die Verwandlung der Welt. Eine Geschichte des 19. Jahrhunderts, Munich: C. H. Beck 2009

Ott, Konrad. "Institutionalizing strong sustainability: A Rawlsian perspective." *Sustainability 2014* (6), 2014: 894–912.

Peters, Glen P., Steven J. Davis, & Robbie Andrew. "A synthesis of carbon in international trade." *Biogeosciences 9*, 2012: 3247–3276.

Randers, Jorgen, & Graeme Maxton. *Reinventing Prosperity: Managing Economic Growth to Reduce Unemployment, Inequality and Climate Change.* Vancouver, Berkeley, CA: Greystone Books, 2016.

Raskin, Paul, Tariq Banuri, Gilberto Gallopin, Pablo Gutman, Al Hammond, et al. *Great Transition. The Promise and Lure of the Times Ahead (A Report of the Global Scenario Group).* Boston, MA: Stockholm Environment Institute – Boston, 2002.

Reusswig, Fritz. "Nicht-nachhaltige Entwicklungen. Zur interdisziplinären Beschreibung und Analyse von Syndromen des Globalen Wandels." In *Nachhaltige Entwicklung*, Karl-Werner Brandt (ed.), 71–90. Opladen: Leske + Budrich, 1997.

Rockström, Johan, Will Steffen, Kevin Noone, Asa Person, F. Stuart Chapin III et al. "A safe operating space for humanity." *Nature 461*, 24. 09 2009: 472–475.

Rosenau, James. "Globalization and governance: Bleak prospects for sustainability." *IPG 3*, 2003: 11–29.

Sachs, J., G. Schmidt-Traub, C. Kroll, G. Lafortune, & G. Fuller. *Sustainable Development Report 2019.* New York: Bertelsmann Stiftung and Sustainable Development Solutions Network (SDSN), 2019.

Sachs, Wolfgang. "Sustainable Development. Zur politischen Anatomie eines internationalen Leitbilds." In *Nachhaltige Entwicklung*, Karl Werner (ed.), Brandt, 93–110. Opladen: Leske + Budrich, 1997.

Savory, Allan. "How to fight desertification and reverse climate change". *TED.com.* 2013. www.ted.com/talks/allan_savory_how_to_green_the_world_s_deserts_and_re verse_climate_change#t-382789 (accessed June 13, 2019).

Scherer, Laura, Paul Behrens, Arjan de Konig, Reinout Heijungs, Benjamin Sprecher, et al. "Trade-offs between social and environmental Sustainable Development Goals." *Environmental Science and Policy 90*, 2018: 65–72.

Spangenberg, J.H. Sustainability science: a review, an analysis and some empirical lessons, in: *Environmental Conservation 38* (3): 275–287.

Steffen, Will, Katherine Richardson, Johan Rockström, Sarah E Cornell, Ingo Fetzer, et al. "Planetary boundaries: Guiding human development on a changing planet." *Science 347* 3.02.2015: 736.

Steffen, Will, Katherine Richardson, Johan Rockström, Sarah E Cornell, Ingo Fetzer, et al, Johan Rockström, Katherine Richardson, Timothy M. Lenton, Carl Folke, et al. "Trajectories of the Earth system in the Anthropocene." *PNAS 115*, 14. 08 2018: 8252–8259.

Stroh, David Peter. *Systems Thinking for Social Change.* White River Junction, VT: Chelsea Green Publishing, 2015.

The Club of Rome. 2019. www.cluboframe.org/ (accessed June 13, 2019).

The Guardian News. 2019. "Rutger Bregman tells Davos to talk about tax: 'This is not rocket science'", https://www.youtube.com/watch?v=P8ijiLqfXP0, accessed Oct 10, 2019.

UN. *Transforming Our World: The 2030 Agenda for Sustainable Development (A/RES/70/1).* New York: United Nations. 2015.

UNDP. *Human Development Indices and Indicators*. New York: United Nations Development Program. 2018.

UNESCO. *Leaving No One Behind. The United Nations World Water Development Report 2019*. Paris: UNESCO. 2019.

UNFCCC. Paris Agreement. Paris: UNFCCC. 2015.

Vos, Marieke. "After shooting 40,000 Elephants, he vowed to find a better way to save two-thirds of the world from becoming desert." *The Epoch Times*. 2014.

WBGU. *World in Transition: A Social Contract for Sustainability*. Berlin: German Advisory Council on Global Change (WBGU). 2011.

WCED. *Our Common Future*. Oxford: Oxford University Press. 1987.

Wittmayer, Julia M., Flor Avelino, Frank van Steenbergen & Derk Loorbach, Actor roles in transition: Insights from sociological perspectives, in: *Environmental Innovation and Societal Transitions,*. 24 (2017), 45–56.

World Ocean Review. *Coasts – A Vital Habitat under Pressure*. Hamburg: maribus gGmbH. 2017.

WWF. *Living Planet Report 2018: Aiming Higher* (M. Grooten & R.E.A. Almond (eds.)). Gland, Switzerland: WWF, 2018 14146.

PART 1

Barriers

There are many kinds of reasons why we are not more sustainable.[1] Without claiming completeness, the following chapters will look at several of these sustainability barriers from the perspective of different disciplines because understanding the barriers comprehensively is a precondition for addressing them effectively and facilitating the sustainability transition.

In order to structure the different kinds of barriers, I will suggest a typology for them. This typology is meant as a conceptual guide, as assistance for comprehension but does not intend to describe any substantial, "ontological" reality.

The first conceptual distinction I suggest relates to the question whether or not a barrier is inevitably tied to the concept of sustainability. For instance, whenever we strive for sustainability, we will always have to deal with trade-offs, we will always have to face complexity, we will always have to realize that people have conflicting interests. We can never hope to achieve more sustainability without addressing these barriers. I will call such barriers *intrinsic barriers* because they are intrinsic to the concept of sustainability as such. Other barriers might be just as difficult to address in practice but could be overcome in principle – they are *extrinsic* to the concept of sustainability. For example, the problem with negative externalities can (at least partly) be addressed by putting a price tag on the consumption of public goods, which might be difficult for practical reasons but is conceivable theoretically.

It should be noted that the mapping of the barriers to certain categories of the typology is somewhat flexible and subjective. For instance, complexity is grouped under nature-related barriers, because it is already inherent in natural systems; it is a feature of our physical reality, especially in ecological or biological systems. However, complexity does certainly also occur in social or economic systems.

PART I
Barriers

Intrinsic Barriers

I suggest grouping the intrinsic barriers into three categories: Barriers related to physical reality, barriers related to human condition, and barriers related to social reality.

2

BARRIERS RELATED TO PHYSICAL REALITY

Contents

Some of the reasons why sustainability is so hard to achieve originate in the nature of our physical reality. The fundamental laws of nature confine our endeavours for sustainability. Complexity, for instance, is an inherent feature of many biological and ecological systems (and, of course, social systems as well). Another is related to energy – the energy return on energy invested.

2.1 The problem of ERoEI, resources and pollution

Energy availability is critical for our civilization. However, providing energy does itself require energy and any energy system only functions in the long run if the yield factor, i.e. the amount of energy harvested compared to the energy invested is sufficiently large. The yield factor for fossil fuels has significantly decreased in recent decades, while it is still growing for some renewable energy forms. This calls into question our current economic system and has consequences for both the availability of resources and for questions of pollution.

Energy availability is critical – for any development and for addressing social and environmental issues. However, energy comes at a price – not only in monetary terms but also in energetic terms. We need to invest energy in order to provide consumable energy, which is measured as the energy return on energy invested (ERoEI or EROI).

Let us set aside the global warming potential (GWP) of hydrocarbons for a moment and just consider their energy content. While the EROI of global oil and gas production peaked in 1999 at 35:1, it has declined to 18:1 in 2006 (Gagnon, Hall & Brinker 2009, 492), and with increasingly depleted conventional sources for oil and gas, this ratio is likely to decline further: "The decline in EROI among major fossil fuels suggests that in the race between technological advances and depletion, depletion is winning" (Hall, Lambert Jessica & Balogh 2014, 151). Hall et al. argue that the argument often used by economists, that increasing prices would give incentives for further exploitation, does not work if we get closer to EROIs of 1:1. Obviously the EROI needs to be larger than 1:1 because otherwise you would have a net negative energy balance, i.e. you lose energy while "providing" it. Now the calculations critically depend on the system's boundaries as Hall illustrates in an interview:

> If you've got an EROI of 1.1:1, you can pump the oil out of the ground and look at it. If you've got 1.2:1, you can refine it and look at it. At 1.3:1, you can move it to where you want it and look at it. We looked at the minimum EROI you need to drive a truck, and you need at least 3:1 at the wellhead. Now, if you want to put anything in the truck, like grain, you need to have an EROI of 5:1. And that includes the depreciation for the truck. But if you want to include the depreciation for the truck driver and the oil worker and the farmer, then you've got to support the families. And then you need an EROI of 7:1. And if you want education, you need 8:1 or 9:1. And if you want health care, you need 10:1 or 11:1.
>
> *(ScientificAmerican 2013)*

In other words, to maintain the current economic system and wealth standard of the global North requires EROIs which are already close to the current values.

Gagnon et al. conclude from their analysis of the historic data on global EROIs of oil and gas production that

> declining EROI of main fuels and low EROI of most alternatives, as well as the many economic difficulties of the past year suggest that we are indeed reaching the "limits to growth" and must adjust our perspectives and economic goals accordingly.
>
> *(Gagnon, Hall & Brinker 2009, 502)*

Similarly, exploring the minimum EROI that a sustainable society must have, Hall et al. indicate a ratio of ten to one as for the "mean for society" (Hall, Balogh & Murphy 2009, 45).

Murphy and Hall describe our situation as an economic growth paradox because the exploitation of unconventional oil resources is only profitable at high prices – but in the current set-up, high energy prices lower economic growth:

> increasing the oil supply to support economic growth will require high oil prices that will undermine that economic growth. From this we conclude that the economic growth of the past 40 years is unlikely to continue in the long term unless there is some remarkable change in how we manage our economy.
>
> *(Murphy & Hall 2011, 52)*

It is important to note that the previous studies have *only looked at the energy balance*. The GWP of hydrocarbons was not even considered – and still the conclusion is that the current growth path will come to an end.

However, considering energy options for the future, the GWP of hydrocarbons can certainly not be neglected. We must clearly rely on renewable energy. Yet renewable energy also has an EROI. The sun does not charge us, to be sure, but the production and installation of solar panels, for instance, requires a lot of energy, resources and labour.

In a systematic meta-analysis of 232 references, Bhandari et al. explored the EROIs for different types of modules of solar panels (e.g. cadmium telluride or copper indium gallium diselenide), and indicate EROIs ranging from 8.7 to 34.2 and energy pay-back times (EPBTs) between 1.0 and 4.1 years (Bhandari et al. 2015, 133). Fraunhofer ISE, one of the leading institutes for applied research on solar energy systems in Germany, numbers the EBPTs for solar panels in the range of 0.7 to 2.0 years. Assuming a lifetime of 30 years, rooftop photovoltaic (PV) systems would therefore produce net clean electricity for approx. 95% of their lifetime (Fraunhofer ISE 2019).

Compared to non-renewable energy forms like nuclear or hydrocarbons, some renewable energy technologies (still) have a much lower EROI, as Weißbach et al. (2013) explored. Solar and wind power certainly depend on natural conditions, and fluctuations in supply need to be matched to fluctuations in demand. What really counts are the "buffered" values, which include the energy needed to provide storage facilities, as Weißbach et al. argue. The authors compared the EROIs and EPBTs of electricity generating power plants for renewable, fossil and nuclear energy sources and found "that nuclear, hydro, coal, and natural gas power systems (in this order) are one order of magnitude more effective than photovoltaics and wind power" (Weißbach et al. 2013, 210). Weißbach et al. indicate numbers for buffered EROI for concentrated solar power (CSP) systems in the Sahara in the range of 8.2 to 9.6, and between 1.5 and 2.3 for solar PV in Germany. Assuming that our societies in its current mode of operation require a minimum EROI of 10:1, there is still some way to go.

The good news is that there is a clear trend towards increasing EROIs or decreasing EPBTs through technological progress. For instance, between 1990 and 2013, the EPBT decreased from 3.3 years to 1.3 years for multicristalline PV rooftop systems installed in Southern Europe (Fraunhofer ISE 2019). The hope is therefore that future progress in production, transmission and storage technologies can still increase the EROIs needed.

The question, however, which the physicist and economist Robert Ayres raised two decades ago still needs to be answered, namely decreasing the flow of physical inputs can generate an increasing flow of final services (Ayres 1996). The "physical inputs" which Ayres refers to are energy (technically speaking: exergy[2]) and material. So far we have only looked at the former, but there are also challenges related to the mass fluxes, which constrain the idea of a circular economy and which have implications for pollution.

If you pour milk into your coffee, it will blend with the coffee immediately. You will never see the opposite happening, i.e. that the milk in the coffee will spontaneously separate itself from the coffee. This reflects a fundamental principle in nature, an asymmetry in time, an "arrow of time", which is a fundamental law of thermodynamics. The first law of thermodynamics is known as energy conservation: the total amount of energy does not change over time in any closed system – which means any system which does not exchange energy or material with its environment. An isolation of the milk in the coffee might be possible energetically. Yet we all know that it does not occur – this is due to the second law of thermodynamics, which introduces the concept of entropy. Entropy measures the degree of irreversibility during energy transformation: "Any real system that goes through a cycle of operations and returns to its initial state must lead to an increase of the entropy of the surroundings" (Schaub & Turek 2016, 17).

Nicholas Georgescu-Roegen claimed that the second law of thermodynamics would even constrain economic processes, i.e. that economic processes themselves are entropic (Georgescu-Roegen 1986). Just as exergy becomes progressively unavailable because of the entropy law, matter would behave similarly, even if sufficient amounts of exergy were available. Ayres has, I submit, convincingly demonstrated that this argument is based on a flawed assumption insofar as it neglects the fact that the earth is not an isolated system (Ayres 1996). If we had sufficient amounts of exergy we could recover any element from any source, regardless how diffusely it occurs – at least energetically. The critical assumption here is, however, that we have enough exergy. Georgescu-Roegen might be wrong in stating that the diffusion of materials on the earth is a problem *in principle* – here I think Ayres is right. But Georgescu-Roegen's argument might be adjusted to say that the diffusion of substances is a problem *in practice* – and in fact, it is a huge one. Once evenly distributed, it costs an enormous amount of exergy to concentrate a substance again.

This has implications both on the side of *sources* and on the side of *sinks*, i.e. with regard to resource supply as well as with regard to pollution.

Regarding *resource supply* it means that we will never reach 100% of circularity in our mass flows. There will always be losses. Even countries with a long tradition of recycling and recovery (e.g. Germany) have not reached anything even close to 100%. In Germany, for instance, which has the highest recycling rate of household trash in the EU, recycling rates are 66% (European Environment Agency 2018), which means that one-third of all household trash is *not* recycled and the resources contained in it are lost. Moreover, these numbers could be called sugar-coated because trash which is actually exported and allegedly recycled abroad is also considered as recycled here.

For certain materials we can already experience a widening gap between the recycling rate and expected future demand. Several rare earth elements, which are necessary in the high-tech industry or for the production of low-carbon energy technologies have a global average post-consumer recycling rate of below 1% (UNEP 2011, 19). At the same time, the demand for some of these elements is likely to increase by several hundred per cent until 2030 (compared to 2012) (EU 2016, 14).

What can be done? Can we not simply increase this ratio? We can increase the ratio, but not simply. We can (and should) improve our collection processes, our separation and recycling technologies and thereby ultimately increase the ratio of recycled input material for our industrial processes. However, this comes at a cost – the cost of energy (more precisely: exergy). Preparing raw materials, processing them to pre-products and then producing them into compound materials consumes a lot of exergy. This conflicts, however, with the limitations of the EROI discussed above because both the extraction of virgin material and the recycling of products consumes substantial among of exergy. It is therefore evident that the wastefulness of resources in our take-make-waste economy must be reduced as soon as possible.

This holds true from the resource point of view but also with regard to *pollution*. Poor recycling rates correlate not only with wastefulness but also with pollution. If rare earths are being recycled at rates of 1% this means that 99% of them are not appropriately processed – and partially scattered into the environment. This will increasingly become an issue of pollution. From a physical point of view, it might be possible to re-collect toxic material or radioactive nuclides – *in principle*. In any realistic scenario, however, this is extremely difficult or practically impossible. Firstly, this would require huge amounts of energy – and the more effort is needed for cleaning up the environment the higher the EROI of our energy sources needs to be. We just saw that our current civilization runs on an EROI of approximately 10:1, which is already a challenge for future energy systems. If we were to increase the effort for environmental clean-up (which we see already starting, for instance, by initiatives to free the oceans from plastic), this ratio would even be higher. In other words, it is likely that we will run into issues of our energy supply. Secondly, decontamination of natural environments is an extremely difficult endeavour because one can often not clean up a landscape, a plant or an animal from capillary distributed toxic substances in a non-destructive way.

Therefore, while decontamination might be possible at small scales, for inanimate matter and with high effort, it is currently not conceivable how this could ever be possible for globally distributed substances. We have so far in many cases benefited from the fact that many toxic substances were degraded by natural processes, although this might "only" take a long time. But there are substances which are hardly degradable or not at all, like persistent organic pollutants (POPs), heavy metals, or nuclear isotopes, for which we cannot rely on the power of nature – maybe we can only hope that natural processes will sooner or later diffuse the substances to a degree which is less harmful.

Moreover, the problem of recycling gets more complicated because we have not yet developed efficient large-scale recycling facilities for many product groups. For instance, there is a trend towards using more compound materials, which worsens the problem. This is simply a side effect of the fact that our knowledge of chemical processes and substances has increased enormously in the last decades. We can tailor substances, products and packaging materials to very specific purposes. Packaging materials, for instance, have to fulfil specific requirements of the packaged good. In case of vegetables, they need to be durable and resistant, light, transparent, often permeable to moisture, and above all: cheap, to name just a few obvious ones. Whether or not they are recyclable has only recently become of interest. In many cases, there is no single substance alone, which can fulfil all these criteria. Therefore, packaging material today often contains several different forms of plastics – which is a huge challenge for recycling and makes incineration ("thermal utilization") often the preferred option.

Similarly, aiming to reduce the fuel consumption of their car fleets, car manufacturers have reduced the share of metals and increased the share of compound materials in today's cars. Whereas cars used to contain mainly steel sheets and a very limited number of materials, which were relatively easy to disassemble and decompose, and melted down to make new steel, it is much more difficult to disassemble today's cars. Finally, we often develop and roll out new technologies without having clarified the end-of-life phase. There are, for instance, no good concepts for the recycling of wind power stations, in particular the rotor blades (Seitz & Kaiser 2014).

Solution perspectives

A great challenge arises from the fact that coal remains cheap, not only financially but also energetically, i.e. it still has a considerably high EROI. However, due to the fact that coal has the highest GWP of all energy sources, we simply cannot afford to burn it any more. Therefore, continued efforts for scalable and cheap solutions for energy storage and transmission technologies as well as further optimization of renewable energy production technologies are needed.

A true circular economy paradigm, which maintains precious resources and does not pollute the environment is needed. There has to be much stronger focus on keeping as many resources as possible within the "industrial metabolism". This demand stems both from the shortage of resources and the reduction of pollution. Much can be done if we have enough exergy – but as we have seen, exergy availability is limited and will also put constraints on both the exploitation of natural resources as well as the decontamination of environmental pollution. We are likely to see a boost in environmental clean-up technologies – but they do consume energy. Financially and energetically it makes much more sense to prevent pollution in the first place. Finally, we need less energy-intensive modes of value creation, for energy efficiency, energy savings and sufficiency wherever possible (see 13.2).

2.2 Complexity

Many biological, ecological or social systems exhibit an enormous complexity. The more complex systems get, the more difficult it is to understand their behaviour. This challenges our struggle for sustainability because the anthropogenic impact on our global ecosystems has increased considerably, engendering unintended consequences. Considered action is needed on all levels but difficult. Consumers can hardly assess the social and ecological impact of purchasing decisions, while policy makers rarely anticipate societal responses of technological innovations or political decisions.

Complexity is a concept which is used in a variety of different disciplines, with partially distinct meanings. As a common denominator one might say that systems are complex insofar as their components interact with each other, and exhibit feedback loops and non-linear, often unpredictable behaviour (see Kauffman 1995; Johnson 2001; Barabási 2002; D. H. Meadows 2008). As humans we mostly experience a direct and linear relation between a cause and an effect, to which we have adapted in our evolutionary history: The harder you kick a ball the faster it hits the goal; the more you deflect a pendulum, the larger is its amplitude. You can always differentiate the effect from its cause (which allows you to address issues by addressing the cause) and the impact of the cause relates to the magnitude of the effect. All this is challenged by complexity. In complex systems with their feedback loops and non-linear behaviour, it is often hardly possible to single out one reason and identify *the* one reason or cause of an issue. System responses to external perturbation are extremely sensitive to changing initial conditions and cannot be predicted after a few phases. They exhibit counter-intuitive effects, sometimes opposite to the expected (or intended). Furthermore, the effect can be much bigger in magnitude than the stimulus, which is the famous "butterfly effect". Tiny causes can have huge effects in other regions of the world, in spatially and temporally distant locations.[3]

As Stuart Kauffman has shown by numerical simulations, Boolean networks[4] become chaotic if the number of connections per node exceeds a certain threshold, which is an astonishingly small number (i.e. 4–5 connections per node) (Kauffman 1995, 82).[5] Our world exhibits a breathtaking complexity of natural, social, economic and technological systems, which are highly fragile and susceptible to error. In the last 200 years, humankind has accelerated this complexity by processes of networking, of trade, globally spread value chains, and global markets in an unprecedented way, increasing the length of causal chains, the number of interacting system components and feedback loops, hence increasing complexity (see Chapter 8). This has repeatedly led to unexpected and unintended consequences of human interaction with nature.

- A classic example of such unintended system response was caused by the World Health Organization (WHO) when fighting malaria in Borneo in the 1950s. WHO sprayed large amounts of DDT to kill the malaria-causing mosquitoes. However, the DDT not only killed the mosquitoes but also tiny parasitic wasps which had previously been controlled by thatch-eating caterpillars. In the absence of the wasps, the caterpillars ate the thatched roofs which caved in. Moreover, the DDT accumulated in the food chain and poisoned cats, which led to increased rodent populations. Eventually, WHO parachuted thousands of live cats into Borneo to stop the rodent plague (O'Shaughnessy 2008).

- Another example is the deliberate introduction of the cane toad (lat. *Rhinella marina*) from Latin America to Australia in the 1930s for the purpose of biological control of insect pests for agriculture (sugar cane). Due to a lack of predators in the new habitat, the toad was able to spread over large parts of Australia and still cannot be stopped. It occupies an area of more than 1,000,000 km^2 and is advancing across north-west Australia at a rate of more than 100 km each year (Tyler, Wassersug & Smith 2007, 11; Wikipedia 2019). This is an extreme example of an intentional intervention – but humans have unintentionally introduced thousands of species into new habitats through ships' ballast (see WWF 2009).

- Finally, a more recent example of unintended consequences is the German Renewable Energy Act. The Renewable Energy Act from 2000 was enacted to support the roll-out of different kinds of renewable energy technologies throughout Germany, such as solar panels, wind turbines, and bio-fuels. Many great achievements and positive developments followed. It stimulated roll-out of wind power and solar panels across Germany. In the latter category, Germany has the highest per capita installation of solar cells globally. However, it also resulted in a massive expansion of the cultivation of energy crops (e.g. maize and rapeseed) in large monocultures. This intensification of agriculture for energy crops has, however, contributed to biodiversity loss. In the last three decades the biomass of insects in natural reserves in Germany has decreased by up to 80% (BfN 2017, 12). There is meanwhile strong evidence of the trade-off between the energetic use of biomass and biodiversity. The expansion of bioenergy cropland may actually offset positive effects of climate change mitigation for global vertebrate diversity (Hof et al. 2018). The main original intent of bioenergy, i.e. mitigating climate change, is likely to negatively impact biodiversity (Hof et al. 2018, 13294). So while biodiversity is on the one side threatened by climate change itself (Bowler et al. 2017), mitigating climate change by bioenergy might have the same devastating effect on biodiversity. Biodiversity is "likely to suffer severely if bioenergy cropland expansion remains a major component of climate change mitigation strategies" (Hof et al. 2018, 13294).

These are all examples of how best-intentioned policies have had devastating effects. These incidents have certainly increased awareness of the need to consider sustainability issues in their complexity. The German Advisory Council Global Change introduced the concept of syndromes to limit the intricacy of the global issues. With the aid of this concept, the great variety of issues can be reduced to a manageable number of syndromes in three categories: sources, sinks, and development (WBGU 1996). A similar, more recent approach is the nexus concept: the idea to study not isolated phenomena but nexuses of related phenomena. For instance the water-energy nexus looks at the correlation between energy and water usage (see GAO 2009; Spang et al. 2014), the water-energy-food nexus adds the food aspect to it, although it seems that joint consideration of the issues on the level of policy debates does not help much in resolving the question of issue prioritization (Al-Saidi & Elagib 2017, 1131).

In sum, complexity is a barrier to sustainability for a number of reasons:

1. Complexity makes long-term simulation and forecasting much more difficult; by its very nature, sustainability requires long-term thinking and careful action which will both become much more difficult, especially in combination with other barriers (e.g. trade-offs, conflicting interests).
2. Human activity increases natural complexity and has a substantial impact on natural ecosystems (e.g. by introducing invasive species).
3. Things are becoming not only more complex in reality but also more diffi-cult to understand, more complicated; it is not easy to grasp the nature of complex systems if you have not studied them yourself (see 3.1). Even worse, sometimes scientific positions change, which increases insecurity. The media contribute to irritation by capitalizing on David-against-Goliath stories (which are the kind of stories people want), so it is easy to find counter-arguments for almost any topic. This causes insecurity and anxiety and prompts people to fall for populist lines. Simple minds would rather believe that global warming is a Chinese invention than try to understand its true background. This is, in fact, a challenge for society: political deci-sions can only be made by elected politicians who need scientific expertise to understand the fundamental issues; they need to draw conclusions and justify their decisions to the public by means of media which lives on sim-plification. Populism is therefore an adjacent problem because it partly responds to challenges caused by complexity (see 4.3).
4. More complex situations and more complicated subject matters lead to uncertainty with regard to the expected future impact of developments and policy measures. However, this is a challenge for politics because political action depends on three preconditions: some public perception of the prob-lem, a politically favourable environment of power and decision making, and the availability of a solution. However, both the public perception and the support of corresponding policy measures are difficult to sustain if there is uncertainty about the likely effects (SRU 2019, 118; Herweg & Zaharia-dis 2017).
5. Finally, sometimes uncertainty is used to hide or neglect a problem. How can one communicate that complex systems cannot be predicted but that the development of the global climate (which exhibits definitely chaotic elements) can nevertheless be forecast with sufficient accuracy? This diffi-culty is exploited by "climate sceptics" who refer to uncertainties in the sci-entific account to invalidate it. Scientific projections always come with some uncertainty, to be sure, especially for highly complex systems in the distant future, but by focussing attention on the possible sources of uncer-tainty these people have tried to delay respective policies (Sala, Ciuffo & Nijkamp 2015, 320).

Solution perspectives

How to deal with this? How can such complexity be addressed or managed?

1. Learning and experiencing complex systems at early age. We first need to better understand the complex interactions in natural systems (ecosystems research), in complex systems (chaos theory, cybernetics, modelling), in global change research, in sociology, socio-ecological research and many more disciplines. This requires cross-disciplinary collaboration and there are many promising developments for this, although the majority of academic disciplines reside in their silos (see 9.1, on specialism). Understanding complexity is particularly important in education to get used to counter-intuitive system reactions at an early stage.
2. Do not expect that complex systems can be governed. By understanding the components and their relationship with each other, we can at best hope to influence the system in the desired direction.
3. Establish and enforce rules for dealing with complexity: apply only small changes, check outcomes and scale up if appropriate (see Chapter 16).
4. Strengthen the resilience of systems – in society, economy, technology and nature – and make them more resistant to perturbations. The precautionary principle needs to be implemented when new technologies are developed and rolled out (see 13.6).

Notes

1 I have so far only published some first rough ideas of the typology suggested here (Berg 2017).
2 Exergy is a thermodynamic concept and describes the amount of energy in a system which can be used for work.
3 The "butterfly effect" is typical for chaotic systems. However, chaotic behaviour is not restricted to complex systems. Non-linearity can certainly occur in situations which are not really complex. A double pendulum, for instance, also exhibits chaotic behaviour but is a pretty simple mechanical construction, which one would rarely, at least as such, call complex.
4 Boolean networks are networks in which the nodes can have only binary statuses and only connections of AND and OR.
5 I have investigated the implications of networking processes for sustainability in some detail elsewhere (see (Berg 2005, 2008)).

Bibliography

Al-Saidi, Mohammad, & Nadir Ahmed Elagib. "Towards understanding the integrative approach of the water, energy and food nexus." *Science of the Total Environment 574*, 2017: 1131–1139.

Ayres, Robert U. *Eco-Thermodynamics: Economics and Second Law*. Fontainebleau: INSEAD, 1996.

Barabási, Albert-László. *Linked. The New Science of Networks*. Cambridge, MA: Perseus Publishing, 2002.

Berg, Christian. *Vernetzung als Syndrom. Risiken und Chancen von Vernetzungsprozessen für eine nachhaltige Entwicklunge.* Frankfurt: Campus, 2005.

———— "Global networks. Notes on their history and their effects." In *Futurology – The Challenges of the XXI Century*, A. Kuklinski and Krzysztof Pawlowski (eds.), 199–209. Novy Sacz: MiasteczkoMultimedialne, 2008.

———— "Shaping the future sustainably – Types of barriers and tentative action principles." In *Future Scenarios for Global Cooperation – Practices and Challenges (Global Dialogues 14)*, Nora Dahlhaus, and Daniela Weißkopf (eds.), 79–92. Duisburg: Käte Hamburger Kolleg/Centre for Global Collaboration Reserach, 2017.

BfN. *Agrar-Report 2017. Biologische Vielfalt in der Agrarlandschaft.* Bonn: Bundesamt für Naturschutz (Federal Office for Nature Conservation), 2017.

Bhandari, Khagendra P., Jennifer M. Collier, Randy J. Ellingson, & Defne S. Apul. "Energy payback time (EPBT) and energy return on energy invested (EROI) of solar photovoltaic systems: A systematic review and meta-analysis." *Renewable and Sustainable Energy Reviews 47*, 2015: 133–141.

Bowler, Diana E, Christian Hof, Peter Haase, Ingrid Kröncke, Oliver Schweiger, et al. "Cross-realm assessment of climate change impacts on species' abundance trends." *Nature Ecology & Evolution 01*, 2017: 0067.

EU. *Raw Materials Scoreboard.* Luxembourg: European Union, 2016.

European Environment Agency. *Recycling of Municipal Waste.* 29. 11 2018. www.eea.europa.eu/airs/2018/resource-efficiency-and-low-carbon-economy/recycling-of-municipal-waste (Zugriff am 15. 07 2019).

Fraunhofer ISE. "Photovoltaics report." 14. 03 2019. www.ise.fraunhofer.de/content/dam/ise/de/documents/publications/studies/Photovoltaics-Report.pdf (Accessed 15. 07 2019).

Gagnon, Nathan, Charles A. S. Hall, & Lysle Brinker. "A preliminary investigation of energy return on energy investment for global oil and gas production." *Energies 2*, 2009: 490–503.

GAO. *Energy-Water Nexus.* Washington, DC: United States Government Accountability Office, 2009.

Georgescu-Roegen, Nicholas. "The entropy law and the economic process in retrospect." *Eastern Economic Journal 12*, 01. 03 1986: 3–25.

Hall, Charles A. S., Jessica G. Lambert, & Steven B. Balogh. "EROI of different fuels and the implications for society." *Energy Policy 64*, 2014: 141–152.

Hall, Charles A. S., Stephen Balogh, & David J. R. Murphy. "What is the minimum eroi that a sustainable society must have?" *Energies 2*, 2009: 25–47.

Herweg, Nicole, and Nikolaos Zahariadis. "The multiple streams approach." In *The Routledge Handbook of European Public Policy*, Nikolaos Zahariadis, Lauri Buonanno (eds.), 54–63. London and New York: Routledge, 2017.

Hof, Christian, Alke Voskamp, Matthias F. Biber, Katrin Böhning-Gaese, Eva Katharina Engelhardt, et al. "Bioenergy cropland expansion may offset positive effects of climate change mitigation for global vertebrate diversity." *PNAS (Proceedings of the National Academy of Science) 115*, 26. 12 2018: 13295–13299.

Johnson, Steven. *Emergence. The Connected Lives of Ants, Brains, Cities, and Software.* New York: Scribner, 2001.

Kauffman, Stuart. *At Home in the Universe: the Search for Laws of Self-Organization and Complexity.* Oxford: Oxford University Press, 1995.

Meadows, Donella H. *Thinking in Systems. A Primer*, ed. by Diana Wright. White River Junction, VT: Chelsea Green Publishing, 2008.

Murphy, David J., & Charles A. S. Hall. "Energy return on investment, peak oil, and the end of economic growth." *Annals of the New York Academy of Sciences* 1219 (1), 2011: 52–72.

O'Shaughnessy, Patrick. "Parachuting cats and crushed eggs. The controversy over the use of DDT to control malaria." *Public Health Then and Now 98 (11),* 2008: 1940–1948.

Sala, Serenella, Biagio Ciuffo, & Peter Nijkamp. "A systemic framework for sustainability assessment." *Ecological Economics 119,* 2015: 314–325.

Schaub, Georg, & Thomas Turek. *Energy Flows, Material Cycles and Global Development.* Springer International Publishing Switzerland, 2016.

ScientificAmerican. "Will fossil fuels be able to maintain economic growth? A Q&A with Charles Hall." *Scientific American.* 01. 04 2013.

Seitz, Heike, & Oliver S. Kaiser. *Ressourceneffizienz von Windenergieanlagen.* Berlin: VDI Technologiezentrum GmbH, 2014.

Spang, E. S., W. R. Momaw, K. S. Gallagher, P. H. Kirshen, & D. H. Marks. "The water consumption of energy production: An international comparison." *Environmental Research Letters* 9 (10), 2014: 1–14.

SRU. *Demokratisch regieren in ökologischen Grenzen – Zur Legitimation von Umweltpolitik.* Berlin: German Advisory Council on the Environment, 2019.

Tyler, Michael J., Richard Wassersug, & Benjamin Smith. "How frogs and humans interact: Influences beyond habitat destruction, epidemics and global warming.." *Applied Herpetology* 4 (1), 2007: 1–18. doi: https://doi.org/10.1163/157075407779766741.

UNEP. *Recycling Rates of Metals – A Status Report of the Working Group on the Global Metal Flows in the International Resource Panel.* Nairobi: United Nations Environment Programme, 2011.

WBGU (The German Advisory Council on Global Change), *World in Transition: The Research Challenge* (Annual Report 1996), Berlin /Heidelberg /New York: Springer, 1996.

Weißbach, Daniel, G. Ruprecht, A. Huke, K. Czerski, S. Gottlieb, & A. Hussein. "Energy intensities, EROIs (energy returned on invested), and energy payback times of electricity generating power plants." *Energy 52,* 2013: 210–221.

Wikipedia. *Cane toad.* 29. 01 2019. https://en.wikipedia.org/wiki/Cane_toad (Accessed 15. 07 2019).

WWF International, "Silent Invasion – *The spread of marine invasive species via ships' ballast water*", Gland (Switzerland) 2009.

3

BARRIERS RELATED TO THE HUMAN CONDITION

Contents

3.1 Cognitive limitations: linear and unconnected thinking

Unless specifically trained, humans have very limited understanding of exponential growth, long-term developments and complex systems. For the longest time of human history this was no problem because most daily phenomena could be addressed with linear, short-term and unconnected thinking. Today this is no longer the case, which is mainly due to our powerful technologies with their significantly increased impact in time and space. In fact, our cognitive limitations pose a serious challenge to sustainability because they incite us to underestimate the problems or even prevent us from understanding them.

Let's do a thought experiment. How often would you need to fold a sheet of paper (80 g/m^2) until it reaches the ceiling in your office? Of course, you might reply that it's impossible to fold it more than 6 or 7 times. However, let's assume it were possible – or let's rephrase the question into: how often would you need to double the layers of paper, which is the equivalent. As a rough estimation you can do this by mental arithmetic. Let's assume one sheet of paper has a thickness of 1/10 millimetre (500 sheets you can buy in a copy shop measure roughly 5 cm). Every time you fold the paper you double the thickness of the pile. After 10 foldings you reach $2^{10} \times \frac{1}{10} mm = 1024 \times \frac{1}{10} mm \cong 10$ cm, with the 11th folding you get to 20 cm and after 15 foldings you reach 3.20 m. Most people who are not used to exponential growth are amazed about this small number. I frequently ask this question in talks to illustrate the nature of exponential growth. Almost all respondents estimate far too high. Can you imagine what I once heard during a meeting of bank managers? "500,000 times!" That is way too much. After 43 foldings you would reach the moon, the pile would be more than 800,000 kilometres high. Folding 300 times, i.e. 2^{300}, would be a number which is larger than the number of atoms in the universe, which can be estimated to be somewhere in order of $10^{80} \cong 2^{265}$.

What does that tell us? As humans we have difficulties in estimating exponential developments. We are used to linear ones, in which things grow day by day the same amount, doubling the input doubles the effect. All other things kept equal, the time it takes to build ten houses is roughly ten times the time it takes to build just one. As a matter of fact, over short periods of time and at small growth rates, exponential growth can well be approximated by a linear function (see Figure 3.1a).

In the range from zero to 30 or 40 the two graphs are almost identical – at least at the current resolution. However, one of them is a linear one ($g(x)$) and one is an exponential one,[1] only for larger values of x do deviations become noticeable.

As humans we are used to operating in linear contexts. Small changes, small effects, larger changes, larger effects. Growth is likely to be neglected if it takes place at small rates or short periods of time. This is illustrated by Figure 3.1 b–d.

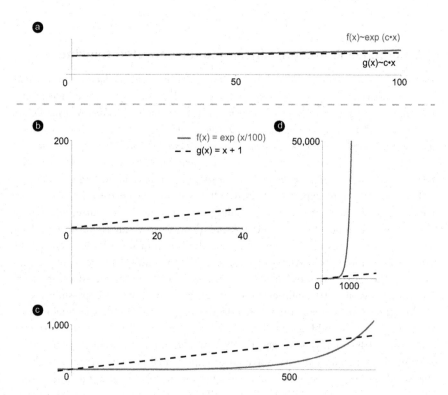

FIGURE 3.1 Linear and exponential growth: a: for small growth rates and limited time frames, linear and exponential developments might appear similar; b–d: all figures show the same two graphs – one linear, the other exponential – with the same scaling of the coordinates, the only difference between the three figures is the represented range (in Figure b the exponential graph is indistinguishable from the x-axis).

(Source: Own illustration)

These three figures apparently exhibit quite different functions. However, Figures b–d display exactly the same two graphs, one linear one and one exponential one, and even the ratio of the axes is the same (i.e. x:y = 1:10). The only difference in the three figures is the intervals shown. In b) the exponential curve is indistinguishable from the x-axis and growth is only visible for the linear function. In c) the exponential curve has picked up and overtakes the linear one, while the linear function is negligible compared to the exponential one for large arguments. Why is all this relevant? Human action is much more likely to be induced by *noticeable changes*, not so much by absolute numbers. This is most evident with much-discussed parameters like economic ones. The wealth of an economy is much less considered than its growth rate (in mathematical terms: the derivative) or even the change in growth rate (second derivative). However, numerous developments in society and the environment today are on paths of exponential growth. Compared to the economic growth rates, they get much less public attention although they are critically important for humanity in the long run. One can easily see the relevance of such exponential growth by looking at some socio-economic and earth system trends which Will Steffen et al. have illustrated (Steffen et al. 2015). Figure 3.2 illustrates some of these trends from 1750 to 2000.

Ever since the first report to the Club of Rome in 1972, *The Limits to Growth* (Meadows et al. 1972), there have been passionate discussions on economic growth, which need to be ignored for the moment. The point to be made here is simply that humans have no intuitive approach to exponential growth, which makes it difficult to argue the urgency of developments whose absolute degree still seems to be manageable.

Since real-world phenomena cannot reach infinity, exponential developments in nature are always confined, which means that different phenomena are correlated and bend the values to finite numbers. A classical example for such a case is the predator-prey model, in which two first-order differential equations are coupled (see Hoppensteadt 2006; Jischa 2018, 59ff.). In the limiting cases where there are either no predators or where there is an infinite pool of prey, each species would grow indefinitely, but the coupling of the two equations keeps both predator and prey within finite limits. The behaviour of many natural systems, like predator-prey or maximum sustainable yield in fisheries, can be modelled similarly.

Under certain conditions, systems with non-linear, mutually reinforcing developments can exhibit chaotic behaviour – which means that their behaviour becomes principally unpredictable and extremely sensitive to the initial conditions. As development within a chaotic regime is by definition not predictable, we have certainly also no intuitional means to assess complexity and chaotic behaviour. Even a simple construct like a double pendulum cannot be predicted in the long run. Given all the great achievements of science and technology, this is something which people can hardly believe. Yet you can say that we are surrounded by chaos everywhere. We prefer to live on islands of order in oceans of chaos.

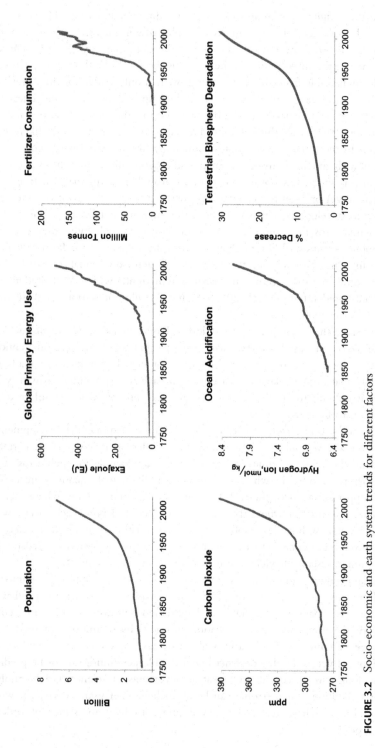

FIGURE 3.2 Socio–economic and earth system trends for different factors

Source: Data taken from Steffen et al., The trajectory of the Anthropocene: The Great Acceleration, January 16, 2015, 81–98 https://doi.org/10.1177/2053019614564785.

One impressive real-world example of the unexpected behaviour of a dynamical system was described in the Millennium Ecosystem Assessment: the collapse of Atlantic cod stocks off the east coast of Newfoundland in 1992, which led to a closure of the fishery after hundreds of years of exploitation (see Figure 3.3).

> Until the late 1950s, the fishery was exploited by migratory seasonal fleets and resident inshore small-scale fishers. From the late 1950s, offshore bottom trawlers began exploiting the deeper part of the stock, leading to a large catch increase and a strong decline in the underlying biomass. Internationally agreed quotas in the early 1970s and, following the declaration by Canada of an Exclusive Fishing Zone in 1977, national quota systems ultimately failed to arrest and reverse the decline. The stock collapsed to extremely low levels in the late 1980s and early 1990s, and a moratorium on commercial fishing was declared in June 1992. A small commercial inshore fishery was reintroduced in 1998, but catch rates declined and the fishery was closed indefinitely in 2003.
>
> *(WRI 2005, 12)*

No biological or ecological system can experience exponential growth over a long period of time. Since all human activities, societies and economies depend upon

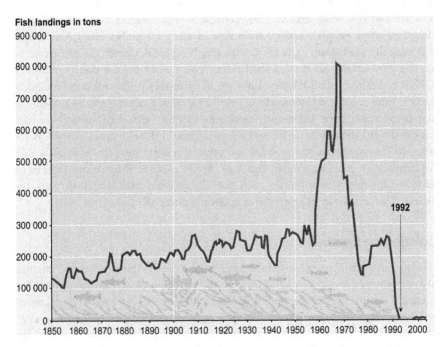

FIGURE 3.3 Fish landings off the Newfoundland coast

Source: Millennium Ecosystem Assessment (2005), *Ecosystems and Human Well-being: Synthesis*, Island Press, Washington, D.C., p. 12.

biological and ecological systems, they cannot experience such growth either. The fact that it was a group of bank managers who suggested such high numbers in the personal anecdote I reported above is somewhat disturbing given that interest rates, however small their percentage might be, lead to exponential growth.

There is yet another aspect of human inability to deal with issues of sustainability which is related to exponential growth: our inability to assess long-term developments, multiple dimensions and interrelations, and feedback loops.

The defining moment of the Anthropocene is that the human influence on the earth has become the most dominant driver for change. This change in the earth's ecosystems, however, is in most cases difficult to experience for an individual; it is difficult to see with your own eyes. Of course, it is possible to draw conclusions regarding changes. For instance, a fisherman can remember that his grandfather told about his catch fifty years ago in which he caught species of fish which he has never caught himself. The natural interpretation of this is that the respective species have disappeared from the specific fishing ground and that this is related to habitat changes – which is, in turn, in some way or another related to human activity. That is all evident and can be (and is) scientifically proven. However, it is a totally different kind of "experience" than escaping from a fire whose heat burns on the naked skin.

Our biological constitution has adapted to the kinds of threats which our ancestors faced in their environments. The phenomena of global change are too abstract and too long-term for the hormones which signal danger to be produced in most people's bodies. The time needed for global changes is simply too long for the human capacity to measure. You cannot *watch* the extinction of species, you cannot *see* the sea level rising, you cannot *feel* the concentration of GHGs increasing. In all relevant cases we need to rely on instruments, measurements, time-sequences, comparisons, etc. That is not a problem in day-to-day life because we have learned to trust our scientists and their results – which makes the situation even more precarious if general distrust in science is sowed. We do not question the validity of science when we buy a medicine in a pharmacy or enter an aircraft. But how do we imagine that we can achieve an emotional, sensory relationship with the changes scientists detect all over the world if we cannot even grasp the simplest phenomena?

Solution perspectives

The key to addressing cognitive limitations is education – this is a truism. Although we might never develop an intuitive sense of the nature of exponential growth and complex systems, we can certainly learn to handle this lack of intuition by substantial knowledge and experience. Getting used to it makes a difference. However, this learning should not start as late as secondary school but as soon as possible. According to Frederic Vester, a German biochemist and a pioneer in cybernetics and networked thinking, it is particularly small children who think in relations and not in classes or categories. Vester argues that relational capabilities are present in children but get lost in a certain school education. In a test series, kindergarten children were asked children questions like: "What is a chair?" and received the answer: " ... when I can sit on it". "What is a house?" – "Where I sleep and mum is there"; summer is "when it's warm and it smells like hay" (Vester 2000, 145). This kind of thinking stops immediately once the kids go to school. A chair becomes furniture, a house becomes a building, and summer is a season. Vester reads this as a transition from a holistic and relational view of the world to a mechanistic one in which disciplinary, linear-causal, often even punctiform thinking becomes cemented.

Forty years ago, a report to the Club of Rome appeared, entitled *No limits to learning. Bridging the human gap* (Botkin, Elmandjra & Malitza 1979). It dealt with the human element in the "predicament of humanity". While the first reports to the Club of Rome had addressed the physical limits to growth, this report dealt with the human capacity to address global challenges, which the authors called the "human gap": "The human gap is the distance between growing complexity and our capacity to cope with it ... Global problems, currently the chief manifestations of complexity, are first and foremost human problems. They are only secondarily attributable to natural causes" (Botkin, Elmandjra & Malitza 1979, 6.7). Of course, many details of this report are time-dependent, but its main intent – to elucidate the need for and the relevance of learning for resolving our major global challenges – remains as true today as it was four decades ago. For Botkin et al., learning goes beyond the narrower literal sense of "gaining knowledge or understanding of or skill in by study, instruction, or experience" (Merriam-Webster 2019); for them it "encompasses the acquisition and practice of new methodologies, new skills, new attitudes, and new values necessary to live in a world of change" (Botkin, Elmandjra & Malitza 1979, 8). Despite the almost ubiquitous availability of information today, our educational practice is far too focused on conveying information, as I see it, and less so on skills, even less on values and orientation capabilities. Learning needs to be understood as "the process of preparing to deal with new situations" (Botkin, Elmandjra und Malitza 1979, 8) – this is more true than ever.

Therefore, in my view, by far the most important measure to address this barrier is education – in a great variety of forms.

a) The basic foundations of our thinking are coined at early age, therefore it is children who should learn about the nature of connected systems as early as possible (Nguyen & Bosch 2014). Vester already developed board games that provide experiential experience with systems behaviour in the 1980s. Later he developed online versions of it, published as "Ecopolicy" (Malik Management 2011). In these games, players can learn to understand regulatory circuits in which social, ecological and economic systems interact. If you are the ruler of the imaginary country Cybernetia ("Kybernetien") you have a certain budget to invest, e.g. in education, production, restoration, quality of life, etc., and you will learn how different policy measures affect different goals.

b) There is so much to improve in our schools, regardless of tight budgets. Only gradually are we realizing the importance of interactive, team-based, collaborative, inter- or transdisciplinary and solution-oriented learning. Interpreting and embedding information, assessing situation and context, drawing conclusions, preparing questions for normative deliberation and seeking solutions acceptable to majorities yet respecting minorities – these are the kinds of skills most readers would not have been trained in at primary or secondary school. However, it is these skills that shape our capability to address the issues of today. Furthermore, the main focus is on the cognitive, much less on the emotional – although we have strong evidence that emotions are especially important in the learning process. An OECD Practitioner guide for education calls emotion the "primary gatekeeper" to learning – for good and for bad (OECD 2012). Given this pivotal role of the learning process, and of the emotions within it, it is quite embarrassing to see the dilapidated state of the average public school or university building, compared with the sales offices of car dealerships or financial institutions. Of course, I know that one can well *explain* these differences (public funding, free education, etc.) – but does that make it any better? What message does our generation bring across to the youth, what image of society do we transmit by that?

c) Education must start with the little ones but should not stop there. Scientists should realize their social responsibility and seek ways to transfer their insights to societal stakeholders. We cannot take it as a matter of course that people are interested in exponential relations and systems thinking. But there are so many fascinating insights to share about science and technology which can help educate broader circles about the correlations on which we all depend.

d) Finally, we must learn to understand and deal with complex systems (see Chapter 16). We must understand that we cannot simply "control" or "govern" them as we can do with linear systems. We need to understand the critical leverage points, the factors that influence several parameters, we need to define criteria for long-term success, and, eventually we will need to learn that complex systems should be handled with care. The best advice in dealing with complex systems is: Hands off!

3.2 Moral limitations: greed, selfishness and ignorance

All religious and cultural traditions have struggled with the limitations of the human condition, its predisposition to selfishness, greed and ignorance, even in light of others' suffering. The moral standards of the great religious and philosophical traditions of antiquity can make one wonder whether progress in the human condition is ever possible. This is and continues to be a challenge for humans, their community and thereby also for any just and sustainable society.

This section deals with humans as moral agents and their deficiencies as a barrier to achieving more sustainability. The activities of NGOs and (environmental) activists do to a large extent address humans as moral agents. Consumers are called to change their purchasing behaviour, assume more frugal lifestyles and celebrate simplicity ("voluntary simplicity"). Everybody knows that there are occasions in which one could have and should have done better. Corruption, greed, personal misconduct and irresponsible behaviour have often contributed to (or even caused) ecological disasters or detrimental social developments (e.g. Exxon Valdez, Enron, financial crises).

Although personal motivation can hardly be assessed by others, there are sufficient cases which suggest morally reprehensible behaviour as a cause for devastating effects on environment and society. When poachers hunt down wildlife only to make a fortune by trafficking the trophies (e.g. Traffic 2017; IAPF 2019), when fishing trawlers stay at sea for months with the crew de facto captive and enslaved, ruining the marine ecosystems, the local fisheries, and the lives of the crew, this is a multiple-party disaster just to benefit influential criminals (EJF 2015); or when electronic waste is illegally shipped from Europe to African or Asian countries to retrieve the contained resources, thereby contaminating large environments and threatening the lives of local people (Huisman et al. 2015) – all this is caused by criminal, illegal and immoral behaviour (apart from wrong incentives, bad governance, default market system and many more systemic issues).

Processes of global influence and public discourse were the circumstances that caused the financial crisis of 2007/2008. The role of greed in this crisis has been intensely discussed (see Reavis 2012). Whereas public opinion might see greed as its main origin, even critics of the prevailing growth paradigm like Tim Jackson argue for a more differentiated view. Discussing the implications and interpretation of the financial crisis in the "age of irresponsibility", Jackson does not see casual oversight or individual greed at its heart:

> The economic crisis is not a consequence of isolated malpractice in selected parts of the banking sector. If there has been irresponsibility, it has been much more systematic, sanctioned from the top, and with one clear aim in mind: the continuation and protection of economic growth.
>
> *(Jackson 2011, 31).*

I do concur with Jackson on the systemic nature of the issue, but this should not prevent us from noticing irresponsible and illegitimate behaviour. The paradigm of "good greed" had been around for too long. As the hero Gordon Gecko in the 1987 movie *Wall Street* states:

> Greed is good. Greed is right. Greed works. Greed clarifies, cuts through and captures the essence of the evolutionary spirit. Greed, in all of its forms – greed for life, for money, for love, knowledge – has marked the upward surges of mankind.

Although it was a movie actor who said this, it is reportedly pretty much in line with the attitude which a real prototype of Gecko, the stock trader Ivan Boesky, displayed in reality. The actions of Boesky and others (including Milken) came to be viewed as emblematic of the greed and excesses often seen as typifying the 1980s on Wall Street. In 1986, prior to his guilty plea, Boesky had given an infamous speech at the University of California extolling the positive aspects of greed, stating that he thought greed was healthy.

The eighteenth-century British aristocrat Baron Rothschild, member of the eponymous family, is credited with saying that "the time to buy is when there's blood in the streets" (Myers 2009). This anti-cyclical behaviour might be common among fund managers – John Templeton was one of the first to exploit this principle of investment at the point of maximum pessimism (Myers 2009) – but it reveals that they literally capitalize on the loss of others. This might be clever in economic terms – but can, and in my view should, be objected to on moral grounds.

Moreover, selfishness and ignorance projected to the national level lead to slogans like "prima Italia" or "America first", wilfully ignoring other peoples' needs and detrimental to any spirit of international, global cooperation which is needed to achieve sustainability (see SDG 17, in UN 2015).

Finally, greed and selfishness also contribute to the perpetuation of unjust, inequitable circumstances, since those in power rather pursue their own interests, which will be explicated later (see 4.5).

Solution perspectives

The question of how to address human deficiencies is as old as humanity. All great religious and philosophical traditions have dealt with it, and their verdict on greed is univocal and unambiguous. Buddha said: "Whatever is not yours: let go of it. Your letting go of it will be for your long-term happiness and benefit" (Saṃyutta Nikāya, 35:101). The Hindu Bhagavad Gita clearly condemns greed as one of three "gates leading to the hell of self-destruction for the soul" (Bhagavad Gita, BG 16.21) (the other two being lust and anger). The Jewish Bible invented the Sabbath year (Deut. 15:1–2), a year in which slaves should be liberated, debts forgiven. The Prophet Jeremiah sharply criticized wrongly acquired profit (Jer. 6:13); the Proverbs of Solomon even say that being "greedy of gain" takes away the owner's life (Prov. 1:19). In the Sermon on the Mount, Jesus talks about the danger of riches and is explicit in his decree on it: You cannot serve both God and mammon (Matt. 6:24). The Quran praises those who transcend greed as those that are successful (Quran, 64v16).

Herman Daly, the father of the Steady-State-Economy, spoke openly about the connection between the requirements of sustainability and religion.

> Sustainable development will require a change of heart, a renewal of the mind, and a healthy dose of repentance. These are all religious terms, and that is no

coincidence, because a change in the fundamental principles we live by is a change so deep that it is essentially religious whether we call it that or not.

(Daly 1996, 201).

Daly points to the linkage between economics and ethics. Often enough it is the system set-up which suggests illegitimate personal behaviour.

Kollmuss and Agyeman (2002) have studied barriers to environmental behaviour. According to them, pro-environmental behaviour is made up of a complex mix of "environmental knowledge, values, and attitudes, together with emotional involvement", embedded in "broader personal values and shaped by personality traits and other internal as well as external factors" (256). The main barriers, on the other hand, are the absence of appropriate incentives as well as old behavioural patterns.

Christian Felber, inventor of the economy for the common good (*Gemeinwohlökonomie*) sees a strong correlation between our economic and our societal values:

> In our friendships and day-to-day relationship we are doing well when we practice human values: building trust, honesty, appreciation, respect, listening, empathy, collaboration, and mutual help and sharing. The "free" market economy is based on system rules, pecuniary reward, and competition. These incentive structures support egoism, greed, parsimony, envy, ruthlessness and irresponsibility … Until today the capitalist market economy has been centrally legitimized by the assumption that the selfishness of individual actors leads to the greatest welfare of all. In my view, however, this assumption is a myth and fundamentally wrong; competition without any doubt incites people to perform … but it does much more harm to society and the relationships among people. If people's ultimate ambition is their own benefit and operating against each other, they learn to obstruct others and to view this as being right and normal. However, by nobbling others we do not treat them as equal value humans. We violate their dignity.
>
> *(Felber 2018, 12, 14).*

3.3 Value-action gap

A particular form of limitation, although closely related to the previous one, is the value-action gap. It deals with the chasm between human aspirations and the banality of everyday life, perfectly described by the Apostle Paul in his letters to the Romans (7,15): "For I don't understand what I am doing. For I do not do what I want – instead, I do what I hate." This phenomenon is closely investigated by psychologists.

The great success of modern science and technology has supported the trust in expert knowledge, which has long dominated also the discourses on environmental and sustainability issues. For quite some time an essentially "top-down" model of communication presumed that deficits in public knowledge and understanding of environmental issues could be addressed by expert knowledge (Burgess, Harrison & Filius 1998, 1446). The perception used to be that the public could be reached and convinced to act in environmentally friendly ways if only the right messaging was used (ibid.). Only by expert knowledge would individuals "accept their own responsibilities and acknowledge the need to change" (ibid.). Such a view, however, is rather pie-in-the-sky: "Hopes are fading of achieving dramatic changes in individual lifestyles simply on the basis of 'more' or 'better' information" (ibid., 1447). Adequate information might be *necessary* to induce behavioural changes within people but it is certainly *not sufficient*.

That people behave differently from what they actually supposedly value most is called the value-action gap or knowledge-action gap. Claude Martin, a Swiss biologist who has dedicated a large part of his life to tropical rainforests, said that he had never met anybody who was indifferent about their future (Martin 2015). Studies have shown that consumers are very much aware of social and environmental issues related to their consumption, for instance "of the issues that workers in developing countries might work under bad working conditions" but do not act accordingly (Joergens 2006, 362; see also Yildiz et al. 2015). Joergens concludes that knowledge apparently does not – at least not as much as is wished – impact action (369). De Groot and Steg argue that short-term individual benefits overshadow long-term collective ones, which is called the "low-cost hypothesis" (de Groot & Steg 2009, 63).

James Blake investigates the value-action gap and concludes that additional information might help in *raising environmental awareness,* but it does *not* suffice to *promote pro-environmental behaviour* (such as lifestyle changes). In that case, social and political institutions would need to be considered as well (Blake 1999). He concludes that "the 'value-action gap' cannot be overcome simply by invoking an 'information deficit' model of participation, informed by a social psychological attitude-behaviour model." Rather, "research suggests that policy must be sensitive to the everyday contexts in which individual intentions and actions are constrained by socioeconomic and political institutions" (Blake 1999, 274).

In its transformation report, the WBGU states that "many people all over the world show a marked sensitivity towards environmental issues" (WBGU 2011, 76), but that the attitudes documented in surveys

> does not mean that a widespread actual rejection of non-sustainable practices has already taken place, and that concrete environmental policy reforms such as, for example, the introduction or increase of eco-taxation, or the introduction of environmental standards, enjoy the unequivocal approval of all citizens.
>
> *(WBGU 2011, 76)*

Survey questions may often be answered hypothetically and the mere perception of problems may not trigger corresponding actions (WBGU 2011, 77).[2]

Drawing on Kollmuss and Agyeman (2002, 242, 247, 249) and Leist (2014), we can summarize the main issues of the value-action gap as follows:

1. People will not exhibit pro-environmental behaviour if they do not have great concern for the environment.
2. People might not feel responsible because they think that their own contribution cannot really influence the situation anyhow: marginality or lack of responsibility (Leist 2014, 398). A special case of this is a position which denies responsibility or the rationalizing of immoral behaviour, which we see, for instance, among "climate sceptics", who downplay the individual's contributions or behaviour, or the credibility of the scientific account (Symmank & Hoffmann 2017).
3. People feel a lack of time, money, or information to adequately respond to the challenges: lack of practicality.
4. People's behaviour depends on the economic consequences of their choices.
5. Direct experiencing is often missing. Seeing a dead fish in a river allows for a stronger correlation between attitude and behaviour than indirect experiencing, such as learning about a topic at school or university. However, most insights about environmental and sustainability issues are only indirectly experienced.
6. There is a normative influence of social norms and cultural traditions which might widen the gap between knowledge and action if these norms propagate unsustainable habits.

It seems evident that the value-action gap contains elements of personal deficiencies as well as lack of institutional support and policy (see Blake 1999; WBGU 2011, 77; Schneidewind & Zahrnt 2014). The transition report of the WBGU calls for a "material and cognitive basis" to achieve an actual change of behaviour:

> The suitable socio-economic and legal framework must also be given ... Frequently, for example, the intention to use environmentally friendly public transport for private journeys and commuting is hampered by a status quo bias, but also by a lack of basic information, (supposed) disadvantages in terms of cost (both monetary and non-monetary), and wrong fiscal incentives, and a lack of infrastructure.
>
> *(WBGU 2011, 77)*

Solution perspectives

How can the value-action gap be addressed? In my view, solution perspectives should take several starting points, since neither reference to values nor price mechanisms nor nudges nor any other single measure will be sufficient alone.

Rather, the greatest chance will lie with approaches that combine a variety of measures, address a variety of different actors and institutional measures. Several of the following concepts will be discussed in other parts of this book.

- Since *price mechanisms* are very effective and efficient ways to trigger people's behaviour, it should be a prime goal to work for the internalization of external social and environmental costs – as we will discuss later (see 5.1). Such measures would not only be effective for most consumers, they would be relatively easy to implement. Moreover, they would relieve the "conscious consumers" of the burden which charges every act of consumption (or abstinence) with a moral imperative. In my view, "ethical consumption" should only be an exception for a limited number of people in a limited number of contexts. One cannot expect that all consumers will be reached by this, and we should not expect that the conscious consumers will change the rules of the game alone. The more sustainable product needs to be the cheaper one.

- The more sustainable products should not only be cheaper than the non-sustainable ones, they should also be *more handy*. Convenience is key – but often more sustainable consumption is more complicated. People's individual behaviour needs to be supported in a way that the more sustainable option is the easier option, e.g. the default option. If you rent an apartment, why should the default power supply be from the standard utility provider instead of a green supplier? Why are office printers not set to duplex printing by default? Such defaults could be demanded by governments. If the government demands that the more environmentally friendly option is the default, they would "stipulate 'good' solutions as the standard, but always include the option of choosing an alternative (opt out)" (WBGU 2011, 78). Such "nudging", which is a strategy of "libertarian paternalism", can help overcome barriers to behavioural changes which are the result of a lack of long-term orientation. When costs and benefits of an action are separated in time, nudges "can help individuals to make their decisions in such a way that the benefit is optimised in the long-term" (WBGU 2011, 78).

- People need to have *positive experiences with or in nature*, ideally in their childhood. This is difficult, since the majority of humankind lives in cities and many kids have had hardly any truly natural experiences. However, empathy, positive experiences and positive feelings are critical for motivating pro-environmental behaviour. You can only protect what you love, and you will only love what you know. Therefore, it is critical to teach the appreciation and understanding of nature very early on (see 13.7).

- Although *education* alone is not sufficient for a change in behaviour, it is nevertheless critical for better understanding of the challenges, and it is critical that this education does not stay on the cognitive level only but also touches the *emotional aspects* (see 13.7). Kollmuss points to a gender difference in this contexts: "Women usually have a less extensive environmental knowledge than men but they are more emotionally engaged, show more concern about environmental destruction, believe less in technological solutions, and are more willing to change" (Kollmuss & Agyeman 2002, 248).

- It is important that people *know how to respond by concrete action*. They need to know what they can do to address the issue observed. This is challenging since not all issues can be addressed by individual action. However, people need to have a guide on what is more and what is less sustainable. Motivation can deteriorate if it turns out that the adopted behaviour is actually not improving (or even worse: exacerbating) the situation. Cynicism and apathy might be the consequence then: "Apathy and resignation are often the result of a person feeling pain, sadness, anger, and helplessness at the same time. If the person has a strong feeling that he or she cannot change the situation (see locus of control), he or she will very likely retreat into apathy, resignation, and sarcasm" (Kollmuss & Agyeman 2002, 255). It is therefore critical that consumers get advice about the right principles for sustainable action, which we will discuss in the second part of the book.

- Furthermore, more sustainable, more pro-environmental behaviour is also a *matter of culture and habits*. Like any other habit, sustainable consumption needs *training and exercise*. We need to develop a culture of sustainable consumption. Just think of how smoking habits have changed in the last two decades in many countries around the globe. While smoking used to be "cool" in the 1970s, the general perception has totally changed. In a similar way, our ways of consumption can change. "Consumer" could become a swear word – because those who consume take away something which is no longer available for use by others (see Chapter 11). What do we need to do? Kollmuss is pretty clear about this: "If we want to establish a new behaviour, we have to practice it" (Kollmuss & Agyeman 2002, 256).

- Last but not least, the *practicing and teaching of moral behaviour* by inspirational leaders will also be critical – however limited the motivational dynamics of ethical considerations might be. As Gadenne et al. (2011) argue, obligation, feelings of guilt, sense of social responsibility could motivate environmental behaviour: "A feeling of moral obligation is a considerable behaviour motivator; this includes pro-environmental behaviour", "highly environmentally orientated consumers were strongly influenced by the belief that there could be control over a situation" (Gadenne et al. 2011, 7686).

3.4 Trade-offs

Trade-offs might be the most severe barrier to sustainability because they will always be there when conflicting goals are to be reached. There are trade-offs between social goals and environmental goals, between poverty reduction and nature conservation, between short-term profit and long-term viability.

These trade-offs can only be addressed by comprehensive views on the respective system as a whole, by understanding the relationships within the system and by transparency on the stakes which the critical stakeholders have. I subsume trade-offs under barriers related to human condition because trade-offs ultimately stem from the multiple demands and wishes we as humans have.

Maybe the most difficult sustainability barrier of all are trade-offs. Trade-off means "a balancing of factors all of which are not attainable at the same time" (Merriam-Webster 2019). Ultimately, trade-offs result from our multiple needs and wants; we aspire to several different things at the same time, we follow conflicting goals. Is it more sustainable to invest in protecting the rainforest or to alleviate the hunger of people in need? How do we balance between short-term and long-term needs – and what *is* long-term, 20 years, 200 years, or 20,000 years? Final storage facilities for nuclear waste will have to be safe for 1,000,000 years – but it will be hopeless to consider the needs of all generations until then. Trade-offs occur everywhere, at various levels, in various fields, for various kinds of actors. How does an employee manage the right balance between private life and business life? How do we as a society balance the interests of the present versus future generations? How do corporations balance different stakeholder expectations and find the suitable path between short-term profitability and long-term success? Due to the nature of the issue, trade-offs imply that there is no well-defined optimum solution, no best choice you can make. There will always be trade-offs between different aspects of sustainability, different timescales, different regions, actors, etc.

Santoyo-Castelazo and Azapagic compare different studies which look at scenarios for electricity generation (Santoyo-Castelazo & Azapagic 2014). The scope of the studies, their aims, the number and kind of indicators considered, the time horizon, and the question whether a life-cycle perspective is included or not – all these questions can be and are being looked at differently. It is not surprising that even in this relatively limited task – compared to the vastness of sustainability demand in general – it is not possible to identify any optimal scenario: "there is no overall 'best' scenario, as each option is better for some sustainability criteria but worse for others" (133). How much more difficult will the situation get if you confront 17 SDGs with 169 targets? The trade-offs among the 17 SDGs are presumably the most difficult barrier to sustainability – for the rationality of the goals or targets can hardly be questioned – *as long as they are considered each on its own.* But this is certainly not possible. There are serious concerns about the trade-offs in the SDGs. Some people even think that the trade-offs within the SDGs prevent their realization. As quoted above, a study by the Institute for Advanced Sustainability Studies does even conclude that the SDGs "cannot be met sustainably" (IASS 2015, 4).

Scherer et al. explored the trade-offs between the social and the environmental goals within the SDGs and found that

> pursuing social goals is, generally, associated with higher environmental impacts. However, interactions differ greatly among countries and depend on the specific goals ... Although efforts by high- and low-income groups are

needed, the rich have a greater leverage to reduce humanity's footprints. Given the importance of both social and environmental sustainability, it is crucial that quantitative interactions between SDGs be well understood so that, where needed, integrative policies can be developed.

(Scherer et al. 2018, 65)

Machingura and Lally investigated the trade-offs among the SDGs and identified three trade-offs as particularly relevant for the overall achievement of the SDGs:

How can ending hunger be reconciled with environmental sustainability? (SDG targets 2.3 and 15.2) How can economic growth be reconciled with environmental sustainability? (SDG targets 9.2 and 9.4) How can income inequality be reconciled with economic growth? (SDG targets 10.1 and 8.1).

(Machingura & Lally 2017, 9)

In a special report on the impacts of global warming of 1.5°C "in the context of strengthening the global response to the threat of climate change, sustainable development, and efforts to eradicate poverty", the IPCC lists several trade-offs as well as synergies between the SDGs and climate mitigation policies. Synergies are, as it were, the good antagonists of trade-offs:

Mitigation options deployed in each sector can be associated with potential positive effects (synergies) or negative effects (trade-offs) with the Sustainable Development Goals (SDGs). The degree to which this potential is realized will depend on the selected portfolio of mitigation options, mitigation policy design, and local circumstances and context. Particularly in the energy-demand sector, the potential for synergies is larger than for trade-offs.

(IPCC 2018, 22)

The good news is that the IPCC report sees even more possible synergies than trade-offs between mitigation options consistent with 1.5°C pathways across the SDGs, but "their net effect will depend on the pace and magnitude of changes, the composition of the mitigation portfolio and the management of the transition" (IPCC 2018, 21) – in other words, it is not certain whether the synergies or the trade-offs will dominate the relationship between mitigation options consistent with 1.5°C pathways and the SDGs.

The question of how trade-offs are to be dealt with is closely related to the underlying concept of sustainability. If human capital, infrastructure, labour and knowledge can function as a substitute for natural capital, biodiversity and other "ecosystem services", in other words: in a weak concept of sustainability, trade-offs are much easier to resolve (Sala, Ciuffo & Nijkamp 2015, 316). However, there are certainly limits to substitutability – extinct species are lost forever and the ecological damage and ultimately the impact on "ecosystem services" can hardly be estimated.

Solution perspectives

Is there any hope of resolving trade-offs? Not really of resolving, but of addressing and mitigating as some of the following suggestions indicate.

1. *Study background of trade-offs*

The first condition for addressing trade-offs is, of course, a *sound understanding* of the great variety of trade-offs between different sustainability goals, e.g. the 17 SDGs and their 169 targets. Machingura and Lally conclude a study on the SDGs and their trade-offs by stating: "Without better knowledge of the nature of trade-offs and the factors that shape them, our understanding of the whole process of sustainable development over to 2030 is bound to be limited" (Machingura & Lally 2017, 63). The "ability to make a wise choice regarding trade-off is one of the most important yet challenging skills for policy-makers" (ibid.).

Such an understanding of the trade-offs is only possible by integrated, systemic views and policies (see Chapters 9 and 16), which is the *second condition* for addressing trade-offs.

2. *Seek integrated, systemic views and policies*

Obersteiner et al. investigated the land resource-food price nexus of the SDGs and see trade-offs between land usage and hunger/poverty at the centre of the SDG trade-offs:

> SDG policy formulation at national, regional, and international scales should be more inclusive: Policy options developed by sectorial and technical specialists must also be subjected to assessments of total system effects outside the bounds of their silos. Based on the results of these assessments, strategies for SDG implementation can be classified as incoherent, neutral, or coherent.
>
> *(Obersteiner et al. 2016, 5)*

Drawing on their results the authors demand more inclusive policies which take an integrated systems perspective. If this does not happen, the result is problem-shifting and "potentially magnifying the challenges facing sustainable development agendas. In the worst cases, incoherent strategies could put many of the SDG objectives out of reach by 2030" (Obersteiner et al. 2016, 5).

3. *Identify critical drivers*

A third condition for addressing trade-offs is to identify the critical drivers. Obersteiner et al. could show in their quantitative assessment that land system interdependencies are more significant determinants of joint environmental and food security outcomes than are population and economic growth scenarios. This suggests that mounting trade-offs are not our demographic destiny but rather the predictable consequence of siloed policies, initiatives, and choices accreting into incoherent SDG strategies.

(Obersteiner et al. 2016, 5)

The authors argue that

coherent SDG strategies are those that minimize trade-offs between the land and food systems. In many countries, future demand for meat and animal products will have a major impact on resource availability and food security trends. In developed economies, shifts away from these land- and water-intensive commodities ... can also reduce the health-related costs of overconsumption, including mortality. At the same time, such a shift would decrease food prices in developing countries, reduce mortality and deforestation, and enable progress toward food security for all (goal 2). In the same way, investments in agricultural resource efficiency, spoilage prevention, and waste mitigation can reduce land system pressure and minimize the overall costs of SDG strategies.

(Obersteiner et al. 2016, 5)

Policies for sustainable consumption and production (SCP) would be a sound basis for coherent SDG strategies. "Even when trade-offs between coequal goals cannot be eliminated entirely, SCP policies allow policymakers to manage competing pressures proactively and create simultaneous solution spaces for the largest possible number of SDGs" (Obersteiner et al. 2016, 6).

4. *Seek synergies*

The fourth condition is: seek synergies and multiple-win (see 15.2). Comprehensive scenarios can help address several SDGs at once, for there are not only trade-offs between goals, there are also synergies:[3] "Enact holistic and integrated policies which cut across sectoral boundaries and exploit synergies" (Machingura & Lally 2017, 64).[4]

Dusseldorp closely investigated trade-offs in sustainability (Dusseldorp 2017). His conclusion is that there is hardly any systematic advice how to address or even resolve these trade-offs. The most-often heard advice is to seek win-win situations, but it would remain unclear whether this will ever be possible (145). Moreover, he argues that you cannot simply compensate deficits in one area by overfulfilling another one. Dusseldorp argues that sustainability norms are not rules for optimizing but rules for satisfying, since certain minimum standards must be met to ensure sustainability (184). Only a weak concept of sustainability would allow that losses in one area can be compensated by gains in another.

In their 2018 special report on the 1.5°C goal, the IPCC highlights the importance of questions of morality and justice for addressing the impacts but also for mitigation and adaptation strategies:

> The consideration of ethics and equity can help address the uneven distribution of adverse impacts associated with 1.5°C and higher levels of global warming, as well as those from mitigation and adaptation, particularly for poor and disadvantaged populations, in all societies (high confidence).
>
> *(IPCC 2018, 20)*

Furthermore, the IPCC sees potential in measures of Carbon Dioxide Removal (CDR) strategies in agriculture, forestry and other land use. Such measures

> as restoration of natural ecosystems and soil carbon sequestration could provide co-benefits such as improved biodiversity, soil quality, and local food security. If deployed at large scale, they would require governance systems enabling sustainable land management to conserve and protect land carbon stocks and other ecosystem functions and services (medium confidence).
>
> *(IPCC 2018, 19)*

The IPCC sees

> robust synergies particularly for the SDGs 3 (health), 7 (clean energy), 11 (cities and communities), 12 (responsible consumption and production) and 14 (oceans) (very high confidence). Some 1.5°C pathways show potential trade-offs with mitigation for SDGs 1 (poverty), 2 (hunger), 6 (water) and 7 (energy access), if not managed carefully (high confidence) … 1.5°C pathways that include low energy demand … low material consumption, and low GHG-intensive food consumption have the most pronounced synergies and the lowest number of trade-offs with respect to sustainable development and the SDGs (high confidence). Such pathways would reduce dependence on CDR. In modelled pathways, sustainable development, eradicating poverty and reducing inequality can support limiting warming to 1.5°C (high confidence).
>
> *(IPCC 2018, 21)*

5. *Involve all relevant stakeholders*

Finally, it is of course, decisive to involve the relevant stakeholders in all steps from problem definition towards solutions. Sala et al. conclude this in their study on sustainability assessments:

> In our opinion, a very specific requirement of sustainability assessment is the stakeholder's involvement (including the 'broad participation' principle). It

should be embedded in all steps ... in a trans-disciplinary setting, leading to a co-production of knowledge from problem definition towards solutions.

(Sala et al. 2015, 318).

This matches well the principle "engage the stakeholders" which we elaborate on below (see 15.5).

In his commendable book, *Systems Thinking for Social Change*, David P. Stroh discusses what can be done in case of competing goals: "Look for a higher goal that encompasses the competing ones. If achievement of both goals is mutually exclusive, commit to one. If not, determine different corrective actions that lead to the accomplishment of both goals" (Stroh 2015, 155). This requires, of course, that you have involved all relevant stakeholders (see Stroh 2015, 79ff.) because only then can you decide whether or not you can "commit to one".

Only if all stakeholders are involved can they engage in an open dialogue, defend their legitimate interests and negotiate a common solution, only then do we have a chance for addressing trade-offs among global sustainability goals in any fair and just manner. This is why I view *increased transparency* as such a fundamental principle for sustainability (see 16.3).

Notes

1 Technically speaking $f(x) = \exp(x/100)$ and $g(x) = x/100$.
2 In the context of the value-action gap, the WBGU report also discusses "lack of long-term orientation, loss aversion, and path dependencies in general" as barriers "which make it more difficult for certain individuals or social environments to make sustainable decisions" (WBGU 2011, 78) – we skip these aspects here since they are addressed in dedicated sections elsewhere in this book.
3 The International Council for Science (ICSU) drafted a framework for the relationship of the SDGs, ranging from direct opposition, i.e. the respective goals would cancel each other out, to the strongest possible interaction, i.e. inextricable linkage (ICSU 2016).
4 In such a way, I outlined a very high-level and preliminary example centred around biomass production earlier (Berg 2015). The proposal contains three elements: 1. Restoring and rehabilitating eroded land and regaining it for biomass production, 2. Utilizing the high solar radiation these areas often exhibit for production of renewable energy, 3. Turning both previous objectives into economic benefits and developing infrastructure and industrial ecology parks. Of course, the devil is in the details and there would certainly be trade-offs at more granular levels. However, the overall scenario combines ecological gains (carbon sink, water cycles, biodiversity ...) and social development (economic development, infrastructure, jobs, education ...). I submit that such a scenario has the potential to directly address 9 of the 17 SDGs, potentially more with secondary effects.

Bibliography

Berg, Christian. "Desert2Eden – Integrated restoration and development of arid regions to address the Sustainable Development Goals (SDGs)." In *GESolutions. Global Economic Symposion 2015*, Kiel Institute for the World Economy (eds.), 89–93. Kiel: Kiel Institute for the World Economy, 2015.

Bhagavad Gita. "Bhagavad Gita – The Song of God." Ed. by Swami Mukundananda. 2019. www.holy-bhagavad-gita.org/chapter/16/verse/21 (Accessed 15. 07 2019).

Blake, James. "Overcoming the 'value-action gap' in environmental policy: Tensions between national policy and local experience." *Local Environment. The International Journal of Justice and Sustainability 4*, 1999: 257–278.

Botkin, James W., Mahdi Elmandjra, & Mircea Malitza. *No Limits to Learning. Bridging the Human Gap*. Oxford, New York, Toronto, Sydney, Paris, Frankfurt: Pergamon Press, 1979.

Burgess, Jacquelin, Carolyn M. Harrison, & Petra Filius. "Environmental communication and the cultural politics of environmental citizenship." *Environment and Planning 30*, 1998: 1445–1460.

de Groot, Judith M., & Linda Steg. "Mean or green: Which values can promote stable pro-environmental behavior?" *Conservation Letters 2*, 2009: 61–66.

Daly, Herman E. *Beyond Growth*. Boston, MA: Beacon Press, 1996.

Dusseldorp, Marc. *Zielkonflikte der Nachhaltigkeit. Zur Methodologie wissenschaftlicher Nachhaltigkeitsbewertungen*. Wiebaden: Springer, 2017.

EJF. *Thailand's Seafood Slaves. Human Trafficking, Slavery and Murder in Kantang's Fishing Industry*. London: Environmental Justice Foundation, 2015.

Felber, Christian. *Gemeinwohlökonomie*. München: Piper, 2018.

Gadenne, David, Bishnu Sharna, Don Kerr, & Tina Smith. "The influence of consumers' environmental beliefs and attitudes on energy saving behaviours." *Energy Policy 39 (12)*, 2011: 7684–7694.

Hoppensteadt, Frank. "*Predator-prey model.*" 2006. www.scholarpedia.org/article/Predator-prey_model (Accessed 15. 07 2019).

Huisman, J., I. Botezatu, L. Herreras, M. Liddane, J. Hintsa et al. *Countering WEEE Illegal Trade (CWIT) Summary Report*. Lyon, 2015.

IAPF. International Anti-Poaching Foundation. 01 2019. www.iapf.org/(Accessed 31. 01 2019).

IASS. *The Role of Biomass in the Sustainable Development Goals: A Reality Check*. Potsdam: Institute for Advanced Sustainability Studies (IASS), 2015.

ICSU. *Working Paper "A Draft Framework for Understanding SDG Interactions"*. Paris: International Council for Science, 2016.

IPCC. *Global Warming of 1.5 °C*. Switzerland: IPCC, 2018.

Jackson, Tim. *Prosperity without Growth. Economics for a Finite Planet*. London: Earthscan, 2011.

Jischa, Michael F. *Dynamik in Natur und Technik*. Munich: oekom, 2018.

Joergens, Catrin, "Ethical fashion: myth or future trend?" *Journal of Fashion Marketing and Management 10 (3)*, 2006: 360–371.

Kollmuss, Anja, & Julian Agyeman. "Mind the gap: Why do people act environmentally and what are the barriers to pro-environmental behavior?" *Environmental Education Research 8 (3)*, 2002: 239–260.

Leist, Anton. "Why participate in pro-environmental action?" *Analyse & Kritik 2*, 2014: 397–416.

Machingura, Fortunate, & Seven Lally. *The Sustainable Dvelopment Goals and their Trade-offs*. London: Overseas Development Institute, 2017.

Malik Management. "Ecopolicy: Learning to think interconnected." 21 01 2011. www.youtube.com/watch?v=mtJCh2xBw1o (accessed 07 15, 2019).

Martin, Claude. *On the Edge. The State and Fate of the World's Tropical Rain Forrests*. Vancouver: Greystone Books, 2015.

Meadows, Donella H., Dennis L. Meadows, Joergen Randers, & William W. III Behrens. *The Limits to Growth. A Report for the Club of Rome's Project on the Predicament of Mankind.* London: Earth Island Ltd., 1972.

Merriam-Webster. Consume. 2019. www.merriam-webster.com/dictionary/consume (Accessed 27. 02 2019).

Myers, Daniel. "Buy when there's blood in the streets." 23. Feb 2009. www.forbes.com/2009/02/23/contrarian-markets-boeing-personal-finance_investopedia.html (Accessed 15. 07 2019).

Nguyen, Nam C., & Ockie J.H. Bosch. "The art of interconnected thinking: Starting with the young." *Challenges 5*, 2014: 239–259.

Obersteiner, Michael, Brian Walsh, Stefan Frank, & Peter Havlík. "Assessing the land resource–food price nexus of the sustainable development goals." *Science Advances 2 (9)*, 1-10 September 2016.

OECD. *The Nature of Learning. Using Research to Inspire Pracitce.* OECD, 2012.

Quran. https://quran.com/. 2016. https://quran.com/64 (Accessed 15. 07 2019).

Reavis, Cate. "The global financial crisis of 2008: The role of greed, fear, and oligarchs." *MIT Sloan Management 16*, March 2012.

Sala, Serenella, Biagio Ciuffo, & Peter Nijkamp. "A systemic framework for sustainability assessment." *Ecological Economics 119*, 2015: 314–325.

Saṁyutta Nikāya. https://www.dhammatalks.org/. 2019. www.dhammatalks.org/suttas/SN/SN35_101.html (Accessed 15. 07 2019).

Santoyo-Castelazo, E., & A. Azapagic. "Sustainability assessment of energy systems: Integrating environmental, economic and social aspects." *Journal of Cleaner Production*, June 2014: 119–138.

Scherer, Laura, Paul Behrens, Arjan de Konig, Reinout Heijungs, Benjamin Sprecher,, et al. "Trade-offs between social and environmental sustainable development goals." *Environmental Science and Policy 90*, 2018: 65–72.

Schneidewind, Uwe, & Angelika Zahrnt. *The Politics of Sufficiency. Making it Easier to Live the Good Life.* Munich: oekom, 2014.

Steffen, Will, Wendy Broadgate, Lisa Deutsch, Owen Gaffney, & Cornelia Ludwig. "The trajectory of the anthropocene: The great acceleration." *The Anthropocene Review 2*, 2015: 81–98.

Stroh, David Peter. *Systems Thinking for Social Change.* White River Junction, VT: Chelsea Green Publishing, 2015.

Symmank, Claudia, & Stefan Hoffmann. "Leugnung und Ablehnung von Verantwortung." In *Handbuch Verantwortung*, Ludger Heidbrink Claus Langbehn & Janina Loh (eds.), 949–972. Wiesbaden: Springer VS, 2017.

Traffic. *Pendants, Power, and Pathways. A Rapid Assessment of Smuggling Routes and Techniques used in the Illicit Trade in African rhino horn.* Pretoria: Traffic (Strategic Alliance of WWF and IUCN), 2017.

Vester, Frederic. *Die Kunst vernetzt zu denken.* Stuttgart: Deutsche Verlags-Anstalt, 2000.

WBGU. *World in Transition: A Social Contract for Sustainability.* Berlin: German Advisory Council on Global Change (WBGU), 2011.

WRI. *MEA – Millenium Ecosystem Assessment.* Washington, DC: World Resources Institute, 2005.

Yildiz, Özlem, Caterina Herrmann-Linß, Katja Friedrich, Carsten Baumgarth. *Warum die Generation Y nicht nachhaltig kauft, Working Paper No 85.* (11/2015) Berlin: Berlin School of Economics and Law, 2015, doi: 10.13140/RG.2.1.5071.8808

4

BARRIERS RELATED TO SOCIAL REALITY

Contents

4.1 System inertness and path dependencies

This section mirrors, as it were, the theme of the book as a whole. While the book asks how a general transition towards sustainability can be facilitated and which barriers need to be overcome, the following section looks at system inertness and path dependencies on a more granular level.

Every system change requires path dependencies to be overcome. These can be construed as positive feedback loops in a systems theoretical perspective (Gößling-Reisemann 2008, 154). The respective systems can be diverse in kind: technological systems, social systems, government bureaucracies, institutions, networks, etc. An interesting example of path dependency is the success of the QWERTY keyboard for type machines (Stamp 2013). Having an alphabetical ordering in mind in the first place, the inventor Christopher L. Sholes realized that this initial set-up caused a lot of jams in the typing process. When he finally fixed the patent for his keyboard in 1878, this was more a result of trial and error than ergonomic consideration. After the initial roll-out of the QWERTY keyboards, it became literally impossible to change the design again. All typists would be trained on those machines which most companies had – a positive feedback loop was established, which becomes harder to change the longer it lasts, as economies of scale come into play (Gößling-Reisemann 2008, 154).

Which of the different alternatives will win the competition among options is often very much dependent on slightly differing initial conditions in this phase, the process is chaotic (155). Once a certain realization has ruled out its competitors, the system has reached a lock-in effect, which means that there is basically only one option left to follow. Such systems can only be changed by radical external measures, since system-internally there is no need for change.

Ayres deliberated the lock-in effects for coal and for combustion engines back in 1994. There is "a tendency for suboptimal choices to get 'locked in' by widespread adoption". Large investments in coal technology determine corresponding cumulative quantities of carbon dioxide:

> The adoption of catalytic converters for automotive engine exhaust is another case in point. This technology is surely not the final answer ... Yet it has deferred the day when internal combustion engines will eventually be replaced by some inherently cleaner automotive propulsion technology. By the time that day comes, the world's automotive fleet will be two or three times bigger than it might have been otherwise, and the cost of substitution will be enormously greater.
>
> *(Ayres 1994)*

Twenty-five years later, it is the same industries which are struggling with decarbonizing: the energy and the transportation sector. In both industries, the dominant players have huge amounts of physical assets which need to be depreciated. The reluctance of the car industry to invest in new propulsion technologies relates to this lock-in effect. The transition from combustion engine to other forms of propulsion requires such heavy investments in infrastructure, factories, technology, etc., that a gradual transformation from one propulsion technology to another seems impossible. Change is difficult and arduous. In his milestone book, *The Innovator's Dilemma*, Clay Christensen explains why great companies can fail despite, or rather *because* they are trying to please their customers. For instance, in order to meet customers' demands and expectations, investments will ensure that incremental (Christensen calls them "sustaining") improvements satisfy the customer but no disruptive technologies can evolve; growth expectations by various stakeholders require that profitability is delivered (which is not possible in early phases of new technologies) and because "products whose features and functionality closely match the market needs today often follow a trajectory of improvement by which they overshoot mainstream market needs of tomorrow" (Christensen 2016, xxvii) – in other words: it is the rigidity of the system, its inertness which hampers change.

Our current social, political, technological and economic systems and institutions cannot claim to be sustainable. The respective systems were established, however, over several decades, if not centuries; they have established routines and processes, they involve a multitude of different actors with specific demands, needs, concerns and wishes. Max Weber already knew that bureaucracy can hardly be changed:

> Once fully established, bureaucracy is among those social structures which are the hardest to destroy ... Where administration has been completely bureaucratized, the resulting system of domination is practically indestructible ... Such an apparatus makes 'revolution', in the sense of the forceful

creation of entirely new formations of authority, more and more impossible – technically, because of its control over the modern means of communication (telegraph, etc.), and also because of its increasingly rationalized inner structure.

(M. Weber 1978, 987, 989)

Anybody who has experienced a large-scale restructuring process within one organization can imagine what Weber wrote about. Organizational change, the merging of two organizations, or changing long-standing habits in traditional institutions often requires enormous efforts – even if it only involves *one* system. Change management requires that the people affected are convinced about the upcoming change and get truly involved ((Marshak 2005, 22), see 15.5).

Solution perspectives

As this section resembles in a way the overall approach of the book, much of what is said here can be utilized for transition processes on smaller scales.

- *Build a vision* of the aspired ideal solution or state.
- *Do not underestimate the complexity of the issue.* Effectively operating systems are difficult to change the larger, more impactful and more established they are. Very carefully considered action is needed on several levels. The German energy transition (*Energiewende*) by which Germany is phasing out nuclear energy and supporting renewable energy technologies, exemplifies the degree of complexity of political, administrative, juridical, technological and societal issues which most people have apparently underestimated.
- *Address the barriers of change and key leverage points.* A profound understanding of the system's key leverage points is needed, which includes the knowledge of barriers and incentives for a transformation, and the critical actors, as well as a provisioning of guidance for those actors so that they can facilitate change.
- *Involve critical stakeholders.* Change requires involving the people affected – and in the case of the necessary sustainability transformation this basically means the whole of society. It sounds trivial but is critical: the more people support the idea, support the transformation, the easier change will become possible. This also requires actors on different levels: governmental organizations as well as non-governmental organizations, individuals as well as organizations. Cases in which systemic change has functioned from the top down are rare. China's turn to the market economy under Deng Xiaoping is probably one of the exceptions. Systemic change will mostly involve many actors, require substantial efforts and will take a long time, backlashes included. However, if people from the bottom up and actors of different kinds realize that change is needed and a sufficiently large number of agents act accordingly, then all of a sudden change becomes possible. A most

impressive example for such trigger points is the fall of the Iron Curtain in the autumn of 1989 in the former German Democratic Republic.[1]

- *Prevent lock-in effects by creating positive incentives for change.* Once a coal plant starts operating, several decades of GHG emissions are guaranteed. Functioning as a carrot for the better is more promising than the stick of proscriptions. Anthony Giddens, in his book, *The Politics of Climate Change*, argues for such positive incentives on the level of global politics: "A crucial lever in world politics will be supplied if it can be shown that those countries, regions or localities that follow a progressive programme gain economic advantage by so doing. Such a programme can be both technological and social" (Giddens 2009, 222).

4.2 "Meeting the needs of the present"

The Brundtland definition calls for "meeting the needs of the present". This is not only a humanitarian demand but also a precondition for future sustainability and for nature preservation.

4.2.1 Demand for sustainability starts with the present needs

Sustainability is not only a concept for the future – the call for sustainability starts here and now, as was emphasized by all major milestones in the discourse. The Brundtland definition begins with the call for meeting the needs of the present. "Needs" are specified to be "in particular the essential needs of the world's poor, to which overriding priority should be given" (WCED 1987). "A world in which poverty and inequity are endemic will always be prone to ecological and other crises" (WCED 1987, ch. 2, Section 4). The Millennium Development Goals (MDGs) set out to improve the situation of the poor have been partly successful, e.g. by halving the proportion of people living on less than US$1.25 a day (the "absolute poor") already five years ahead of time (PEP 2016, 6). However, "extreme poverty is actually increasing again in sub-Saharan Africa" (Oxfam 2019, 11):

> Poverty is on the rise in several countries in Sub-Saharan Africa, as well as in fragile and conflict-affected situations. In many countries, the bottom 40 percent of the population is getting left behind; in some countries, the living standard of the poorest 40 percent is actually declining.
>
> *(Worldbank 2018, xi)*

A new phenomenon is that "most of the world's poor now live in middle-income countries" (Worldbank 2018, xi). Finally, the SDGs also belong to this tradition and highlight the importance of eradicating poverty and hunger in their first two goals.

Therefore, any development which claims to be sustainable will have to include the present social issues of poverty, hunger, social instabilities and military conflicts. Addressing these issues is a humanitarian demand and does not need any reference to sustainability. However, at the same time it is a prerequisite for providing the basis on which the needs of future generations can also be considered and taken care of. It is a precondition for ecological integrity, for the preservation of natural environments. People in the global North need to realize that they cannot protect the earth's ecosystems for their grandchildren if billions of people in the global South struggle to feed their children today.

In his encyclical "*Laudato Si*", Pope Francis pronounced "that a true ecological approach *always* becomes a social approach; it must integrate questions of justice in debates on the environment, so as to hear *both the cry of the earth and the cry of the poor*" (Pope Francis 2015, 35; original emphasis). This commitment to the "cry of the poor" must not be sacrificed lest the very idea of sustainability become cynical.

4.2.2 Poverty as multi-dimensional phenomenon

The question is, of course, what the needs of the poor are and how poverty can be measured. There is general consensus that poverty is a multi-dimensional issue. Jeffrey Sachs lists six forms of capital that the absolute poor lack: human capital, business capital, infrastructure, natural capital, public institutional capital, and knowledge capital (Sachs 2005, 244f.). Extreme poverty should be understood as "the inability to meet basic human needs for food, water, sanitation, safe energy, and a livelihood" (Sachs 2015, 30). The UN Development Program issued an updated multi-dimensional poverty index (MPI), which gives consideration to the importance of the multi-dimensional approach to poverty eradication within the 2030 Agenda for Sustainable Development. The 2018 MPI will help to better measure progress against SDG 1 (UNDP 2018). It contains ten indicators, which cover, for instance, nutrition and child mortality, school attendance, sanitation, electricity access, or drinking water.

The poor have much worse starting conditions, worse nutrition, worse education, more exposure to pollution, less access to fresh water and sanitation, to health care, balanced diets, etc. There are many vicious cycles for the poor, regardless which social system they exhibit (Stroh 2015, 61). Poverty is self-perpetuating. Answers for how to escape these vicious cycles are legion and they are being passionately discussed – but it is clear that any answer has to consider the variety of aspects of poverty.

The places of poverty have significantly shifted in recent decades. While in 1990 the vast majority (93%) of the world's absolute poor[2] lived in low-income countries, nearly twenty years later (2008) three-quarters of the poor – almost a billion in sum – were living in middle-income countries[3] (Sumner 2010, 14). Sumner deduces from this development "that poverty is increasingly turning from an international to a national distribution problem, and that governance and domestic taxation and redistribution policies become of more importance than ODA" (Sumner 2010, 26) (ODA = Official development assistance). On the other hand, those poor living in LICs (approx. 370 million in 2009) are concentrated in sub-Saharan Africa.

4.2.3 The poor suffer most – environmental injustice

The main contributors to environmental degradation and global change are predominantly people in the global North and emerging economies. Much of their footprint is not visible domestically because pollution and low labour standards are externalized to the global South (Lessenich 2017). On the other hand, it is the poor and the people in the global South who suffer most from pollution, environmental degradation and non-sustainability, from contaminated soil, worsening droughts, soil erosion, sinking groundwater levels and climate change. The poor depend upon the natural environments much more than the affluent, they suffer most from their deterioration, they have fewer options to protect themselves against pollution, water scarcity or droughts, and they often lack knowledge due to limited education. Moreover, arid regions – which are particularly affected by global warming – have a high share of the poor: "Unabated climate change will also prevent the achievement of other development goals, particularly combating poverty" (WBGU 2011, 63).

The paramount example of the poor's harmful exposure to environmental pollution are the landfills in developing countries, primarily those for e-waste. Agbogloshie, a landfill in Ghana's capital Accra, gained notoriety for being a hot-spot of pollution, crime, rape and prostitution; it is called "Sodom and Gomorrah", attracting mostly poor people from rural areas in Ghana's north (Safo 2002). Seeking to extract valuable metals from e-waste, which is partly imported, partly domestic, tens of thousands of poor people try to make a living off the waste of middle-class people throughout the world. Abgogloshie has made it into the top ten of the world's most polluted sites (Pure Earth 2013).

Reasons for this e-waste disaster are complex. Starting with Moore's Law and the rapid development cycles of IT hardware, product designs which do not consider the products' end of life, criminal gangs which illegally export electric and electronic equipment from the mature markets of the global North, rural exodus and migration to cities aspiring to a better life are just some of them.

Migration and rapid urbanization cause additional problems. Conflicts around mineral resources, tribal conflicts, gangs, drug dealing, surrogate wars, military conflicts, or human trafficking – these phenomena are often both cause and effect of poverty and social instability. They all threaten or preclude the fulfilment of basic human needs, of security, food, shelter and human development. However, they are not only symptoms of non-sustainability, they also exacerbate it.

Huge environmental issues also go along with humanitarian hot-spots of migration, turmoil, war and hunger. Not only military action, but also migration pose challenges to the natural ecosystems, as a 2018 UNDP report stresses with regards to the Rohingya refugees in Bangladesh (UNDP Bangladesh 2018).

4.2.4 High ecological footprint or high development – is there no alternative?

Fighting poverty is a humanitarian demand. However, if not done correctly, fighting poverty can also threaten ecological integrity because of the trade-offs between preserving nature and developing society. To be sure, poor countries have the smallest environmental footprint of all and they need more development, economic growth, and yes, also more resource consumption. SDG 8.1 demands at least 7% of GDP for the least developed countries. At the same time, however, in the current market set-up, with externalizing costs, harmful policies of the global North with regard to agricultural subsidies, and (non-tariff) trade-barriers, such high growth rates are likely to strain the ecosystems even further. Moreover, there are hardly any promising blueprints for combining human development and ecological integrity. On the contrary, looking at the relationship of the ecological footprint (measured in hectares per person) versus the Human Development Index (HDI),[4] one can see that not a single country has both an HDI above 0.8 and is still within the ecological limits (see Figure 4.1). In other words, no single country manages to have a relatively high degree of development and still keep a small ecological footprint.

Poor countries are within the ecological limits of the planet but have a poor development status (which means short life expectancy, little income and low educational status), while all countries with an HDI >0.8 have a footprint which

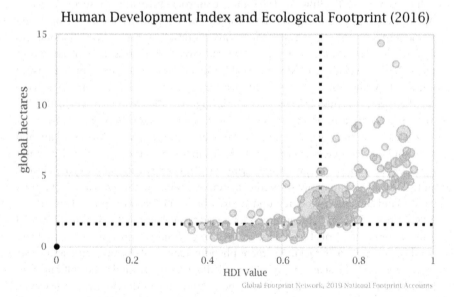

FIGURE 4.1 Ecological Footprint vs. HDI for all countries. No single country has an ecological footprint compatible to the earth's bearing capacity (i.e. <1.7 hectare per person) and a good level of development (i.e. HDI >0.8) at the same time. All SDSN leaders are in the upper right section.

Source: Global Footprint Network (2019), www.footprintnetwork.org.

is higher (often considerably) than the sustainable value of 1.7 hectare per person, which means that the corresponding resource demand would not be replicable globally (Global Footprint Network 2019).

The UN Development Report therefore summarizes the situation thus: "The degradation of the environment and atmosphere, coupled with significant declines in biodiversity, threatens the human development of current and future generations" (UNDP 2018, 11).

But what about the 2030 Agenda or the Paris Agreement – are their results not visible? Unfortunately not. The Sustainable Development Solutions Network (SDSN) ranks countries according to their performance with regard to the SDGs. All the top-performing SDSN countries lie far above the globally sustainable footprint level, implying that they would require several earths. Therefore, the Global Footprint Network concludes

> that fulfilling the Sustainable Development Goals is no guarantee for sustainability. The top-ranked nations on the SDG index all have high Ecological Footprints. If everyone in the world lived like them, we would need more than three planets. In fact, it seems that there may be a tension as material development achievements are far more prominent in the SDGs than the need to preserve the underlying natural capital.
>
> *(Global Footprint Network 2019)*

Looking at these data, there can be no doubt that it is the responsibility of the industrialized countries to push the transformation towards a low-carbon, circular economy and that they reconsider their consumption patterns.

Solution perspectives

What can, what should be done to fight global poverty?[5] There are libraries full of books about poverty, its origins, its dimensions, as well as mitigation strategies. Nobody can claim to know the answer – and there is no single answer. Having studied the nature of poverty for several years, Banerjee and Duflo conclude their book *Poor Economics* by acknowledging that "[e]conomists (and other experts) seem to have very little useful to say about why some countries grow and others do not." It would always be possible to construct a rationale for what happened in each place retrospectively but "we are largely incapable of predicting where growth will happen" (Banerjee and Duflo 2011, 267). Nevertheless, there are many things one can say about global poverty. Here are only a few snapshots at the interface of poverty and sustainability.

Banerjee and Duflo (2011) argue two extreme positions, those of Jeffrey Sachs and William Easterly. While the former sees the poor caught in the "poverty trap" (Sachs 2005) – they live in hot, infertile, and often landlocked countries and have no chance to escape this status without large initial investments from the outside, the latter sees ODA as contributing to this misery rather than reducing it, because people would not search for their own solutions when they rely on external help (Easterly 2006). They would become lethargic, corruption would blossom and economic development be prevented. While Easterly is rightly pointing at challenges for ODA, he seems to be too optimistic about the curative effects of the market system. As Maxwell points out regarding Easterly's book, it is "instructive that there is no discussion of market failure in this book" (Maxwell 2006).

The important point here is, however, that all quoted authors concur that a sustainable development will only be possible if we resolve the most burning social issues of the present as well.

Following the 2002 UN Summit on Sustainable Development ("Rio+10"), a group of organizations formed which are committed to ending extreme poverty while sustaining the environment (PEP 2016, 2) and of which the UNEP-UNDP Poverty-Environment Initiative (PEI) is a member. In their 2016 paper "Getting to Zero", this partnership aims at the triple vision of "zero extreme poverty, zero net greenhouse gas emissions, zero net natural asset loss" which is "proposed to guide the structural reform that will enable poor groups and countries to achieve the SDGs at scale" (PEP 2016, 11).

They call for

a) a greater involvement and empowerment of the poor and marginalized groups into the respective development processes,
b) integrated institutions capable of considering poverty, environment and climate issues in a comprehensive approach,
c) a "whole of society approach" which systematically engages civil society and especially small businesses, and
d) an environmental fiscal reform initiative, by which taxation and subsidy policies incorporate environment and climate actions into fiscal systems (PEP 2016, 42ff.).

In addition to this, in my view, global North countries also need to realize the impact their domestic policies has on other countries and change their practices and policies accordingly: subsided agricultural products enter markets of developing countries, but, in turn, de facto non-tariff trade barriers prevent the products from the developing world entering markets of the North; e-waste is shipped from North to South, fish illegally caught off the coast of Africa enters European markets, and the elites from the South leave their countries to aspire to a better standard of living in the North. The industrialized countries need to develop more coherent strategies for poverty reduction and green development and reconsider their consumerist lifestyles as if there were no tomorrow. Owen Barder calls for the rich countries to tackle the causes of poverty by changing their policies and by investing in global public goods. Income and consumption should, as a matter of global social justice, be transferred from the world's rich to the world's poor to enable them to live better lives (Barder 2009, 2).

For low-income countries, two things which I find particularly important are education and energy – both are only possible through investment.

Invest in education

Education is certainly one of those actions which has multiple benefits for fighting global poverty, especially for girls and women. If girls stay in school longer, they are less likely to marry early, they have a better chance of finding work, they are more likely to be emancipated, they will be better informed about family planning and contraception (see Sachs 2015, 213). For this reason I think fostering education is also a principle of sustainable action – we will return to this below (see 15.6).

Invest in renewable energy infrastructure

Fighting poverty often means fighting *energy poverty*. There is a strong increase in HDI if energy consumption per capita increases from zero to some small amount (see Gapminder 2019). However, it is not only energy consumption per se but apparently electricity consumption which is particularly relevant in the long term. Nadia Ouedraogo (Ouedraogo 2013) investigated the relationship between energy consumption as well as electricity consumption and HDI for twelve sub-Saharan African countries and found a positive correlation between electricity consumption and HDI and a negative one for other forms of energy, which was mainly biomass. She concludes:

> there is an urgent need not only to greatly scale up support for energy access but also to link this support more closely to the climate agenda, to the revitalization of rural areas, and to better management of the urban and peri-urban development that has dominated the changing energy landscape of recent decades. Thus, improving access to adequate energy services, affordable, reliable, effective and sustainable in environmental terms is crucial for economic growth and human development of our sample and to contribute to the fight against climate change
>
> *(Ouedraogo 2013, 39)*[6]

A few years ago, the residential energy consumption of Africa in its entirety was equivalent to that of New York City (WBGU 2011, 293). Since energy is critical for development, and Africa is to a large extent off-grid, a fast expansion of renewable energy supply would be much needed in Africa: "The overcoming of energy poverty is considered a fundamental requirement of successful poverty reduction strategies" (WBGU 2011, 292). Especially in those countries which have little energy availability at the moment, all efforts should be concentrated to build up a low-carbon energy infrastructure.

The following benefits would be related to renewable energy infrastructure in LICs:

a) Replacing traditional energy use with low-carbon technologies "can simultaneously lead to poverty reduction, prevent damage to health, and mitigate the utilisation pressure on natural ecosystems … Efficiency improvements in the area of existing bioenergy use and the switch to modern forms of energy such as electric power and gas are an important precondition for overcoming energy poverty and covering the basic essential need in developing countries" (WGBU 2011, 293).

b) There is a strong correlation between energy availability and *human development*, simply because energy is so critical for our modern civilizations. Renewable energy infrastructure would lessen the burden of biomass or fossil fuel combustion (with negative health effects and less GWP), it would facilitate machinery usage, lighting, productivity gains, safe cooking, etc., and eventually increase income. Electricity availability for residents can, for instance, reduce indoor pollution, which causes 3 million deaths per year in East Asia, South Asia and Africa mainly due to traditional ways of cooking and heating (Our World in Data 2019). Sachs praises the benefits of electricity, which would not only provide "access to lights at night or electricity for home activities, there is a critical lack of power for pumping water for irrigation; for refrigeration; for preservation of agricultural outputs; for industrial processes of food, textiles, and apparel; and for every other kind of industrial activity" (Sachs 2015, 155).

c) Increased income, however, will most likely lead to reduced *fertility rates*. "The negative association of fertility with economic and social development has therefore become one of the most solidly established and generally accepted empirical regularities in the social sciences" (Myrskylä, Kohler & Billari 2009, 741). This is much needed, since especially the least developed countries in sub-Saharan Africa have the highest population growth globally.

d) "As a general trend, energy intensities of economies decrease with time and with increasing income" (Rao & Baer 2012, 673).

e) After some increase in demand for natural resources and worsening pollution and waste problems, the environmental footprint will, according to the Environmental Kuznets Curve, decline.

In light of the fact that Africa is expecting the highest rates of population growth in the next decades, and that a large fraction of people in sub-Saharan Africa still do not have access to electricity and depend on fossil fuels to a substantial extent, the provisioning of an infrastructure for renewable energy is potentially the most critical factor for addressing both several dimensions of poverty and mitigating the climate crisis at the same time.

Banerjee and Duflo (2011, 346) conclude their report on the situation of the poor by highlighting:

1.   that the poor are often lacking critical information (e.g. about vaccination, contraception). This points to the ultimate importance of education and information campaigns (see 15.6);

2.   that the poor have to struggle for many more everyday life things than affluent people. They need preconfigured standard solutions (for water purification, for savings accounts, etc.). Sometimes there is nothing even available for them because there is no market for them, e.g. in life insurance. Halme et al. argue in a similar direction by pointing to the Base of the Pyramid (BOP) approach: "The BOP approach suggests that it is possible to design products, services, and business models that can make life easier for poor people and bring more profits for businesses. To this end, the BOP literature says that, businesses should learn about the needs of low-income, often informal target markets, and apply user-oriented design methods ... Despite this rhetoric, it is not common among businesses to start innovating on the basis of the needs and practices of the poor" (Halme et al. 2016, 114).

At the end of their book, *Poor Economics*, Banerjee and Duflo confess that they

> also have no lever guaranteed to eradicate poverty, but once we accept that, time is on our side. Poverty has been with us for many thousands of years; if we have to wait another fifty or hundred years for the end of poverty, so be it. At least we can stop pretending that there is some solution at hand and instead join hands with millions of well-intentioned people across the world – elected officials and bureaucrats, teachers and NGO workers, academics and entrepreneurs – in the quest for the many ideas, big and small, that will eventually take us to that world where no one has to live on 99 cents per day.
>
> *(Banerjee & Duflo 2011, 273)*

The final word on poverty should come from a genuine fighter against it, Pope Francis:

> The land of the southern poor is rich and mostly unpolluted, yet access to ownership of goods and resources for meeting vital needs is inhibited by a system of commercial relations and ownership which is structurally perverse. The developed countries ought to help pay this debt by significantly limiting their consumption of non-renewable energy and by assisting poorer countries to support policies and programmes of sustainable development.
>
> *(Pope Francis 2015, Section 52)*

4.3 Populism and fundamentalism

Populism and fundamentalism can pose barriers to sustainability by their underlying values on the one hand and by an explicit agitation against political goals in the context of sustainability on the other. I submit that both phenomena can be construed as reactions against the complexities and intricacies of (post-)modern societies with their rapidly changing living environments, the withering of long-established certainties and the emergence of new risks and threats.

Despite overwhelming evidence for the anthropogenic cause of the climate crisis, climate sceptics still find their audience. Their positions have seemingly become even more in vogue lately, coinciding with the upswing in right-wing populism in many regions of the globe. The sceptics still find their disciples, they still manage to get media presence with their noxious thinking, and they still provoke us as we are working on sustainability to explain the state of the science in this field. However, we need to realize that their position cannot be countered by empirical evidence. It is rather to be understood in sociological or psychological, if not religious, terms (see, e.g. Hobson & Niemeyer 2012; Jaspal, Nerlich & van Vuure 2016).[7] We should resist their attempt to relate climate science (or any other) in any respect to some kind of worldview or ideology – and always remain open to substantiated arguments which try to falsify established theories.[8]

Any organization, movement, or direction of thought which runs counter to the values of sustainability – be it implicitly or explicitly – which calls into question the very need for such a concept, or disagrees on its presuppositions will certainly be a barrier for sustainability. There is a distressing increase of corresponding societal developments all around the globe. This is most evident for the increase of populism. Populism is being discussed quite heatedly both in public and in academic discourse. Due to the variety of different concepts, some scholars have even argued that populism could not be a meaningful concept for the social sciences at all (Mudde & Kaltwasser 2017, 5).

However, there seems to be consensus about some of the main features of populism (see Mudde & Kaltwasser 2013; Müller 2016; Mudde & Kaltwasser 2017; Fraune & Knodt 2018). Populists use the framing of "us versus them", they claim to speak for the "true people" versus the "corrupt elite". They highlight issues which "the elites", in their view, have not sufficiently addressed.

Mudde and Kaltwasser describe it as a "thin-centered ideology that considers society to be ultimately separated into two homogeneous and antagonistic camps, 'the pure people' versus 'the corrupt elite', and which argues that politics should be an expression of the volunté générale (general will) of the people" (Mudde & Kaltwasser 2017, 6). Lockwood argues that "nationalism and authoritarianism combine with anti-elitism to construct a world view in which 'the people' are ruled by a corrupt and illegitimate liberal, cosmopolitan elite" (Lockwood 2018, 726). Furthermore, the ideological nature of right-wing populism would also create an attraction to conspiracy theories, which "is a consistent facet of climate scepticism" (Lockwood 2018, 726).

So while populism itself is not characterized by a strong common ideology of its own – but rather by a common legitimating framework as well as a political style and mood (Fraune & Knodt 2018, 2) – it is mostly "attached to other ideological elements, which are crucial for the promotion of political projects which are appealing to a broader public" (Mudde & Kaltwasser 2017, 6). As a consequence, populism cannot deliver any complex or comprehensive answers to the questions modern societies generate (Mudde & Kaltwasser 2017, 6) – although this is exactly what they claim.

Populism is not per se irrational. Ernesto Laclau argues that populism has its own distinctive rationality. Laclau even goes so far to claim that "all democratic politics are, in fact, populist" because under the assumption "that society is inherently heterogeneous, politics must entail the hegemonic articulation of a multiplicity of political demands in a manner that is always provisional and open to revision" (Gandesha 2018). Similarly, populism, which is "most fundamentally juxtaposed to liberal democracy rather than to democracy per se" can also have a positive role for liberal democracies, for instance by politicizing issues which the ruling "elite" neglects (Mudde & Kaltwasser 2017, 1f.; 84).

However, the prevalent populism which is on the rise in many regions around the globe altogether threatens the global pursuit of sustainability. There is ample evidence that especially right-wing populist politicians agitate against political goals in the areas of environmental protection, climate politics, migration politics, etc. Not only national presidents like those in the USA or in Brazil, but also right-wing parties in many other regions of the world serve their supporters with simplistic messages, often illiberal and discriminating against minorities.

These populists often follow the messaging of "climate sceptics" and question the overwhelming congruence within the scientific community of the anthropogenic nature of global warming. The most ridiculous nonsense is spread, like "[t]he concept of global warming was created by and for the Chinese in order to make U.S. manufacturing non-competitive" (Trump 2012), but this apparently does not frighten off the populists' followers. However, questioning well-proven scientific results in such a way is a more foundational challenge to open and liberal societies than being "just" "anti-environmentalist" agitation. It challenges the achievements of the modern constitutional democracy, personal freedom, equality, non-discrimination, freedom of science. It certainly also challenges the very quest for sustainability.[9]

Political decisions in favour of sustainable development always have to balance short-term and long-term goals. Provisions for the future and long-term planning are much harder to argue for than presenting campaign goodies. This has always been the case for politicians (Machiavelli already recommended committing the necessary cruelties at the beginning of a reign), but it is a special challenge for liberal-democratic parties confronted with populist positioning. Populists instead concentrate on the short term, addressing their specific target group, demanding concrete improvements, asking for immediate cures which the "lazy elite" would not be willing to apply. Why should normal people not be taken in by such pied pipers if the alternative is an extra price to prevent an evil which needs a lot of explanation and which will most likely occur after one's own death? Populists make it particularly hard for politicians to argue for sustainability policies: "Their interest is to prioritize short-term programs which favour their select group of the people, rather than longer-term mitigating policies which have widespread economic and environmental benefits" (Dibley 2018). That's why populism and "post-truth politics" increase political polarization also in specific sustainability context like sustainable energy transformations, which Fraune and Knodt have investigated (see Fraune & Knodt 2018, 6).

Populism may be construed as a channel through which people's concern is operationalized in the political discourse, but it is rather defined by the way societal topics are framed, political agendas are set and the public discourse is led. In short, populism responds to the insecurity and anxiety people feel and provides answers in the political arena.

There is another phenomenon which is also defined rather by form than by content and which shares structural similarities with populism: fundamentalism. Populism and fundamentalism offer structurally similar kinds of answers: binary logic, clear speech, exclusivity in claims, rigid moral attitudes, an "us versus them" logic, and strong leadership. Fundamentalism can be found in almost any school of thought, be it religious tradition, political program, or philosophical worldview. In my view, it is not just by chance that populist leaders are especially applauded among groups which share fundamentalist attitudes. A large fraction of Trump's voters, for instance, are from the evangelical camp, often adhering to fundamentalist views. Eighty per cent of Christian evangelicals in the USA voted for Trump in the presidential election in 2016 (Washington Post 2016). Leading evangelicals explicitly attribute to Trump a religious role, seeing "Trump as a latter-day King Cyrus, the sixth-century BC Persian emperor who liberated the Jews from Babylonian captivity" (Borger 2019):

> The comparison is made explicitly in *The Trump Prophecy*, a religious film screened in 1,200 cinemas around the USA, depicting a retired firefighter who claims to have heard God's voice, saying: 'I've chosen this man, Donald Trump, for such a time as this.'
>
> *(Borger 2019)*

Both populism and fundamentalism provide clear, simple and easy answers and can thus be understood as reactions to the increasing complexity of (post-) modern times, to the new challenges arising rapidly over the last few decades and to the anxiety people have against the changes in their living environments.[10] The sociologist Thomas Meyer sees fundamentalism as an anti-modern phenomenon, a reaction against modernity. Following Jürgen Habermas, Meyer depicts modernity as the "first cultural formation in the history of humankind which is condemned to create its self-confidence and its norm out of itself". According to Meyer, this "self-assured jump into the bottomless is an unprecedented cultural, social and political revolution, an open experiment"

(Meyer 1989, 21). Meyer recognizes fundamentalism as a deeply anti-modern, in fact, anti-Enlightenment movement. While the Enlightenment dwelled on the power of reason – for good and for bad – fundamentalism withdraws from the disgrace of having to use one's own reason. Life has become so complex, traditional certainties no longer hold, established institutions wither away, globalized markets threaten reliable jobs, migration exposes people to new value and belief systems – these kind of feelings are the background against which fundamentalism can blossom. Fundamentalism is a pull-back from the disgrace of using reason in one's own right (Meyer 1989, 157).[11]

Apparently both populism and fundamentalism offer certainties and security of expectations by providing simple answers, clear solutions, rigid moral standards, and a feeling of togetherness – one in the socio-political domain, the other in the domain of worldviews.

Why is all this relevant to sustainability?

a) First and foremost, of course the horror and misery militant fundamentalists such as Al-Qaida or ISIS has brought the world in the last decades has not only caused an endless number of innocent fatalities, it also impedes peaceful cooperation among the peoples of the world.

b) Fundamentalist agendas have influenced populist politics. Populist politics can often be understood as the attempt to please fundamentalist voters, for instance, by trying to disqualify transgender people from military service, by acknowledging Jerusalem as Israel's capital, or revoking the nuclear deal with Iran. This course of action – irritating and incomprehensible for most people outside this school of thought – contributes to increasing tension in international politics.

c) More subtly, fundamentalism and populism create an environment which is hostile to the values of sustainable development. For instance, "fundamentalism and populism pose deepening threat to women human rights defenders", as these women are increasingly confronted with discrimination, harassment and violence (UN 2016).

Solution perspectives

What needs to be done to address this barrier?

a) First, we need to acknowledge that we might have underestimated critical dimensions in the debate about sustainability. Whether or not the Nationally Determined Contributions (NDCs) are sufficient to meet the 2-degree limit (within the Paris Agreement) is totally irrelevant if we lose the people, if we fail to get societies around the world behind the respective vision, if populists take over and dominate national and internal agendas with their clientele politics. As Hempel puts it: "If it is true that Trump's most significant and insidious impacts on environmental policy, management, and science will ultimately be traced to his low regard for truth, his cultural assault on political legitimacy and trust, and the morally destructive behavior he models for impressionable minds, then it is incumbent on environmental educators to devote more of our attention to questions of cultural identity, democratic deliberation, and ethics in our treatment of environmental issues" (Hempel 2018, 187).

b) We might need to learn some lessons in humility. There is good reason to think that the dogmatic and patronizing style of some people considered "elitist", sadly enough sometimes combined with personal misconduct, is the soil in which the populist defamation of "corrupt elites" can sprout.

c) We should realize people's quest for guidance. When propagating the need for a sustainable development, it is important to understand the drivers for populism and fundamentalism and acknowledge that they apparently serve people's need for guidance and security. All proponents of sustainability need to find solutions to communicate ways which speak to this demand – but neither by over-simplifying ourselves on the one hand nor by simply stating that the world simply is that complex and simplification would not be possible. We can and should learn, as I see it, from the rise of populism that we cannot afford to avoid the hard questions: "The best way to deal with populism is to engage – as difficult as it is – in an open dialogue with populist actors and supporters … Most importantly, given that populism often asks the right questions but provides the wrong answers, the ultimate answer should not just be the destruction of populist supply, but also the weakening of populist demand. Only the latter will actually strengthen liberal democracy" (Mudde & Kaltwasser 2017, 118).

d) Choose the right framing. When arguing for climate mitigation by renewable energy, for instance, it will be important to emphasize renewables' direct, short-term benefits: "Rather than reducing emissions and tackling global climate change, it may be better to frame mitigation as part of a large-scale effort towards modernization; that is, modernizing energy systems, transportation systems and infrastructure" (Dibley 2018).

e) Religious leaders should seek appropriate messages. Finally, since important streams of fundamentalism have a religious background, leaders in these traditions should ask themselves how they can speak to the hearts of their believers in a way that matches the respect of human rights and the values of our globalized world. Having a background in Christian theology myself I must admit that my tradition has still a long way to go in finding the right translation of traditional belief in our world today. In my view, however, the topic of sustainability bears huge potential not only for revitalizing Christianity's own traditions but also for seeking alliances across belief systems and confessions. There is so much which our religions have in common (and, by the way, which they share with non-religious traditions as well) and which should be explored.

4.4 Inequalities

Although global inequality has altogether decreased due to a rapidly growing middle class in Asia, it remains of great concern in Africa and South Asia. Furthermore, inequality within countries has significantly increased in many countries. Inequalities pose a challenge for sustainable development not only because of its immediate troubling implication of injustice but also because inequalities correlate with several unwanted societal phenomena like poverty, distrust, or corruption.

Inequality is a topic of which people are increasingly aware and concerned about. In his book, *Inequality: What can be Done?*, Atkinson quotes a study according to which inequality was considered "the greatest danger in the world" (Atkinson 2015, 1). It is especially the extremes and the intrigues of the super-rich, which catch people's attention. This small group of people is estimated to hide US$7.6 trillion from the tax authorities (Oxfam 2019, 13). This unbelievable wealth of a tiny but growing global elite contrasts with the billions of people at the base of the pyramid. However, are inequalities are always harmful or unjust? Do market-based economies not always exhibit some inequality? (see IMF 2017, ix). What kind of inequality are we talking about – national inequality or global inequality?

To begin with, it is not true what barroom clichés sometimes suggest: that inequality across the board has increased in the last few decades. On the contrary, inequality at the global level – which means if one is abstracting from any national boundaries – has substantially declined in the last thirty years (IMF 2017, 1). The main reason for this decline is the convergence of the income between developing and advanced economies (IMF 2017, 1), and the rise of a "global middle class", mainly in Asia, especially China (Milanovic 2016, 3). While this global middle class benefited from globalization, the middle- or lower-middle class in the advanced economies mainly lost in this process (Milanovic 2016, 3).

At the same time, however, in many countries, particularly in the advanced economies, inequality has significantly increased *within national boundaries* (IMF 2017, 1) – and it is this income inequality which matters much more for many societal developments than the inequality between countries (see below) (Wilkinson & Pickett 2011, 229). Especially the top shares of income and wealth have increased in nearly all countries in recent decades (Alvaredo et al. 2017, 9).

Why is inequality an issue, a barrier to sustainability, of such concern that fighting inequality has even become one of the 17 SDGs (#10)?

Let us first look at inequality within countries:

1. Inequality is, at least in its extreme forms, simply a *matter of justice*. In many cases, the rich have much better access to education, health services, job opportunities, and they have less exposure to pollution, etc. For instance, it is wealth, not talent, which "dictates a child's educational destiny" in many countries, as an Oxfam study documents. A Kenyan boy from a rich family

has more than a 30% chance to continue studying beyond secondary school, while for a girl from a poor family this chance is just 0.4% – 75 times less (Oxfam 2019, 16). This inequality is not specific to African countries. In the USA, according to Oxfam "the American Dream has become a myth, with social mobility lower today than it has been in decades" (Oxfam 2019, 16). As Oxfam reports, both in the UK and Brazil, the poorest 10% of population have a higher effective tax rate than the richest 10%. Similarly in other countries: Warren Buffet, the US-billionaire, said in an interview that he would probably be paying less tax than his secretary (Isidor 2013). Something similar is occurring in the corporate world. In Europe, large multinational enterprises (MNEs) actually pay less tax than smaller companies: "most countries appear to tax MNEs regressively: the larger the MNE, the lower the effective tax rate" (Janský 2019, 3). In addition, due to international tax competition (capital mobility is particularly high), corporate income taxes have substantially decreased in the last few decades. Average corporate income tax rates decreased by close to 40% in all markets between 1990 and 2015 (IMF 2017, 15).[12] Furthermore, wealth can even help acquire another identity, in the literal sense of the meaning. A 2018 study of Transparency International accused the European Union (EU) that EU citizenship and residency rights can be purchased just as luxury goods can be. An EU passport costs between €250,000 and €10 million (Transparency International; Global Witness 2018).

2. In the current system, inequalities have the *tendency to be perpetuated* if they are not forcefully mitigated by political measures (which is, in turn, ever more difficult in light of rising populism). Better education means much better starting positions.

3. Inequality is positively *correlated to several unwanted societal attributes*:

   - Income inequality, particularly at the bottom of the distribution pyramid, *lowers the trust level* in a society, but trust is an "important determinant of the macro-economic performance" which implies that "increasing inequality may be adversely affecting a country's growth and development over time" (Gould & Hijzen 2016, 21). Inequality affects trust, but there is "no direct effect of trust on inequality; rather, the causal direction starts with inequality" (Wilkinson & Pickett 2011, 55); Wilkinson and Pickett "think of trust as an important marker of the ways in which greater material equality can help to create a cohesive, co-operative community, to the benefit of all" (62).

   - Wilkinson and Pickett view it as an important signpost that there is a "contrast between the material success and social failure of many rich countries". This "suggests that, if we are to gain further improvements in the real quality of life, we need to shift attention from material standards and economic growth to ways of improving the psychological and social wellbeing of whole societies" (Wilkinson & Pickett

2011, 4). In fact, they have convincingly shown that *it is not the income level* of a country which decides on its problems but the *level of inequality!* For instance, *health and social problems are positively correlated to inequality – but not to living standards!*

- Inequality correlates to consumerism since "inequality increases the pressure to consume" (Wilkinson & Pickett 2011, 228).
- The IMF states that "excessive inequality *can erode social cohesion,* lead to political polarization, and ultimately lower economic growth" (IMF 2017, ix; emphasis added).
- *Corruption and inequality seem to reinforce each other.* "The results also indicate that increased corruption is positively correlated with income inequality. The combined effects of decreased income growth and increased inequality suggests that corruption hurts the poor more than the rich in African countries" (Gyimah-Brempong 2002, 183). "The impact of corruption on income inequality and poverty is considerable" (IMF 1998, 29). This holds for countries with difference growth experiences and at different stages of development (ibid.).
- A 2019 Oxfam study summarizes the multiple negative impact of inequality: "In more unequal countries, trust is lower and crime higher. Unequal societies are more stressed, less happy and have higher levels of mental illness" (Oxfam 2019, 14). Or positively phrased, "greater equality is the material foundation on which better social relations are built" (Wilkinson & Pickett 2011, 272).

According to a 2019 Oxfam study there is

a growing consensus that the wealth of individuals and corporations is not being adequately taxed, and instead taxes are falling disproportionately on working people. For every dollar of tax revenue, on average just 4 cents are made up of revenue from wealth taxes.

(Oxfam 2019)

This is in stark contrast to SDG 10, which asks that inequality within and among countries be reduced.

What about *global inequality?* When statistics say that global inequality has decreased, this certainly does not mean that it does not remain a serious issue. The convergence of income in large emerging economies (especially China) has lessened the global gap, but absolute poverty is still a severe issue particularly in sub-Saharan Africa and South Asia.

Globalization comes with increased risk for the developing world (see Stiglitz 2002; 2006, ch. 1). Stiglitz calls for more global solidarity. Why should jobs be protected at home when they could serve a much greater good in another place? Stiglitz recommends that we put ourselves in the position of others – to really feel their need and come to a just shape of laws and regulations – where he also refers to John Rawls and his theory of justice.

Global inequalities will be continue to be the engine for migration. The outlook of a more prosperous future will continue to lure people from the global South towards the industrialized countries. The greater the differences and the more depressing the poorer condition, the more likely this engine will continue running.

Solution perspectives

How can a fair distribution of goods be achieved? This question has troubled great minds for millennia. Many suggestions have been made. For instance, fair is when everybody gets the same – but this neglects the fact that due to natural circumstances, fate, chance, etc., some are more privileged than others, which most people perceive as unjust. Maybe fair is to give to everybody according to her or his merit? That leaves open what would be counted as "merit". Again, the very fact that talents and gifts are also not evenly distributed among people would not be accounted for either. Would it be fair to give everybody according to her or his needs? This sounds like a compassionate solution, but it does not answer what precisely a "need" would be in contrast to a want and whether or not this would be justified. Furthermore, it neglects that people do have an influence (despite uneven natural preconditions) on their own situation – both on what they need and on what they have earned. Furthermore, the utilitarian maxim to consider as just what maximizes the greatest good for the greatest number is confronted with the criticism that it does not (sufficiently) bother about how this greatest good is actually distributed among society (see, e.g. (Schroth 2006)). Finally, a legal positivist solution is simply to state that fair would be that which everybody gets according to law. However, this would not help in situations where law cannot be applied and it does not answer how respective laws would need to be shaped.

In his seminal work, *A Theory of Justice*, John Rawls develops a procedural approach to the concept of justice which can clarify how a fair distribution can be found (Rawls 1971). The starting point is his liberal conviction that all people are free and equal, nobody is privileged. Rawls conceives of a group of people for which talents, goods, privileges, gender, circumstances, etc., are unevenly distributed – as in real life. These people are then tasked to deliberate according to which principles goods shall be distributed, considering the spread of advantages due to natural predisposition, personal inheritance, historic circumstances, etc. It is assumed that everybody takes part in the best neutral manner since nobody knows which position she or he will hold later in the game ("veil of ignorance").

According to Rawls, every rational actor would be able to agree on two fundamental principles. The first principle guarantees every person the greatest possible set of equal basic liberties as long as they do not limit the respective rights of others – which would need to be codified in the constitution. The second principle, more relevant in our context, is a principle of distributional justice. Social and economic inequalities would be acceptable if they are combined with the greatest benefit for the least privileged (principle of difference) as well as with positions and posts which are open to everybody under fair conditions (principle of equal opportunities) (Rawls 1971; 1978).

Of course, modern constitutional democracies have already institutionalized concepts of fairness with a similar intention – also prior to and independent of Rawls' theory. However, their concrete realization often lags far behind the intended result.

a) At least nominally the rich usually have to carry a higher tax burden than the poor. We saw that the opposite is often the case (both for individuals and for corporations), but this is rather a result of, among others, exemptions, better starting positions (better consultants, lawyers, …), carelessness, inter-state competition, lobby-influence (e.g. taxing work higher than capital income), corruption, etc., than any conceptual privilege of the rich. This makes it even more important that this defect, which runs against basic and commonly agreed moral standards, will be tackled by politics.

b) Systems of social security which provide special support for the needy for housing, living expenses, medical care, etc. There are certainly ongoing debates on the adequacy of the degree of support but the general concept of special support for the needy seems to be common – both in most national contexts as well as internationally.

c) Special agreements within the WTO to support the least developed countries are one example which shows that at least in principle this is being tried on an international level as well (despite the failure of the Doha Round, see below). However, while the OECD countries fail by far to meet their target of spending 0.7 % of GDP for ODA, spending just 0.3%, there are significant interest payments of the global South to the global North. In fact the "[n]et transfer of resources from developing countries continues to be negative, which means that capital has been flowing out of these countries" (UN-DESA 2017).

Rawls' theory can be used in many context of sustainability in which a just distribution of goods is sought. It can not only assess the fairness of laws and regulations on a national level, it can also support similar discussions at the international level. Moreover, if future generations are also seated at the imaginary table, we also get criteria for intergenerational justice. Is it fair, for instance, that one generation benefits from a certain technology but a thousand generations bear the cost – as in the case of nuclear energy? Is it fair that future costs of climate change are discounted, which basically lessens the pressure to build up corresponding reserves although we do not know whether we will actually experience historic growth rates in the future? (Current trends question this.) Is it fair that countries which had huge gains from their colonial domination and have massively contributed to the environmental issues of today receive more interest payments from the ex-colonial states than they spend on ODA?

Moreover, Rawls' main argument can directly be linked to other traditions and to common-sense creeds. Rawls emphasizes that his theory does not rest on any particular worldview, ideology, or religious tradition. In a democratic constitutional state, the official concept of justice should be independent of controversial religious or philosophical doctrines as much as possible.

But this does not exclude seeking parallels in other schools of thought. We cannot expect that everybody will be able and willing to follow philosophical arguments. Arguments also reach only rationality but hardly touch people emotionally. I think it is therefore important to point to traditions which exhibit parallels to Rawls and utilize them for current considerations on a fair distribution of (common) goods.

Important religious traditions have also emphasized the importance of impartiality, reciprocity and compassion for the poor.

a) The Hebrew Bible reminds the ancient Israelites to be friendly to the stranger: "But the stranger that dwelleth with you shall be unto you as one born among you, and thou shalt love him as thyself; for ye were strangers in the land of Egypt" (Leviticus 19:34); "Love ye therefore the stranger: for ye were strangers in the land of Egypt" (Deuteronomy 10:19). This core meaning of the principle of reciprocity is critical for Rawls' theory as well.

b) Charity (*zakat*) is one of the pillars of Islam. *Zakat* is obligatory to all Muslims: "Whatever you spend of good is [to be] for parents and relatives and orphans and the needy and the traveler. And whatever you do of good – indeed, Allah is Knowing of it" (Quran, Sura 2.15).

c) In the Gospel of Matthew, Jesus tells his disciples that the help, with which they support the needy, the poor, the hungry, will lastly be considered as a support of Jesus himself (Matthew 25:40).

"Don't criticize a man until you've walked a mile in his moccasins" says a Native American proverb. People think differently when they have been personally touched, when they have personally interacted with people in need, they tend to have much greater sympathy for others in the same calamity. People who have always lived in their home country can hardly imagine how it feels to be a stranger. Europeans often cannot imagine how present the colonial heritage is for many people in Africa.

Concrete action?

a) "According to the IMF, adequate taxation of capital income is needed to protect the overall progressivity of the income tax system by reducing incentives to reclassify labor income as capital income and through a more uniform treatment of different types of capital income" (IMF 2017, x). The

World Inequality Report 2018 argues in the same line: "Tax progressivity is a proven tool to combat rising income and wealth inequality at the top" (WID 2018, 19).

b) The same report also favors more democratic transparency about the dynamics of income and wealth: "A global financial register recording the ownership of financial assets would deal severe blows to tax evasion, money laundering, and rising inequality" (ibid.).

c) Moreover, the 'sluggish income growth rates of the poorest half of the population' need to be addressed by 'more equal access to education and well-paying jobs' (20). According to the WID 2018 report, the inequality within countries affects also the inequality between them: "Within-country inequality dynamics have a tremendous impact on the eradication of global poverty" (18).

d) "[P]olicies that reduce corruption will also lower income inequality and poverty" (IMF 1998, 1).

e) "We need to create more equal societies able to meet our real social needs. Instead of policies to deal with global warming being experienced simply as imposing limits on the possibilities of material satisfaction, they need to be coupled with egalitarian policies which steer us to new and more funda-mental ways of improving the quality of our lives. The change is about a historic shift in the sources of human satisfaction from economic growth to a more sociable society" (Wilkinson & Pickett 2011, 231). More equal societies even recycle a larger part of their waste (232)!

Every political measure comes with winners and losers. The more foundational the societal changes for a transformation are needed (and there is good reason that we do need substantial change), the more society needs to consider how cost and benefits are fairly distributed.

> Economic and political inequality have long-lasting implications for gov-ernance both within and between states. Inequality in either form contrib-utes to a rise in extremism and social unrest, and it also raises the questions of what responsibility the international community should bear for human development beyond just satisfying basic needs, that is, security, food and shelter.
>
> *(Jang, McSparren & Rashchupkina 2016, 3)*

4.5 Conflicting interests

People have conflicting interests. In most cases, these conflicts can be resolved within estab-lished legal and political frames. There are circumstances, however, in which this is not the case because either such frames do not exist (e.g. on the international level), the conflicting interests are hidden (by lobbyism or corruption), or the conflicting parties differ greatly in their negotiating power (due to socio-economic inequalities). Counter-measures include first and foremost clear and undisputed transparency (about dependencies, payments and privileges, etc.), strict regulation (e.g. for mandatory disclosures) and good governance.

In any social community different actors have different, often conflicting interests. The needs and wants of the youth differ from those of elderly people, small and medium-sized enterprises have different concerns than large multinationals, the interests of employees differ from that of employers, etc. Conflicting interests simply reflect the diversity of demands and wishes among the members of any society. It is ultimately the role of politics, insofar as it deals with the "total complex of relations between people living in society" (Merriam-Webster 2019) to mediate opposing interests and seek a solution acceptable to the parties involved. Conflicting interests are therefore at the heart of any political discussion and certainly not specific to the sustainability discourse – but they do impede progress in sustainability. The interests of coal workers, for instance, to maintain their employment conflicts with the interest of environmentalists to reduce GHGs. The interests of the general public to sustain important environmental qualities like biodiversity and the quality of ground water conflict with farmers' interest in increasing yields – which often coincides with the interest of fertilizer-producing chemical companies.

Resolving such conflicts is, as just said, a paramount responsibility of politics, which can organize discourses in which every party expresses their concerns and an acceptable (i.e. fair) resolution of the conflict will be negotiated. This is just the regular procedure to resolve conflicting interests in the public sphere, taking place thousands of times every day all around the globe. However, there are a few qualifications which complicate conflicting interests.

4.5.1 No framework for resolving conflicting interests on international level

In democratic constitutional states, the resolution of conflicting interests is legally codified and can ultimately be decided by law courts. This is much more difficult on the international level, however, where all too often conflicting interests are resolved by the rule of power or military conflicts, which is related to issues of global governance (see Chapter 6).

The difficulties of multilateral international negotiations became most evident by the failure of the Doha Round of the World Trade Organization, which was centred around the question of how the less developed regions can benefit more from global trade (WTO, The Doha Round 2019). The Doha Round was semi-officially also called the "Doha Development Agenda". The industrialized countries (especially the US and EU) had long been accused that their policies harm the developing world. In particular, trade barriers (tariff or non-tariff barriers) as well as subsidies for export and for agricultural products were (are) under attack, which made it hard for developing countries to benefit from global trade. These concerns were particularly relevant for developing countries since agricultural products often contribute substantially to the national income. Therefore, the Doha Round intended to lower trade barriers and revise trade rules.

At the beginning of the discussions in 2001, there seemed to be willingness of the industrialized countries to reach a trade agreement which would not have required that the developing countries also needed to lower their import barriers, similar to the industrialized countries.[13] But since China reached a point at which it exported more than it imported, the rich countries demanded that China also needed to lower their import barriers and reduce agricultural subsidies. However, China and India were not willing to change the system (NewYorkTimes 2016).

So why die Doha fail? One can well argue that it failed because "the United States and EU weren't willing to give up their agricultural subsidies" (Amadeo 2019). On the other hand, China and India were not willing to make the concessions needed. Both positions have a point, though: "because neither developed economies like the United States and the European Union nor developing countries like China and India were willing or able to make fundamental concessions" (New York Times 2016):

> The United States knew that its compromise lay in offering more farm subsidy cuts; the European Union knew it would be required to cut agricultural tariffs; and the larger emerging countries knew they would have to offer deeper industrial and agricultural tariff cuts. Yet after more than five years of preparations, when the deal was there for the taking, none of the key players stepped up to make it happen
>
> *(Charlton 2006)*

After years with several rounds of negotiation, in July 2008, following another negotiation marathon, which managed convergence for 18 out of a to-do list of 20 topics, Pascal Lamy, Director-General of the WTO, expressed his frustration over non-agreement. The developing countries would have been the primary beneficiaries of an agreement "with a rebalancing of the rules of the trading system in favour of developing countries" (Lamy 2008).

4.5.2 Conflicting interests are not always visible

Resolving conflicting interests assumes, of course, that these interests are detectable and known to all parties involved, which is not always the case. Sometimes this is due to ignorance, sometimes one party does not want everybody involved to have full transparency about its own relationships. Lobbyists, for instance, do first of all represent and explain the interests of their clients to policy makers. They support understanding the respective clients' situation, providing background information and facilitating the understanding of technical contexts. Professional lobbyists, however, do often not have an interest in excessive publicity because their arguments in their clients' interest might otherwise be torpedoed by antagonists.

There are numerous examples of political decisions (and sometimes also, inaction) which cannot be explained other than by the effectivity of corporate lobby groups. Of course, it is usually very difficult to prove the exact efficacy of lobbyists and their influence as the origin of political decisions – but that is exactly the problem. However, political analysts do refer to this explanation as being frequently very plausible.

For example, the German Federal Government (at that time, a coalition of Social Democrats and Greens) issued the German Sustainability Strategy in 2002. Within this comprehensive document, they set the objective to reduce surplus of agricultural nitrogen to 80 kg/ha by 2010 (Bundesregierung 2002, 115), which was repeated by the new government (Christian Democrats and Social Democrats) in the strategy's progress report in 2008 (Statistisches Bundesamt 2008, 36). In 2016, when the target for 2010 had long been missed, the new target was marginally reduced to 70 kg/ha but the period substantially extended by twenty years to 2030 (needless to say, this happened almost unperceived by the public) (Bundesregierung, German Sustainable Development Strategy 2016). In 2018, the European Court of Justice of the European Union delivered the judgment that "the Federal Republic of Germany failed to fulfil its obligations … concerning the protection of waters against pollution caused by nitrates from agricultural sources" (Court of Justice of the European Union 2018). The "new" federal agricultural minister, being in office less than a year, was already accused of being driven by the agrarian lobby (Der Spiegel 2019).[14]

Lobby groups will also "support" the implementation process of the UN 2030 Agenda, which evokes sharp criticism. The "Reflection Group on the 2030 Agenda" – a group of civil society organizations from Uruguay, Malaysia, Lebanon, Fiji, the US and Germany, and supported by the German Friedrich-Ebert-Foundation – finds it

> irritating that the International Chamber of Commerce (ICC) as coordinator of the Global Business Alliance for 2030 (an umbrella group of major global industry associations and business organizations) can claim to play a key role in implementing the 2030 Agenda. Corporate lobby groups such as the ICC have been advocating for exactly those trade, investment and financial rules that have destabilized the global economy and exacerbated inequalities in both the global North and the global South.
>
> *(Martens 2016, 12)*

Of course, it is in their legitimate interest that societal actors express their views and have contact with politicians. However, the critical criterion is that this must be as transparent as possible and fully compliant with international standards of compliance and code of conduct. Especially politicians and public decision makers need to reveal their potentially conflicting interests. There need to be strict rules for side-line jobs, and discretionary earnings and dependencies need to be publicized.

A special case of conflicting interest is a conflict of interest in the more technical use of the term. A "conflict of interest" normally refers to conflicting interests embodies within the same actor (either individual or organization). Conflict of interest describes "a situation where an individual or the entity for which they work – whether a government, business, media outlet or civil society organization – is confronted with choosing between the duties and demands of their position and their own private interests" (Transparency International 2017a, 4). Transparency International (TI) issue a Corruption Perceptions Index (CPI), which "measures the perceived levels of public sector corruption worldwide" (Transparency International 2017b, 3). According to TI, more than 6 billion people live in countries which achieve less than 50% of the optimal results, i.e. they "live in countries that are corrupt" (3). Particularly worrying is the correlation between corruption and freedom of expression.

> Freedom of expression is vital for exposing corruption and the injustices that it causes. Top performers in the CPI – those with lower levels of corruption – do far better in protecting the rights of journalists and activists. Conversely, bottom performers – those countries with higher levels of corruption – are more likely to stifle the voices of their citizens and media … CPI results correlate not only with the attacks on press freedom and the reduction of space for civil society organizations. High levels of corruption also correlate with weak rule of law, lack of access to information, governmental control over social media and reduced citizens' participation. In fact, what is at stake is the very essence of democracy and freedom.
>
> *(Transparency International 2017b, 3)*

In 2018, TI reported that "of all journalists who were killed in the last six years, more than 9 out of 10 were killed in countries that score 45 or less on the index" (Transparency International 2018b).

4.5.3 Inequalities imply uneven negotiation powers and impede settlements

Conflicting interests among variably powerful protagonists are particularly difficult to resolve. This is very often related to financial status. Talking about the widening gap between the extreme rich and the extreme poor, the IMF acknowledges that increasing progressivity of income taxation would not hurt economic growth: "However, this could be difficult to implement politically, because better-off individuals tend to have more political influence, for example, through lobbying, access to media, and greater political engagement" (IMF 2017, 13).

Of course, such an imbalance of power exists on all levels, from individual relations to global politics. The US's unilateral terminations of the Iran Nuclear Deal, the INF Treaty, or the COP 21 agreement – which followed populist logic – is only possible because of the US's economic and military power. There is ample evidence that cases of the David-against-Goliath type are only very rarely won by the weaker party. The failure of the Doha Round discussed above might also be attributed to this imbalance. Consumers have little power in lawsuits against large corporations if class actions are not possible. NGOs have much smaller budgets than large corporations and cannot employ as many professional lobbyists and lawyers. On the other hand, the financial elites have means to protect and increase their wealth. Tax havens, the Panama Papers (or Paradise Papers) have revealed, sequester the huge volume of funds which are hidden away from the public. In many such cases, which range from legal tax avoidance to human trafficking and everything in between, the missing taxes are only the smallest damage to the global community.

4.5.4 Leadership and power structures

A subtle form of "conflicting interests" is given when a certain group of people with relatively uniform interests is in power and they influence the way leadership and power structures are shaped. Change is very hard to trigger if those in power dictate the rules of the game. At least in the public sphere democratic constitutional states should control such self-service of the elites by corresponding institutional measures, foremost among them the separation of powers and the corresponding checks and balances. But as we have just discussed, such institutional control measures might be bypassed by lobbyism or corruption. Moreover, there is a wide arena of informal, subtle power structures which are not captured by codified written laws and regulations. There are dominant power structures in societal areas which are not, or only partially regulated by the state.[15] Naturally, such power structures are not obvious – at least not to the uninvolved – and precisely because of this they are so difficult to detect, to discuss and to change. However, quite often they are related to a misuse of power, unfair treatment and discrimination, sometimes even to sexual abuse. The fact that scandals of misuse of power, of sexual abuse and discrimination have occurred in diverse organizations – in the Catholic Church as well as the movie industry, among musicians and pop singers as well as in university contexts, in large corporations as well as in public organizations, among athletes as well as in youth work – provides evidence of the fact that power and leadership structures must be very carefully managed and controlled.

The public discourse on incidents of misuse of power has increased awareness of this topic – but sensitivity to power abuse or experience discrimination is presumably still much more prevalent among the sufferers of discrimination than in the general public. Women are much more aware of male dominance, abuse and harassment than men. Average believers are much more critical about power structures in religious institutions than the officials. People in the global South are often much more aware of their colonial background than people in the countries of the former colonial rulers.

Solution perspectives

The conflicting interests discussed above have in common that established codified norms do not facilitate fairness because laws are not existent or their application is hindered by obscuration or because strong inequalities produce unfair situations. How can this be changed? I submit that apart from applying the fairness principle we discussed above, there is primarily one strategy which is important: *increase transparency*. It is crucial that all stakeholders involved in conflicting interests have access to the same (relevant) information. How can you negotiate if your counterpart has relevant information which you do not have? This must be considered a minimum requirement for any resolution of conflicting interests. Only then can the negotiation partners openly discuss and negotiate a fair solution. For issues of public concern it is particularly important that all relevant information is made public. Public sensitivity to issues of conflicts of interest, lobbyism and corruption has increased in recent years but more still needs to be done towards transparency. Lobbyism is a grey zone. The examples given above indicate that much more transparency is needed to really make the role of lobbyists transparent and fair.

1. Much more needs to be done to fight corruption. Although the OECD recommends enacting a dedicated whistle-blower protection law, applying across public and private sectors, this is not implemented throughout: "Only about a quarter of world exports come from countries with active law enforcement against companies bribing abroad" (Transparency International 2018a, 4). 53% of the G20 score below 50 (out of 100) in the Corruption Perception Index 2017. Although corruption is meanwhile punishable in many countries, it can be quite challenging to prove in concrete cases. The German Criminal Code, for instance, places that "[w]hosoever undertakes to buy or sell a vote for an election ... shall be liable to imprisonment" (Federal Law Gazette (Bundesgesetzblatt) 2013, 108e). However, it is not easy to prove that somebody wanted to buy or sell a vote. Therefore, in my view the best way to drain the swamps of illegitimate influence is full transparency on payments. Where is the thin line between legitimate advice and illegitimate influence? Precisely because this cannot be clarified in general, the best mitigation strategy towards conflicting interest is transparency. Of course, for the negotiation partners involved, transparency makes things more complicated. New stakeholders might enter the scene, ask critical questions, explanations need be given, etc. This costs time and energy, makes the process longer and potentially increases cost. However, this is the price we need to pay for fairness and justice.

2. The control of leadership and power structures requires a governance system which defines, among others, ways to manage and control the exercise of power. Every structure of power enforcement needs some system of checks and balances which limit power enforcement.

3. Diversity needs to be fostered. Diversity is one of the sustainable action principles in the second part of the book (see 16.2). Diversity makes companies more innovative and profitable. It is the role of an ideal diversity management process to challenge and to change existing cultures of organization, structures and conditions of power (Özdemir 2019).

4. Closely related to new cultures of organization, of management, is also a new culture of leadership. Künkel sees a disconnect between the discourses on global transformation and on leadership: "Hardly any scholars writing on sustainability conceptualize global sustainability challenges as leadership challenges for governments, businesses, civil society organizations, and citizens. But there is a growing body of research and literature on leadership approaches that move beyond individualistic notions of leadership limited to organizational contexts" (Künkel 2019, 45). Künkel founded the Collective Leadership Institute whose vision is "to empower future-oriented people to lead collectively towards a sustainable future" (CLI 2019). Core ideas of this approach are that leadership should shift from the concept of an individual focus, siloed and hierarchical approaches, in which dialogue and collaboration are mere side-issues to a new kind of "collective leadership", in which non-hierarchical structures and collaboration become the norm, and contribution to the common good an inherent goal of leadership (Künkel 2019, 48).

Notes

1 Of course, this was only possible because of Mikhail Gorbachev's policy of glasnost and peristroika in the USSR.
2 Absolute poor is commonly defined as living on less than US$ 1.90 per day (World Bank 2015).
3 This is largely caused by the fact that the very populous countries China, India, Pakistan, Indonesia, and Nigeria have moved into the MIC-category (Sumner 2010a, 14).
4 HDI is a measure normalized between zero (no human development) and one (full development), to which educational status, life expectancy and GDP p.c. in PPP contribute, each by one-third.
5 In the following we concentrate on global poverty, i.e. the poor living in LICs because in the MICs it would primarily be the respective domestic government who is in charge of this.
6 Similarly Niu et al. have investigated electricity consumption and human development level in 50 countries from 1990–2009 and state: "the higher the income of a country, the greater is its electricity consumption and the higher is its level of human development" (Niu et al. 2013, 338). They measure human development by five indicators: per-capita GDP, consumption expenditure, urbanization rate, life expectancy at birth and the adult literacy rate.
7 Hobson investigated the values and beliefs of climate sceptics and how entrenched these are in other beliefs. She concludes that rational arguments on the conceptual level will not have any effect. "In short, if 2 hours seeing (at times quite challenging) climate scenarios for your local region, and then 3 days spent deliberating cannot dispel the myriad of forms of climate scepticism, what will?" (Hobson & Niemeyer 2012, 409).

8 How can a climate sceptic ever take any medicine, sit in an airplane, or undergo surgery but not trust what the vast majority of scientists say? It has been shown that "the use of religious labels and metaphors in the construction of climate identities performs important rhetorical functions. Use of such metaphors serves to delegitimize climate science and to separate out 'good' scientists from 'bad' scientists, given the presence of scientists in the sceptics camp, but in the process it runs the risk of delegitimizing science itself" (Jaspal, Nerlich & van Vuure 2016, 821). We should realize that these people will not be convinced by further empirical data.

9 The precise correlation between right-wing populism and climate scepticism is not clear yet. Lockwood speaks of "a surprising dearth of academic research" that investigates the nature and causes of the right-wing populism and "climate skepticism" (Lockwood 2018, 713).

10 In their study "Right-wing populism and market-fundamentalism", Ötsch and Pühringer argue that right-wing populism is also reinforcing "market fundamentalism" and vice versa, which would be threatening democracy in the twenty-first century (Ötsch & Pühringer 2017).

11 Peter Beyer has a slightly different perspective on this and emphasizes the relation between fundamentalism and globalization. Fundamentalisms would be "rather modern developments in a globalized world" (Beyer 2010, 283). He sees fundamentalism as "reaction against" but not as "reactionary" as in any prior situation, an "*ancien régime*" would be sought to establish (Beyer 2010, 272, 283).

12 The IMF study calculated average statutory corporate income tax rates for balanced samples of 37 advanced economies, 92 emerging markets, and 59 low-income developing countries.

13 At the outset of the Doha Round in 2001, ministers committed "to comprehensive negotiations aimed at: substantial improvements in market access; reductions of, with a view to phasing out, all forms of export subsidies; and substantial reductions in trade-distorting domestic support. We agree that special and differential treatment for developing countries shall be an integral part of all elements of the negotiations and shall be embodied in the schedules of concessions and commitments and as appropriate in the rules and disciplines to be negotiated, so as to be operationally effective and to enable developing countries to effectively take account of their development needs, including food security and rural development" (WTO, Ministerial Declaration 2001, Section 13).

14 Another example was German Chancellor Merkel's blockade of an EU deal for emissions reduction of European cars in 2013. Many people in Germany were irritated by the "similarity between the position taken by Merkel and that represented by the automobile industry" as *Der Spiegel* wrote (Spiegel Online 2013). According to German press agency DPA, EU diplomats "had voiced anger with German efforts to torpedo the deal. 'It is a scandal,' one unnamed diplomat told DPA" (Spiegel Online 2013).

15 In the German corporate world, for instance, it will take another forty years until the top executives of the top 100 German companies are equitably distributed among men and women – if the current trend continues (BCG 2019).

Bibliography

Alvaredo, Facundo, Lucas Chancel, Thomas Piketty, Emmanuel Saez, & Gabriel Zucman. "Global inequality dynamics: New findings from WID.world." *NBER Working Paper Series*, Cambridge, MA: National Bureau of Economic Research, 2017.

Amadeo, Kimberly. "Doha round of trade talks. The real reason why it failed." *The Balance* 25, 2019: 06.

Atkinson, Anthony B. *Inequality: What can be Done?* Cambridge, MA: Harvard University Press, 2015.

Ayres, Robert U. "Industrial metabolism: Theory and policy." In *The Greening of Industrial Ecosystems*, Braden R. Allenby and Deanna J. Richards, (eds.), 23–37. Washington, DC: The National Academies Press, 1994.

Banerjee, Abhijit V. & Esther Duflo. *Poor Economics. A Radical Rethinking of the Way to Fight Global Poverty*. New York: Public Affairs, 2011.

Barder, Owen. "What is poverty reduction?" Working Paper, Center for Global Development, 2009.

BCG. "(Em)Power Women. Wo Chefetagen in Sachen Vielfalt stehen." Boston Consulting Group, Boston Consulting Henderson Institute, Technical University of Munich, 2019.

Beyer, Peter. "Religion out of place? The globalization of fundamentalism." In *The Routledge International Handbook on Globalization Studies*, Bryan S. Turner, (ed.), 269–286. Abingdon, Oxon: Routledge, 2010.

Borger, Julian. "'Brought to Jesus': the evangelical grip on the Trump administration." *theguardian.com*. 11. 01 2019. www.theguardian.com/us-news/2019/jan/11/trump-administration-evangelical-influence-support (Accessed 17. 01 2019).

Bundesregierung. *Perspectives for Germany. Our Strategy for Sustainable Development*. Berlin: Federal Government Publication Office, 2002.

———. *German Sustainable Development Strategy*. Berlin: Federal Government Publication Office, 2016.

Charlton, Andrew. "The collapse of the Doha trade round." *CentrePiece – The Magazine for Economic Performance 210*, 2006 (Autum): 21.

Christensen, Clayton M. *The Innovator's Dilemma. Why New Technologies Cause Great Firms to Fail*. Boston, MA: Harvard Business Review Press, 2016.

CLI. Collective Leadership Institute. 2019. www.collectiveleadership.de/blog/article/cli-history/(Accessed 24. 04 2019).

Court of Justice of the European Union. "Judgment of the court (Ninth chamber) of 21 June 2018 – Commission v Germany." *Case C-543/16*, Luxembourg, 21. 06 2018.

Dibley, Arjuna. "Why the rise of populist nationalist leaders rewrites global climate talks." *Nature: Climate Change 5*, 2018: 12.

Easterly, William. *The White Man's Burden. Why the West's Efforts to Aid the Rest have done so much Ill and so Little Good*. New York: Penguin, 2006.

Federal Law Gazette (Bundesgesetzblatt). "StGB (German criminal code)." 2013.

Francis, Pope. *Laudato Si*. Vatican, 2015. http://w2.vatican.va/content/francesco/en/encyclicals/documents/papa-francesco_20150524_enciclica-laudato-si.html

Fraune, Cornelia, & Michèle Knodt. "Sustainable energy transformations in an age of populism, post-truth politics, and local resistance." *Energy Research & Social Science 43*, 2018: 1–7.

Gandesha, Samir, "Can Europe make it? Understanding right and left populisms." 23 May 2018 www.opendemocracy.net/can-europe-make-it-samir-gandesha/understanding-right-and-left-populisms [accessed 9. Sept. 2019].

Gapminder. 2019. www.gapminder.org (Accessed 15. 07 2019).

Giddens, Anthony. *The Politics of Climate Change*. Cambridge: Polity Press, 2009.

Global Footprint Network. globalfootprintnetwork.org. 20. July 2019. www.globalfootprintnetwork.org (Accessed 11. 01 2019).

Gößling-Reisemann, Stefan. "Pfadwechsel – schwierig aber notwendig." In *Industrial Ecology*, von Armin von Gleich & Gößling-Reisemann Stefan (eds.) 154–161. Wiesbaden: Vieweg + Teubner, 2008.

Gould, Eric D, & Alexander Hijzen, "Growing Apart, Losing Trust? The Impact of Inequality on Social Capital." IMF Working Paper (WP/16/176), August 2016, Washington, DC.

Gyimah-Brempong, Kwabena, "Corruption, economic growth, and income inequality in Africa." in: *Economics of Governance 3*, 2002: 183–209.

Halme, Minna, Galina Kallio, Lindeman Sara, Arno Kourula, Maria Lima-Toivanen, & Angelina Korsunova. "Sustainability innovation at the base of the pyramid through multi-sited rapid ethnography."*Corporate Social Responsibility and Environmental Management 23 (2)*, 01 2016: 113–128.

Hempel, Monty. "AnthropoTrumpism: Trump and the politics of environmental disruption." *Journal of Environmental Studies and Sciences 8 (2)*, 2018: 183–188.

Hobson, Kersty, & Simon Niemeyer. "'What sceptics believe': The effects of information and deliberation on climate change scepticism." *Public Understanding of Science 22 (4)*, 2012: 396–412.

IMF. *Does Corruption Affect Income Inequality and Poverty? Working Paper of the Internationally Monetary Fund WP 98/76*, Washington, DC: International Monetary Fund, 1998.

———. *Fiscal Monitor: Fiscal Monitor:Tackling Inequality*. Washington, DC: International Monetary Fund (IMF), 2017.

Isidor, Chris. "Buffett says he's still paying lower tax rate than his secretary." *CNN.com*. 04. 03 2013. https://money.cnn.com/2013/03/04/news/economy/buffett-secretary-taxes/ index.html (Accessed 08. 07 2019).

Jang, Jinseop, Jason McSparren, & Yuliya Rashchupkina. "Global governance: Present and future." *Palgrave Communications 2* 19. 01 2016: 1–5.

Janský, Petr. "Effective tax rates of multinational enterprises in the EU." Herausgeber: www.greens-efa.eu/. 22. 01 2019. www.greens-efa.eu/files/doc/docs/ 356b0cd66f625b24e7407b50432bf54d.pdf (Accessed 15. 07 2019).

Jaspal, Rusi, Brigitte Nerlich, & Kitty van Vuure. "Embracing and resisting climate identities in the Australian press: Sceptics, scientists and politics." *Public Understanding of Science 10*, 2016: 807–824.

Künkel, Petra. *Stewarding Sustainability Transformation. An Emerging Theory and Practice of SDG Implementation*. Springer Nature, 2019.

Lamy, Pascal. "Day 9: Talks collapse despite progress on a list of issues." 29. 07 2008. www.wto.org/english/news_e/news08_e/meet08_summary_29july_e.htm (Accessed 22. 01 2019).

Lessenich, Stephan. *Neben uns die Sintflut. Die Externalisierungsgesellschaft und ihr Preis*. Berlin: Hanser, 2017.

Lockwood, Matthew. "Right-wing populism and the climate change agenda: Exploring the linkages." *Environmental Politics 27 (4)*, 2018: 721–732.

Marshak, Robert. "Contemporary challenges to the philosophy and practice of organization development." In *Reinventing Organization Development. New Approaches to Change in Organzations*, David L. Bradford & Warner W. Burke. (eds.), 19–41. San Francisco, CA: Pfeiffer, 2005.

Martens, Jens. *The 2030 Agenda – A New Start Towards Global Sustainability?* Beirut, Bonn, Ferney-Voltaire, Montevideo, New York, Penang, Rome, Suva: Reflection Group on the 2030 Agenda for Sustainable Development, 2016.

Maxwell, Simon. "Review: The white man's burden." *Overseas Development Institute*, 2006. 21 September 2006.

Merriam-Webster. "Politics." 2019. www.merriam-webster.com/dictionary/politics (Accessed 15. 07 2019).

Meyer, Thomas. *Fundementalismus. Aufstand gegen die Moderne*. Reinbek bei Hamburg: Rowohlt, 1989.

Milanovic, Branko. *Global Inequality: A New Approach for the Age of Globalization.* Cambridge, MA: Harvard University Press, 2016.

Mudde, Cas, & Cristóbal Rovira Kaltwasser. "Exclusionary vs. inclusionary populism: Comparing contemporary Europe and Latin America." *Government and Opposition 48 (2),* 01 2013: 147–174.

Mudde, Cas, & Cristóba Rovira Kaltwasser. *Populism: A Very Short Introduction.* Oxford: Oxford University Press, 2017.

Müller, Jan-Werner. *Was ist Populismus?* Berlin: Suhrkamp, 2016.

Myrskylä, Mikko, Hans-Peter Kohler, & Francesco C. Billari. "Advances in development reverse fertility declines." *Nature 460,* 08 2009: 741–743.

New York Times. "Global trade after the failure of the Doha round." *New York Times,* 2016.

Niu, Shuwen, Yanqin Jia, Wendie Wang, He Renfei, Hu Lili, & Liu Yan. "Electricity consumption and human development level: A comparative analysis based on panel data for 50 countries." *Electrical Power and Energy Systems,* 2013: 338–347. doi: 10.1016/j.ijepes.2013.05.024.

Ötsch, Walter O., & Pühringer. Stephan. "Right-wing populism and market-fundamentalism." *ICAE Working Paper Series – No. 59,* Linz: Institute for Comprehensive Analysis of the Economy, Linz University, 2017.

Ouedraogo, Nadia S. "Energy consumption and human development: Evidence from a panel cointegration and error correction model." *Energy, 31.* 10 2013: 28–41.

Our World in Data. *https://ourworldindata.org/.* 2019. https://ourworldindata.org/(Accessed 15. 07 2019).

Oxfam. *Public Good or Private Wealth?* Oxford: Oxfam, 2019.

Özdemir, Feriha. *Managing Capability. Ein Ansatz zur Neubestimmung von Diversity Management.* Wiesbaden: Springer Fachmedien, 2019.

PEP. "Getting to zero." *The Poverty-Environment Partnership (PEP),* 2016.

Pure Earth. "Agbogbloshie dumpsite." Ghana. *http://worstpolluted.org/.* 2013. http://worstpolluted.org/projects_reports/display/107 (Accessed 11. 01 2019).

Quran. "https://quran.com." 2016. https://quran.com/2/215 (Accessed 08. 07 2019).

Rao, Narasimha D., & Paul Baer. "'Decent living' emissions: A conceptual framework." *Sustainability* 4 (4), 2012: 656–681.

Rawls, John. *A Theory of Justice.* Cambridge, MA: Belknap, 1971.

Rawls, John. "The basic structure as a subject." In *Values and Morals,* Goldman, A. I., J. Kim & D. Reidel (eds.), 47–71. Dordrecht: D. Reidel, 1978.

Sachs, Jeffrey D. *The End of Povery. Economic Possibilities for Our Time.* New York: The Penguin Press, 2005.

———. *The Age of Sustainable Development.* New York: Columbia University Press, 2015.

Safo, Amos. "www.newsfromaftrica.org." Oct 2002. web .archive.org/web/20110718110241/www.newsfromafrica.org/newsfromafrica/articles/art_827.html (Accessed 11. 01 2019).

Schroth, Jörg. "Utilitarismus und Verteilungsgerechtigkeit." *Zeitschrift für philosophische Forschung 60,* 2006: 37–58.

Spiegel, Der. "Miss Ernte". *Der Spiegel 03/2019.* Hamburg, 11. 01 2019.

Spiegel Online. *Germany blocks CO2 reduction deal.* 27. 06 2013. www.spiegel.de/international/europe/germany-delays-eu-decision-on-lower-co2-emissions-for-cars-a-908176.html (Accessed 23. 01 2019).

Stamp, Jimmy. "Fact of fiction? The legend of the Qwerty keyboard." *Smithsonian.com.* 03. 05 2013. www.smithsonianmag.com/arts-culture/fact-of-fiction-the-legend-of-the-qwerty-keyboard-49863249/ (Accessed 25. 05 2019).

Statistisches Bundesamt. *Nachhaltige Entwicklung in Deutschland. Indikatorenbericht.* Wiesbaden: Statistisches Bundesamt, 2008.

Stiglitz, Joseph. *Globalization and Its Discontents.* New York, London: W.W. Norton, 2002.

———. *Making Globalization Work.* New York: W.W. Norton, 2006.

Stroh, David Peter. *Systems Thinking for Social Change.* White River Junction, VT: Chelsea Green Publishing, 2015.

Sumner, Andy. "Global poverty and the new bottom billion: What if three-quarters of the world's poor live in middle-income countries?" Working Paper, IDS – Institute for Development Studies, Brighton, UK, 2010.

Transparency International. *Access All Areas. When EU Politicians Become Lobbyists.* Brussels: Transparency International EU, 2017a.

———. *Corruption Perceptions Index 2017.* Berlin: Transparency International, 2017b.

———. *Exporting Corruption. Progress Report 2018: Assessing enforcement of the OECD Anti-Bribery Convention.* Berlin: Transparency International, 2018a.

———. *Transparency International.* 21. 02 2018b. www.transparency.org/news/feature/corruption_perceptions_index_2017 (Accessed 24. 01 2019).Date 21 February 2018 https://www.transparency.org/news/pressrelease/corruption_perceptions_index_2017_shows_high_corruption_burden_in_more_than

———; Global Witness. *European Gateway. Inside the Murky World of Golden Visas.* Transparency International; Global Witness, 2018.

Trump, Donald. Twitter, 06. 11 2012.

UN. "Fundamentalism and populism pose deepening threat to women human rights defenders, UN experts warn." *News.un.org.* 25. 11 2016. https://news.un.org/en/story/2016/11/546242-fundamentalism-and-populism-pose-deepening-threat-women-human-rights-defenders (Accessed 17. 01 2019).

UN-DESA. "Global context for achieving the 2030 Agenda for sustainable development international financial flows and external debt (Development Issues No. 10)." *UN Development Strategy and Policy Analysis Unit - Development Policy and Analysis Division,* 2017. New York.

UNDP. *Human Development Indices and Indicators 2018 Statistical Update.* New York: United Nations Development Program, 2018.

UNDP Bangladesh. *Environmental Impact of Rohingya Influx.* Dhaka: UNDP Bangladesh, 2018.

Washington Post. "White evangelicals voted overwhelmingly for Donald Trump, exit polls show." *Washingtonpost.com.* 09. 11 2016. www.washingtonpost.com/news/acts-of-faith/wp/2016/11/09/exit-polls-show-white-evangelicals-voted-overwhelmingly-for-donald-trump/?utm_term=.e726110d2d65 (Accessed 17. 01 2019).

WBGU. *World in Transition: A Social Contract for Sustainability.* Berlin: German Advisory Council on Global Change (WBGU), 2011.

WCED. *Our Common Future.* Oxford: Oxford University Press, 1987.

Weber, Max. *Economy and Society. An Outline of Interpretetive Sociology. 1,* Guenther Roth and Claus Wittich (eds.), Berkeley, Los Angeles, London: University of California Press, 1978.

WID (Facundo Alvaredo, Lucas Chancel, Thomas Piketty, Emmanuel Saez &, Gabriel Zucman). *World Inequality Report 2018,* edited by the World Inequality Lab, 2018.

Wilkinson, Richard, & Kate Pickett. *The Spirit Level. Why Greater Equality Makes Societies Stronger.* New York, Berlin, London, Sydney: Bloomsbury Press, 2011.

Worldbank (International Bank for Reconstruction and Development), *Poverty and Shared Prosperity 2018: Piecing Together the Poverty Puzzle.* Washington, DC: World Bank, 2018.

World Bank. "World bank forecasts global poverty to fall below 10% for first time. Major hurdles remain in goal to end poverty by 2030." 04. 10 2015. www.worldbank.org/en/news/press-release/2015/10/04/world-bank-forecasts-global-poverty-to-fall-below-10-for-first-time-major-hurdles-remain-in-goal-to-end-poverty-by-2030 (Accessed 27. 04 2019).

WTO. "Ministerial declaration." *Doha WTO Ministerial 2001: Ministerial Declaration.* Geneva: WTO, 20. 11 2001.

———. "The Doha round." *Wto.org.* 2019. www.wto.org/english/tratop_e/dda_e/dda_e.htm (Accessed 22. 01 2019).

Extrinsic barriers
1– Institutional deficiencies

While the barriers considered thus far are intrinsic to the concept of sustainability, the following are not necessarily tied to the concept of sustainability – they are extrinsic. Extrinsic barriers are grouped into those which relate to institutional deficiencies and those which are zeitgeist-dependent. The institutional-deficiency barriers include our dominant social institutional frameworks: the systems of the market, of politics, of law, of technology and of our institutions in general. I call zeitgeist-dependent those barriers which express attitudes and values of our current era but might potentially be different in other times. Zeitgeist, a German loan word in English, describes the "general intellectual, moral, and cultural climate of an era" (Merriam-Webster 2019). Accordingly, I see some sustainability barriers as zeitgeist-dependent, namely short-term orientation and consumerism.

Extrinsic barriers
I – Institutional deficiencies

5

ECONOMY

Faulty market system

Contents

Although the sustainability barriers related to the economy are considerable in practice, it might nevertheless be possible to capture them here in essence, since they are relatively easy to describe on a conceptual level. Under Market failure we will first discuss the tragedy of the commons, free-riding and externalization (5.1) before deliberating on the pervasiveness of economic thinking (5.2).

5.1 Market failure

Out of the vast topic of market failures we can only look at three closely related aspects which I consider critical for our non-sustainability: the problem of the public goods or the "tragedy of the commons", the problem of free-riding, and the problem of externalizing environmental and social costs.

The "belief in the efficacy of the market mechanism is a fundamental organising principle of the policy prescriptions of modern economics" (Perman et al. (2011/1996), 5). Adam Smith assumed that the profit-seeking of "any man" renders societal benefit:

> He generally, indeed, neither intends to promote the public interest, nor knows how much he is promoting it … he is, in this as in many other cases, led by an invisible hand to promote an end which was no part of his intention … By pursuing his own interest he frequently promotes that of society more effectively than when he really intends to promote it.
>
> *(quoted from Perman et al. 2011, 5)*

However, the functioning of the market mechanism rests on preconditions which are often not given under real world situations, in particular when public goods are involved. "A market failure occurs when the market does not allocate scarce resources to generate the greatest social welfare" (Hanley, Shogren & White 2006, 42). Habitat destruction or biodiversity loss are results of such market failure. To be sure, what is considered "market failure" here is often related to corresponding policy failures, since the market framework is ultimately set by policy decisions.[1] We will discuss some of the challenges of such policy failures at the global level in the next chapter (see Chapter 6).

Of the several market failures being discussed in environmental economics (e.g. externalities, non-exclusion, non-rival consumption, and asymmetric information (Hanley, Shogren & White 2006, 75)), we focus on the most obvious and most harmful cases: public goods and their non-excludability and externalities.[2]

5.1.1 Public goods and the tragedy of the commons

Economists categorize goods according to two important criteria: rivalry and excludability. The former is given when a good which is consumed by one agent cannot be consumed by another one. The ice cream in a kiosk is both rivalrous and excludable: the more ice cream one person consumes the less remains for others (rival); and consumers can be excluded from consumption (e.g. via the price). (Perman et al. 2011, 113f.). If both excludability and rivalry are given, goods are pure private goods. Here the market mechanisms work fine (although other preconditions of the ideal market might be absent as well, such as perfect competitiveness or perfectly informed agents).

Environmental problems often relate to public goods (or public bads). Open-access resources like fishing grounds are definitely rivalrous, since the amount of catch of one agent reduces the potential catch of the others. However, they are non-excludable since nobody is being prevented from fishing outside territorial waters. Pure public goods are neither excludable nor rivalrous – at least not in certain limits. The navigation supported by a lighthouse does not reduce other ships' navigation possibilities. It is clear that really consistent examples of pure public goods are rare because in extreme cases they become rivalrous. The clear air and clean water a person enjoys does not reduce the possibility of others to do the same, as long as the related activity does not inflict damages on the good. However, of course in our highly consumptive society this is often the case: the more people swim in clean water the dirtier it gets (e.g. by sunscreen). Therefore, it is widely accepted that the market mechanisms heralded by Adam Smith faces serious conceptual and operational challenges for environmental problems: "Nobody who has seriously studied the issues believes that the economy's relationship to the natural environment can be left entirely to market forces" (Perman et al. 2011, 15). The main reason is that "markets cannot supply public goods … The supply of public goods is (part of) the business of government. The existence of public goods is one of the reasons why all economists see a role for government in economic activity" (118).[3]

In short: pure public goods do not have a price, and because they are considered free, they cannot be allocated by market mechanisms. It is the role of the government to preserve the public goods by different kinds of measures: bans, thresholds, protective measures, price mechanisms, etc. We cannot (and need not) discuss this here – there is no end of literature on the proper handling of public goods.

In his famous article, "The Tragedy of the Commons", Garrett Hardin, already some fifty years ago, pointed out that "as the human population has increased, the commons has had to be abandoned in one aspect after another" (Hardin 1968, 1248). This was said when the human population was roughly 3.5 billion – not even half of today's figure. However, Elinor Ostrom, who won the 2009 Nobel Memorial Prize in Economics for her work on the commons, challenges Hardin's conclusion:

> Garrett Hardin ... earlier argued that users were trapped in accelerated over-use and would never invest time and energy to extract themselves. If that answer were supported by research, the SES [social-ecological systems] framework would not be needed to analyze this question. Extensive empirical studies by scholars in diverse disciplines have found that the users of many (but not all) resources have invested in designing and implementing costly governance systems to increase the likelihood of sustaining them.
>
> *(Ostrom 2009b, 420)*

The technical details regarding how to protect public goods, and which coordinating measures will show best results, etc., are technical economic problems which we cannot discuss here. On a national level, government intervention for the sake of the public good has partly been successful. Strict environmental regulation, price incentives, bans and fines, limits and thresholds, taxes and subsidies have all helped to protect public goods, to reduce pollution and preserve the natural environment. On an international level, however, this is only just beginning to take place. There have been promising examples of establishing mechanisms on the global scale (e.g. the Montreal Protocol, the Kyoto Protocol), but there are substantial difficulties at the global level to come to agreements which are binding, enforceable, sufficient, and operational (see Chapter 6), which has to do with two other issues related to public goods: free-riding and externalities.

Solution perspectives

Donella Meadows indicates "three ways to avoid the tragedy of the commons:

- *Educate and exhort.* Help people to see the consequences of unrestrained use of the commons. Appeal to their morality. Persuade them to be temperate. Threaten transgressors with social disapproval or eternal hellfire.
- *Privatize the commons.* Divide it up, so that each person reaps the consequences of his or her own actions. If some people lack the self-control to

stay below the carrying capacity of their own private resource, those people will harm only themselves and not others.
- *Regulate the commons.* Garret Hardin calls this option, bluntly, 'mutual coercion, mutually agreed upon.' Regulation can take many forms, from outright bans on certain behaviours to quotas, permits, taxes, incentives. To be effective, regulation must be enforced by policing and penalties" (Meadows 2008, 119).

The challenge, however, is of course how to regulate the *global* commons with regulation which is to the largest extent national only – up to now. We will return to this from a policy point of view in the next chapter. From an economic point of view, Ostrom's work can show the way, for she closely investigated this very point: how commons can be regulated. Ostrom would presumably agree with Meadows that regulation needs to be enforced. However, this enforcement need not come from an external authority, it can be established by self-governance. This will be one result of our discussion of global governance below (see Chapter 6), but it is also supported by Ostrom's work on self-governance. Ostrom et al. constructed an empirical common-pool resources game and explored to what extent individuals would be able to self-govern the use of such resources (Ostrom, Walker & Gardner 1992). Whereas Thomas Hobbes said that "Covenants, without the sword, are but words, and of no strength to secure a man at all", Ostrom et al. argue that individuals may well come to agreements and covenants which are effective for governing the use of common-pool resources even in the absence of one external force if certain conditions are met (Ostrom et al. 1992):

> Individuals may be able to arrive at joint strategies to manage these resources more efficiently. To accomplish this task, they must have sufficient information to pose and solve the allocation problems they face. They must also have an arena where they can discuss joint strategies and perhaps implement monitoring and sanctioning. In other words, when individuals are given an opportunity to restructure their own situation, they frequently – but not always – use this opportunity to make credible commitments and achieve higher joint outcomes without an external enforcer.
>
> *(Ostrom et al. 1992, 414)*

This would not mean that cooperation is always happening but it does "challenge the Hobbesian conclusion that the constitution of order is only possible by creating sovereigns who then must govern by being above subjects, by monitoring them, and by imposing sanctions on all who would otherwise not comply." The authors conclude: "these experiments suggest that covenants, even without a sword, have some force" (Ostrom et al. 1992, 414).

Whereas "isolated, anonymous individuals overharvest from common-pool resources", it makes a huge difference if people can interact: "Simply allowing communication, or 'cheap talk,' enables participants to reduce overharvesting and increase joint payoffs contrary to game theoretical predictions" (Ostrom 2010, 641).

Unlike the medieval use of common land which functioned well for centuries, common-pool resources today are overcultivated because the different agents do not know each other and cannot directly communicate with one another. According to Ostrom, groups in which participants are frequently communicating achieve almost optimal usage of common goods without overexploitation (Ostrom 2009a, 220).

Regarding the role of government and administration offices, one can say that the key success factor for administration concepts lies in the relationships between the actors having an interest in successful resource management. The social capital which humans can create by networking on different levels, with NGOs and with government actors, substantially influences an effective feedback, the learning processes and ultimately the development of new and better solutions (Ostrom 2009a, 227f.; Ostrom 2009b, 419).

In any case, the communication and networking processes envisioned here will call for common values, prosocial values in particular, which might be challenging in a global context. This is one of many reasons why the establishing of a global civil society is needed to support the protection of the global commons.

5.1.2 Free-riding

The costs of establishing a public good like climate protection should, of course, be covered by all who benefit from the results – which is, in case of global public goods, all of humanity. However, since people experience the climate regardless of any contribution of their own to mitigate its change, there is the problem of free-riding, which means that agents enjoy a good, but do not want to participate in the costs. Regardless of what other countries are doing, everybody would benefit from climate protection (Hanley et al. 2006, 78).

Free-riding occurs on different levels, from the personal to the international; on the latter for instance in the competition on location advantages: "In recent decades, international tax competition … has led to a steady downward trend in corporate income tax rates", which reduces overall tax progressivity (IMF 2017, 15).

In cases where there are sanctioning mechanisms (e.g. strict environmental administrative law), the attraction of free-riding for potential wrongdoers is being reduced by high risk of discovery and severe punishments. In many cases of environmental pollution the risk of discovery is relatively low. Similarly, the severity of the penalty for violating environmental regulation is also quite often pretty moderate. Sometimes it might even be more "rational" for agents to just accept the risk of paying a penalty upon detection rather than adjust their own behaviour accordingly. For instance, environmental regulation was long said to be so harmless "that it was cheaper for big corporations to pay the fine than obey the law" (Stateimpact 2013). According to the philosopher Vittorio Hösle, it is a slap in the face of justice that a shoplifter is more severely punished than someone who assaults human livelihood for rapid profit (Hösle 1994, 126).

Solution perspectives

There will always be free-riding – but there are different ways to keep it within limits. Sanctioning mechanisms do have a discouraging effect. Since risk is related to the product of likelihood and impact, a low probability of detection can be countered with higher fines – so that a "rational agent" would rather comply. Consequently, officials all over the world call for higher fines for environmental pollution, be it in Texas (Stateimpact 2013), Taipei (Taipeitimes 2017), or the UK (Out-Law 2016; Wateractive 2009). In a global context there is, of course, not a comparable enforcement authority as on the national level, thus making free-riding particularly problematic for global public goods.

However, despite the absence of one single enforcement authority at the global level, sanctioning mechanisms might still work, as just discussed around Ostrom's work. Grechenig, Nicklisch & Thöni (2010) also argue that sanctioning mechanisms need not come from the state to be effective – as long as effective punishment can take place. In the case where several agents need to collaborate in provisioning a public good, the "option to punish others substantially improves cooperation" (ibid.). As we will also see in Chapter 6, on global governance, modes of horizontal regulation can be quite effective if there is sufficient self-interest for countries to collaborate: "The international policy objective is clear but elusive: find incentives to motivate nations with strong and diverse self-interests to move voluntarily toward a collective goal of reduced carbon emissions" (Hanley et al. 2006, 79).

In case of climate change one can argue that it is particularly the rich, "developed" countries that need to start addressing the global issues:

a) They not only have the power to do so – because of their economic and social strength;
b) They not only have responsibility to do so – because of their unsustainable growth path, seen with hindsight from a historical perspective;
c) They should also realize that free-riding is not an option for them (neither is the excuse that other countries emit more) because their free-riding could function as an excuse for others to also free-ride, which must be avoided by all means. There are no winners on a sinking ship.
d) Taking action against global issues does not need to imply a reduction in the quality of life. Wilkinson and Pickett point out that the severe limitation of economic growth in the rich countries, which is needed to cut carbon emissions, "does not mean sacrificing improvements in the real quality of life – in the quality of life as measured by health, happiness, friendship and community life, which really matters" (Wilkinson & Pickett 2011, 231).

The problem of free-riding is, in my view, particularly problematic in the rational choice model, which sees self-interest as the only motivation for "rational actors".[4] If the ruling paradigm of action is self-interest, there is no way to escape the free-rider problem. One might object that the problem of free-riding is real and not dependent on any specific economic model. That is certainly true. However, I do think that our dominant theories about economic transactions and the underlying view of human beings has a considerable effect on behaviour. Empirical studies have looked at the socio-economic and societal backgrounds of people in free-riding contexts. The results support the idea that people behave according to their beliefs. Marwell and Ames corroborated the "weak" free-rider hypothesis during a study conducted among first-year graduate students and confirmed that "free-riding does exist", although they observed more contribution to the public good than the individual decision theory would suggest (Marwell & Ames 1981, 307f.). Moreover, they found that free-riding seems to be, by trend, more prevalent among economists. Deliberating about potential explanations, they wonder that maybe economists "start behaving according to the general tenets of the theories they study. Confronted with a situation where others may not behave rationally, they nevertheless behave the way good economic theory predicts" (ibid., 309).

Frank, Gilovich & Regan (1993) support this interpretation. They state that students of economics "are much more likely than others to free-ride in experiments that called for private contributions to public goods" (ibid., 159). As suggested by "a variety of evidence", there is "a large difference in the extent to which economists and noneconomists behave self-interestedly. We believe our survey of charitable giving and our prisoner's dilemma results lend additional support to the hypothesis that economists are more likely than others to free-ride" (ibid., 170). The authors tie this conclusion to the recommendation:

> in an ever more interdependent world, social cooperation has become increasingly important – and yet increasingly fragile. With an eye toward both the social good and the well-being of their own students, economists may wish to stress a broader view of human motivation in their teaching.
>
> *(Frank et al. 1993, 170f.)*

Luckily there is meanwhile sound empirical evidence that the rational choice model is erroneous. In a paper summarizing her speech on the occasion of receiving the Nobel Prize in Economics, which she received especially for her work on the commons, Ostrom gives an account of the academic discussion around rational choice and concludes: "The most important lesson for public policy analysis derived from the intellectual journey I have outlined here is that humans have a more complex motivational structure and more capability to solve social dilemmas than posited in earlier rational-choice theory" (Ostrom 2010, 664). However, on the same occasion she also bemoans that some "policy analysts, public officials, and scholars ... have not yet absorbed the central lessons articulated here" (664).

The learning from this for our current purpose is, I propose, that our understanding of humans as moral and social agents does very well matter. We are by far not the "rational" – which is often equated to "self-interested" – beings that economists of the past used to suggest.

Bechtel et al. argue that both reciprocity and altruism are effective means to combat free-riding. Reciprocity – presumably a core element of *any* ethics[5] – mitigates free-riding, because "it stabilizes expectations about others' co-operative behavior which fosters the evolution of co-operation." Altruism, in turn, "defined as a general concern for the well-being of other individuals "could well "increase an individual's willingness to support global climate co-operation" (Bechtel, Genovese & Scheve 2017, 5f.).

5.1.3 Externalization societies: shift costs to the weak, to nature, and to the future

People in the global North tend to think that their own societies are doing relatively well in addressing environmental pollution, and it is true that the environmental policies of the last decades have contributed to a much cleaner environment – locally. What many people do not realize, however, is that this is only possible because costs are constantly externalized to other regions, to nature, and to the future. An "externality exists where a consumption or production activity has unintended effects on others for which no compensation is paid" (Perman et al. 2011, 10). Externalities are often related to public goods or bads. Acid rain caused by sulphur emissions of a coal-burning plant is one example. Due to the structural change which many economies of the global North have undergone in the past decades, pollution is not so visible any more in these countries – but it has not disappeared. To some extent, the dirty, polluting activities or industries have just been relocated to other regions. China, for instance, has become the world's workshop in the last few decades and has experienced significant environmental issues because of this. The World Bank estimated that the cost of environmental pollution in China in the last decades of the twentieth century ranged between 3–8% of GDP (World Bank 1997, 71). In the first decade of our century, the cost related to health losses due to air pollution alone would be in the range of 3–4% of GDP (World Bank 2007, 15).[6]

At the same time, China has shipped products to Europe and North America and with that an invisible carbon load. There are significant carbon flows embodied in the trade between regions – and it is the poor countries who pay the price for the outsourcing of the "developed world":

> A general finding supported by all studies is that there is a large and growing flow of embodied carbon from poor and emerging to developed

countries. This is important to understand regional emission drivers and
may have a variety of applications in policy.

(Peters, Davis & Andrew 2012, 3273)

However, it is not only carbon emissions that are being externalized – it is the
overall environmental and social cost. The water contained in melons which
grew in arid regions and are exported to the North, the fertilizers used to grow
oranges, the GHGs emitted by flying tropical fruits or flowers to market, the
cotton grown in dry regions which require several thousand litres of water per
kilogram produced. Above all, it is the labour standards and social rights of the
people in the developing world upon which the global production system
capitalizes.

Some of these circumstances might exceed the strict economic sense of
external cost – in the sense that *one* transaction produces an effect to those
not involved in the transaction. However, from a systems perspective,
I presume one can well argue that the global North is externalizing costs to
developing and emerging economies.[7] In this respect, externalizing is certainly no new phenomenon – it has been going on for centuries and is (at
least partly) rooted in colonial history. Two hundred and fifty years ago,
Rousseau wondered about an occupier's right to claim ownership over land.
He named three conditions, which must be met that a "first occupier" can
rightly do so: "the ground wasn't already occupied by someone else; he
occupies only as much as he needs for his subsistence; he takes possession of
this ground not by an empty ceremony but by labour and cultivation"
(Rousseau 2017/1762, Book 1, §9). In the inglorious history of Europe's
colonialism, for the largest part none of these conditions were met – even if
we disregard the shameful chapter on slavery. The ground had been occupied, it was mostly taken for greed and glory, and often not taken by labour
and cultivation.

The Western world has pushed externalizing in such a way that the German
sociologist Stephan Lessenich calls these societies "externalization societies" (Lessenich 2017). Regardless of whether one sees external cost as a weakness of the
market or a failure of the state, somebody has to internalize the cost, somebody
has to pay the price (Lessenich 2017, 46). According to Lessenich, people in the
externalization society do not live beyond their own means, they live beyond
the means of *others*: "People in the externalization society are doing well because
others tighten their belt, because abstinence is practiced *elsewhere* – and this per-
manently and lasting, so that the externalization society may prosper not only
today but tomorrow and in future" (ibid., 65; original emphasis).

Moreover, costs are not only externalized to the environment and to poorer
regions but also to the future. We are constantly piling up debt, be it as states,
corporations, or individual consumers. The credit card symbolizes this motto:
enjoy now, pay later. We are so used to consuming by increasing our debts that
we do not even realize that this has not always been the case.

To some extent it is normal that we value the present more than the future. Economists call this "positive time preference", i.e. people have a preference for receiving benefits now rather than later (Conrad 2010, 11). As long as there is a positive interest rate (i.e. discount rate), for economists this is not a problem because the one who lends money to others will get the interest in return. For society as a whole, the discount rate would

> reflect its collective 'sense of immediacy', which, in turn, may reflect a society's level of development ... higher discount rates tend to favour more rapid depletion of non-renewable resources and lower stock levels for renewable resources. High discount rates can make investments to improve or protect environmental quality unattractive when compared with alternative investments in the private sector ... Such a situation could lead to the current generation throwing one long, extravagant, resource-depleting party that left subsequent generations with an impoverished inventory of natural resources, a polluted environment, and very few options to change their economic destiny.
>
> *(Conrad 2010, 15)*

The question of time preference and discount rate is relevant for calculating future costs of climate change and pollution. Economists discount future costs (e.g. a future damage) in order to estimate the savings which are needed today to compensate future costs. Given the characteristic of exponential growth (as is all growth with constant interest rates), two parameters are absolutely critical: the discount rate and the time (see Figure 3.1). If one expects a cost of € 1000 to be due in 100 years, it is sufficient to save € 1.15 today – at a discount rate of 7%. If the cost is only due in 200 years, the savings would be as little as € 0.1, or 10 cents. However, if the discount rate is zero, then the total future cost has to be saved today. Therefore, the

> choice of the social discount rate plays a critical role in cost-benefit analysis and project evaluation, and has been a subject of intense debate for the last several decades. In a perfectly competitive world without market distortions, the market interest rate is the appropriate social discount rate.
>
> *(Zhuang et al. 2007, 21)*

The market interest rate, however, has varied considerably in the past – with peaks of more than 15% in the USA in the early 1980s and close to zero in the present decade.

An appropriate discount rate is also important to calculate the financial return on investments, for instance in education. The higher the assumed discount rate, the lower the financial return of education. For the OECD average, the net financial returns for a man attaining tertiary education as compared to a man attaining upper secondary education, in equivalent US$ converted using PPPs for GDP is US$ 267,000 at a discount rate of 2% but only US$ 44,000 at a discount rate of 8% (data for 2015) (OECD 2018, 109).

Similarly, the future cost of climate change and environmental destruction varies greatly depending on the discount rate and is, of course, heatedly discussed among economists. The Stern Report, which called for immediate action for climate protection to avoid high cost in the future, was criticized because it would have assumed too low a discount rate. For instance, the economist William Nordhaus from Yale University challenges the report, among other reasons, because the discount rate of just 1.4%, which Stern assumes, would be far too low compared to historical values of discount rates (Nordhaus 2007). Others, like the renowned economist Kenneth Arrow, do also challenge Stern's values of the discount rate but do not find the general policy implications changed by that (Arrow 2007).

The critical question is how realistic it is to extrapolate from past developments. If it is true that climate change, to stay with this argument, is an unprecedented phenomenon in human history – and we have the best possible evidence for that – then we should at least be very careful with extrapolations from the past.

The jurist Joseph Guth raises even a more fundamental concern about discounting. He challenges the very idea of cost-benefit analysis (which is the basis for discounting) if the entire future of humankind is at stake. Guth calls this a "paradox of discounting":

> while each small part of the ecologically-functioning biosphere may seem dispensable for some finite gain, the entire biosphere, though finite and composed only of these small parts, is nevertheless indispensable ... We can sacrifice any of the individual parts, but we cannot sacrifice the whole ... An economy that sells off bits and pieces of the earth without means for recognizing they are parts of an invaluable whole cannot be projected into a future in which that economy is assumed to grow forever.
>
> *(Guth 2008, 112)*

Solution perspectives

The problems of environmental externalities have been discussed for ages. Almost half a century ago, the OECD recommended: "In other words, the cost of these measures should be reflected in the cost of goods and services which cause pollution in production and/or consumption" (OECD 1972, Annex A. a) 4). So in principle and on a generic level, the solution for externalities is rather simple. Pollution must become expensive, public goods must get a price tag. And since it is particularly global public goods which are at stake, these public goods should ideally get a global price. Only then will we truly reduce both the environmental and social burden – otherwise pollution will just be exported. How this would be implemented in detail is a matter of efficiency, practicability and enforcement.

A first step would be to phase out environmentally harmful subsidies. Subsidies for fossil fuels in the order of 5% of global GDP systematically discriminate against renewable energy. Of course, this would not need to take place overnight. Subsidies could be phased out over ten years or so.

The next step could be to revise the taxation system in a way that resource-friendly activities are supported while consumption is taxed. This goes back to an idea proposed by Arthur Pigou in 1920.[8] Taxes on labour could be reduced while those on the consumption of energy or raw materials, or surface sealing should be introduced (see Repetto et al. 1992; von Weizsäcker, Lovins & Lovins 1995; Berg et al. 2012; Sachs 2015, 216). Repetto et al. suggested to switch

> some of the revenue burden from taxes on income, employment, and profits to environmental charges on resource waste, collection, and pollution would yield double economic benefits. Reducing tax rates on income and profits would reduce the marginal excess burden by $0.40 to $0.60 per dollar of reduced tax revenue.
>
> *(Repetto et al. 1992, 11)*

If this is done gradually, long-term and tax-revenue neutrally, this could boost energy and resource efficiency, support new business models of access (instead of possession), and facilitate the transitions towards service economies. We cannot and need not discuss here which of the several policy measures would be most appropriate: taxes, subsidies, cap-and-trade-mechanisms, or others. The instruments are available in principle – and some have already been proven to be effective in practice. They need to be rolled out and to reflect the true ecological and social costs, on a global scale.

With regards to the externalization to the future, some economists from the "prescriptive school" "argue that society should not adopt the preference of individuals" (Perman et al. 2011, 78). Drawing on an argument of Amartya Sen, Perman et al. distinguish between the

> individuals' roles as consumer and citizen. This can be applied in the intertemporal context. As citizens exhibiting commitment toward others, future generations in this case, individuals would not necessarily wish to discount the future at the same rate as they do when considering the distribution of their own utility over time.
>
> *(Perman et al. 2011, 78)*

However, while Perman et al. refer to the prescriptive school of economists who would imply a zero rate of utility discounting (Perman et al. 2011, 78),

Guth even wants to abandon the concept of discounting for environmental decisions entirely: "No rate of discounting, whether positive, negative, zero, or variable, can mold that structure into a form that can manage large-scale ecological degradation" (Guth 2008, 112).

It is here where meaning and purpose become important, as does our understanding of ultimate reality. If individual self-interest, one's own well-being, is the ultimate measure of things, arguing in favour of true care for future generations will be as hard as arguing for fellow humans in remote regions of the globe. Seeing oneself, however, in close communion with the whole of creation, being indebted to one's ancestors, being responsible for one's children and grandchildren and wishing them a life as fulfilling and rich as the one we are privileged to live, leads to a completely different perspective on discounting the utility function. This leads us to the next section: the pervasiveness of economic thinking.

5.2 The pervasiveness of economic thinking

Apart from their failures and limitations, the market mechanisms can still claim to allow very efficient allocation of goods and services within their legitimate domain. However, economic thinking has become pervasive today and permeates all sectors of human and social life way beyond what is appropriate.

"The life of money-making is one undertaken under compulsion, and wealth is evidently not the good we are seeking; for it is merely useful and for the sake of something else" (Aristotle, EN I.3. 1096a). Aristotle's verdict on money is outspoken. Money and wealth are not goods in themselves, they are mere means for another, ultimate goal. For Aristotle, good in itself was *eudaimonia*, the happy, successful life. I presume many people would agree with Aristotle's judgement even today – at least in theory. But reality is different. Money makes the world go round and is often the dominant paradigm. While Aristotle considered wealth as "merely useful", the usefulness of things has become their ultimate rationality in modern thinking. It is therefore not surprising that "[n]ormative resource and environmental economics is predominantly founded in utilitarian ethics" (Perman et al. 2011, 59).

Bosselmann elaborates on the relationship between the concepts of ecology and economy. When Ernst Haeckel introduced the term "Oecologie" (ecology) in 1866, he used Herder's image of housekeeping. Ecology would describe "the fundamental principle of housekeeping against which 'economy' appears as a mere subdiscipline of efficient housekeeping" (Bosselmann 2017, 19). What used to be the servant has evolved into the master, economic logic has become pervasive, everything is subsumed under the logic of the market. Following this very logic and the ideas of the Washington Consensus, public services which used to be provisioned by the state (e.g. health care, public transport, utilities) were privatized in the 1990s, implying a huge shift of power from the public to the private sector (E. U. von Weizsäcker 2007).

The pervasiveness of market logic is closely related to the rational choice model of decision making. If humans are basically driven by their self-interest as the rational choice model suggests, it would only be natural to apply that kind of logic to all human activity. However, as discussed above, the rational choice model has come under attack from several sides. Tversky and Kahneman, pioneers in behavioural economics, had already noticed in the 1980s that "the logic of choice does not provide an adequate foundation for a descriptive theory of decision making." (Tversky & Kahneman 1986, 252). Hodgson points to the self-immunizing character of rational choice: "This utility-maximising version of rational choice theory has the character of a universal 'explanation' that can be made to 'fit' any set of events" (Hodgson 2012, 94). Since "any manifest behavior" could be fitted into the rational choice framework, the rational choice theory would be unfalsifiable according to Hodgson (104). McKinnon argues in a similar direction and points to the religious character of rational choice and its corresponding perception of the market:

> If we think about how the market is commonly talked about by economists and in the business pages of the newspaper, we find that it is often talked about in the terms that Jews, Christians and Muslims use to speak of God. The market is not only viewed as a powerful force, but a personified entity: the market acts, reacts, punishes and rewards. The market is often seen as omniscient and omnipotent, and well on its way to becoming omnipresent: who can listen to Adam Smith's oft repeated turn of phrase about the 'invisible hand', and not hear the echo of god language? When the god makes demands, what can we do but obey? ... Rational choice theory's literalized market metaphor naturalizes a view of human nature that makes the one-dimensional society of neo-liberal politics seem inevitable: the selfish god of rational choice theory, given the chance, will create human beings in its own image ... the popularity of rational choice thinking stems in part from its consonance with the 'new common sense' neo-liberal politics has created. That the theory reflects this new socio-political reality is not surprising, but that its god metaphor threatens to recreate human beings in its image is a good deal more worrying. No longer is there theoretical scope for human action that is not calculatingly self-serving. Rational choice theory suggests that not only is the capitalist market natural, even the gods are subject to its laws. We, too, must obey.
>
> *(McKinnon 2013, 540)*

The pervasiveness of market logic becomes most evident by the fact that it also influences moral convictions and value systems. Vohs et al. investigated the "psychological consequences of money" (Vohs, Mead & Goode 2006). In nine experiments, the authors "tested whether activating the concept of money leads people to behave self-sufficiently, which we define as an insulated state wherein people put forth effort to attain personal goals and prefer to be separate from others" (Vohs et al. 2006, 1154). They found that "Relative to participants primed with neutral concepts, participants primed with money preferred to play alone, work alone, and put more physical distance between themselves and a new acquaintance" (ibid.). Compared to people not reminded of money, "people reminded of money reliably performed independent but socially insensitive actions" (ibid., 1156). The authors were surprised by the magnitude of the effect, since their participants were all highly familiar with money. They see their own results resonating with the work of Frank et al. (1993) cited above, who

> reported that university students majoring in economics made self-interested moves in social dilemma games more often than students of other disciplines. Economics students were also more convinced than non-economists that their competitors would make self-interested moves, a result which echoes the present thesis that money evokes a view whereby everyone fends for him- or herself. The self-sufficient pattern helps explain why people view money as both the greatest good and evil. As countries and cultures developed, money may have allowed people to acquire goods and services that enabled the pursuit of cherished goals, which in turn diminished reliance on friends and family. In this way, money enhanced individualism but diminished communal motivations, an effect that is still apparent in people's responses to money today.
>
> *(Vohs et al. 2006, 1156)*

In an investigation of "morals and markets", Falk and Szech present "controlled experimental evidence on how market interaction changes how human subjects value harm and damage done to third parties" (Falk & Szech 2013). In their experiment, "subjects decide between either saving the life of a mouse or receiving money", and they conclude that

> market interaction displays a tendency to lower moral values, relative to individually stated preferences. This phenomenon is pervasive. Many people express objections against child labor, other forms of exploitation of the workforce, detrimental conditions for animals in meat production, or environmental damage. At the same time, they seem to ignore their moral standards when acting as market participants, searching and buying the cheapest electronics, fashion, or food, and thereby consciously or subconsciously creating the undesired negative consequences to which they generally object. We have shown that this tendency is prevalent already in very simple bilateral trading where both market sides are fully pivotal in that if they refuse to trade, the mouse will stay alive. In markets with many buyers and sellers, diffusion of being pivotal for outcomes adds to moral decay. This 'replacement' logic is a common feature of markets, and

it is therefore not surprising that the rhetoric of traders often appeals to the phrase that 'if I don't buy or sell, someone else will'.

(Falk & Szech 2013, 710)

Moreover, apparently people can sense whether others behave egoistically or altruistically. Oda et al. studied social exchanges involving money and found that people behave differently toward altruists and non-altruists: "This represents strong evidence for the evolution of cognitive adaptations providing counter-strategies to subtle cheating by identifying altruists and engaging in exchanges with this group exclusively" (Oda et al. 2009).

Falk and Szech have shown that "market interaction causally affects the willingness to accept severe, negative consequences for a third party" (Falk & Szech 2013, 707). The authors do see the tremendous virtues of markets in generating information about scarcity and allocating resources efficiently, and they concede that other organizational forms of allocation and price determination do not generically place higher value on moral outcomes:

However, focusing on the causal effects of institutions, we show that for a given population, markets erode moral values. We therefore agree with the statement … that we as a society have to think about where markets are appropriate—and where they are not.

(Falk & Szech 2013, 707, 713)

Solution perspectives

Fighting the prevalence of economic reasoning will first and foremost require that the existence of areas of life which follow a different logic is acknowledged. It is not the market which is to blame *if we allow it* to control our life. If people lose sense for those aspects of life that defy economic rationality, we need not wonder about its prevalence. If we do not manage to uphold those things which we appreciate as ends in themselves, it is only natural that everything will become a means, which is ultimately subdued under the logic of the market. Children's play, human relationships, social collaboration, empathy and trust, love and compassion – none of these can truly be grasped by the logic of preference.

Criticism of the prevailing economic rationality comes from quite different angles. According to Erich Fromm, psychologist and philosopher, economic activity used to be a means toward an end, and the end was life itself. In capitalism, however, economic activity, success and material profit became an end in itself. It has become the fate of humans that they have to contribute to the success of the economy, that they have to accumulate capital, not for their own happiness or salvation but as end in itself. Man became a cog in the great business machine – a cog the purpose of which was outside himself (Fromm 1942, 95).

The German economist Wolfgang Sachs criticized economists' fixation on GDP, by which nature and community got out of focus, despite 30–50% of social work being performed informally, mostly by women (Sachs 1993). Joseph Stiglitz criticizes the "market fundamentalism" of the Washington Consensus, which was imposed by international financial institutions (Stiglitz 2002, 221). Schneidewind and Zahrndt question the cumulation logic of our society and call for the right measure for time and space, for possessions and market. This would not necessarily imply less, but also different, better, nicer (Schneidewind & Zahrnt 2014, 51).

Christian Felber sees a strong correlation between our economic and our societal values. While the capitalist market economy would be legitimized by the assumption that the selfishness of individual actors leads to the greatest welfare of all, Felber views this assumption is "a myth and fundamentally wrong". Competition would to some extent incite people to perform. However it would do even greater harm to society and the relationships among people.

> If people's ultimate goal is to maximise their own benefit and therefore operate against each other, they learn to cheat others and to view this as being right and normal. However, by cheating others we do not treat them as humans of equal value; we violate their dignity.
>
> *(Felber 2018, 14)*

Dignity is, as I see it, the decisive concept to counter the prevalence of market logic. By taking up a distinction made by Seneca, Immanuel Kant distinguished between price and dignity.

> In the realm of ends everything has either a price or a dignity. What has a price is such that something else can also be put in its place as its equivalent; by contrast, that which is elevated above all price, and admits of no equivalent, has a dignity.
>
> *(Kant 2002, BA 78)*

Dignity defies the calculating logic of economic rationality – it is "elevated above all price". We need to rediscover the importance of those aspects of life which escape economic rationality and are to be dignified in this sense. Culture is one of them. Culture and cult have the same origin – both etymologically and historically. Religious cults have functioned as communal and formalized experience in which the divine could reveal itself.[9] Since their earliest beginnings, religious cults have been fountains of cultural creativity. Many great pieces of art, of literature, of music have their "setting in life" (*Sitz im Leben*) in religious cults.

It is here, in cult and culture, where life is taking place, where amazement is happening and awe expressed. I consider amazement as the origin of all great human endeavours, of religion, philosophy, music, art, literature, science … However, amazement is, as it were, the opposite of the calculus of cost-benefit analysis. How can you be amazed by something if you calculate cost and benefit? How can you build up trust if you suspect that your counterpart is just trying to calculate or even manipulate your behaviour (e.g. by generous presents)? Love is, in its essence, vulnerable because the loving person gives him- or herself but can never be sure that this love will be answered. The beloved one cannot be controlled, is inaccessible. This has been an insight of large parts of the occidental philosophy and the Judaeo-Christian tradition.

However, there is an interesting intersection with the current ideas of a German sociologist – Hartmut Rosa. Rosa states that a "world which were fully known, planned and controlled, would be a dead world" (Rosa 2019, 8). A successful relation to the world, which Rosa calls an experience of *resonance*, is characterized (among others) by a moment of inaccessibility. Things over which we have full control lose their quality of resonance (ibid., 52). According to Rosa, vitality, contact and true experience can only arise from the inaccessible.

Therefore, one can argue from different lines of thought that we need to make sure not to subdue everything under the calculating logic of the market and its cost-benefit analyses. A fully known, a fully calculated world would be boring, would not inspire us, would not give rise to amazement and awe. We need to foster the cultural, the communal, the social aspects of life. As we will discuss below, frugality and simplicity are less to be seen as moral obligations but as chances for living with little burden (see 14.4).

Notes

1 I'm indebted to Rolf J. Langhammer for pointing me to this correlation as well as for several other valuable comments on this current chapter.
2 Perman et al. summarize the market failure as such: "Actual economies do not satisfy the conditions of the ideal competitive economy. Agents do not have perfect information, markets are incomplete, markets are often not perfectly competitive, markets cannot supply public goods, and much of consumption and production behaviour generates external effects. These 'failures' will result in inefficient allocation of resources" (Perman et al. 2011/1996, 134).
3 Of course one cannot say that *all* economists see it like this, given the striking differences among them. For instance, German economist C. Christian von Weizsäcker frankly states: "The global problems are being resolved by surrendering the leading role to the economy over politics" (C. C. von Weizsäcker 2000, 166). Only if markets are de-politicized could true democracy take place (ibid.).
4 This description of rational choice is a simplification, to be sure. Ostrom summarized the rational agent as follows: "Fully rational individuals are presumed to know (i) all possible strategies available in a particular situation, (ii) which outcomes are linked to each strategy given the likely behavior of others in a situation, and (iii) a rank order for each of these outcomes in terms of the individual's own preferences as measured by utility. The rational strategy for such an individual in every situation is to maximize expected utility" (Ostrom 2010, 643).
5 I am always impressed about such reciprocity in the exhortation by which ancient Israel was called to care for the stranger: "Do not oppress a foreigner; you yourselves know how it feels to be foreigners, because you were foreigners in Egypt" (Exod. 23:9).

6 "Evaluation of the health losses due to ambient air pollution using willingness-to-pay measures raises the cost to 3.8 percent of GDP" (World Bank 2007, 15) (xv).

7 The International Resource Panel states: "Even in the two countries which arguably have made the most explicit efforts towards decoupling, Japan and Germany, and where at first glance domestic resource consumption shows stabilization or even a modest decline, deeper analysis shows that many goods contain parts that have been produced abroad using major amounts of energy, water and minerals. Thus some of the advanced countries are managing the problem of high resource intensity by 'exporting' it elsewhere" (UNEP 2011, ix).

8 He had quite up-to-date ideas, for instance that our impact on the environment has lasting impact on the future: "the environment of one generation can produce a lasting result, because it can affect the environment of future generations. Environments, in short, as well as people, have children" (Pigou1932/1920, 96).

9 In the Jewish tradition it is not, as often stated, humans which are the culmination, the pride of creation. Rather it is the Sabbath, the place in which all of creation comes to rest in the presence of God.

Bibliography

Aristotle Nicomachean Ethics ed. Web Atomics The Internet Classics Archive Daniel C. Stevenson Übersetzung: W. D. Ross. Kitchener. Batoche Books, 1999.

Arrow, Kenneth J. "Global climate change: A challenge to policy." *Economists' Voice* 6, 2007: 1–5.

Bechtel, Michael M., Federica Genovese, and Kenneth F. Scheve. "Interests, norms and support for the provision of global public goods: The case of climate co-operation." *British Journal of Political Science* 2017: 1–23.

Berg, Christian, Christian Calliess, Christa Liedtke, Georg Meran, Ortwin Renn, Wolfgang Schmalz, & Miranda Schreurs. "Sustainable economic activity and growth." In *Dialogue on Germany's Future*, Bundespresseamt, (ed.), 86–95. Berlin: Federal Government Publication Office, 2012.

Bosselmann, Klaus. *The Principle of Sustainability. Transforming Law and Governance.* London, New York: Routledge, 2017.

Conrad, Jon M. *Resource Economics.* Cambridge: Cambridge University Press, 2010.

Falk, Armin, & Nora Szech. "Morals and markets." *Science 340*, 10.05 2013: 707–711.

Felber, Christian. *Gemeinwohlökonomie.* München: Piper, 2018.

Frank, Robert H., Thomas Gilovich, & Dennis T. Regan. "Does studying economics inhibit cooperation?" *Journal of Economic Perspectives* 7(2), 1993: 159–171.

Fromm, Erich. *The Fear of Freedom.* London: Kegan Paul, 1942.

Grechenig, Kristoffel, Andreas Nicklisch, & Christian Thöni. "Punishment despite reasonable doubt – a public goods experiment with sanctions under uncertainty." *Journal of Empirical Legal Studies 7(4)*, 18. 11, 2010: 847–867.

Guth, Josef H. "Resolving the paradoxes of discounting in environmental decisions." *Transnational Law & Contemporary Problems Nr 18(1)*, Oct., 2008: 95–114.

Hanley, Nick, Jason F. Shogren, & Ben White. *Environmental Economics.* London: Palgrave Macmillan, 2006.

Hardin, Garrett. "The tragedy of the commons." *Science 162*, 13. 12 1968: 1243–1248.

Hodgson, 2012" to the references list: Hodgson, Geoffrey M. On the Limits of Rational Choice Theory, *Economic Thought* 1:94–108.

Hösle, Vittorio. *Philosophie der ökologischen Krise.* Munich: Beck, 1994.

IMF. *Fiscal Monitor: Tackling Inequality*. Washington, DC: International Monetary Fund (IMF), 2017.

Kant, Immanuel. *Groundwork for the Metaphysics of Morals* (ed. Allan W. Wood). New Haven, CT, London: Yale University Press, 2002 (1785).

Lessenich, Stephan. *Neben uns die Sintflut. Die Externalisierungsgesellschaft und ihr Preis*. Berlin: Hanser, 2017.

Marwell, Gerald, & Ruth E. Ames. "Economists free ride, does anyone else?" *Journal of Public Economics 15*, 1981: 295–310.

McKinnon, Andrew M. "Ideology and the market metaphor in rational choice theory of religion: A rhetorical critique of 'religious economies." *Critical Sociology 39 (4)*, 2013: 529–543.

Meadows, Donella H. *Thinking in Systems. A Primer* (ed. Diana Wright). White River Junction, VT: Chelsea Green Publishing, 2008.

Merriam-Webster. www.merriam-webster.com/dictionary/. 2019. www.merriam-webster.com/dictionary/zeitgeist (accessed 27. 02 2019).

Nordhaus, William D. "A review of the Stern Review on the economics of climate change." *Journal of Economic Literature XLV 9* 2007: 686–702.

Oda, Ryo, Takuya Naganawa, Shinsaku Yamauchi, Noriko Yamagata, & Akiko Matsumoto-Oda. "Altruists are trusted based on non-verbal cues." *Biological Letters 5 (6)*, 23.12 2009: 752–754.

OECD. *Recommendation of the Council on Guiding Principles concerning International Economic Aspects of Environmental Policies*. Paris: OECD, 1972.

———. *Education at a Glance 2018: OECD Indicators*. Paris: OECD Publishing, 2018.

Ostrom, Elinor. „Gemeingütermanagement – Perspektive für bürgerschaftilches Engagement." In *Wem gehört die Welt? Zur Wiederentdeckung der Gemeingüter*, Silke von Helferich & Heinrich-Böll-Stiftung (eds.), 218–228. Munich: oekom, 2009a.

———. "A general framework for analyzing sustainability of social-ecological systems." *Science 325*, 24. 07 2009b: 419–422.

———. "Beyond markets and states: Polycentric governance of complex economic systems." *The American Economic Review 100*, 06 2010: 641–672.

———, James Walker, and Roy Gardner. "Covenants with and without a sword: Self-governance is possible." *American Political Science Review 86*, 06 1992: 404–417.

Out-Law. "Record Scottish waste penalty shows high fines becoming normal in environmental cases, says expert." *Out-law.com*. 23. 02 2016. www.out-law.com/en/articles/2016/february/record-scottish-waste-penalty-shows-high-fines-becoming-normal-in-environmental-cases-says-expert/(Accessed 27. 04 2019).

Perman, Roger, Yue Ma, Michael Common, David Maddison, & James McGilvray. *Natural Resource and Environmental Economics*. Harlow, Essex: Pearson Education Ltd., 2011 (1996).

Peters, G. P., S. J. Davis, and R. Andrew. "A synthesis of carbon in international trade." *Biogeosciences 9*, 2012: 3247–3276.

Pigou, Arthur Cecile. *The Economics of Welfare*. London: Macmillan, 1932 (1920).

Repetto, Robert, Roger C. Dower, Robin Jenkins, & Jacqueline Geoghean. *Green Fees: How a Tax Shift Can Work for the Environment and the Economy*. Washington, DC: World Resources Institute, 1992.

Rosa, Hartmut. *Unverfügbarkeit*. Wien, Salzburg: Residenz-Verlag, 2019.

Rousseau, Jean-Jacques. *The Social Contract*. Jonathan Bennett, 2017 (1762).

Sachs, Jeffrey D. *The Age of Sustainable Development*. New York: Columbia University Press 2015.

Sachs, Wolfgang. „Die vier E's: Merkposten für einen maß-vollen Wirtschaftsstil." *Politische Ökologie 11 (33)*, 1993: 69–72.

Schneidewind, Uwe, and Angelika Zahrnt. *The Politics of Sufficiency. Making It Easier to Live the Good Life.* Munich: oekom, 2014.

Stateimpact. *Stateimpact.* 2013. https://stateimpact.npr.org/texas/2013/02/11/polluters-and-penalties-will-higher-fines-make-a-difference-in-texas/(Zugriff am 27. 04 2019).

Stiglitz, Joseph. *Globalization and Its Discontents.* New York, London: W.W. Norton, 2002.

Taipeitimes. *taipeitimes.com.* 14. 03 2017. www.taipeitimes.com/News/taiwan/archives/2017/03/14/2003666738 (Zugriff am 27. 04 2019).

Tversky, Amos, & Daniel Kahneman. "Rational choice and the framing of decisions." *The Journal of Business 59(4), Part 2: The Behavioral Foundations of Economic Theory,* 1986: 251–278.

UNEP. *Decoupling Natural Resource Use and Environmental Impacts from Economic Growth. A Report of the Working Group on Decoupling to the International Resource Panel.* Paris: UNEP, 2011.

Vohs, Kathleen D., Nicole L. Mead, & Miranda R. Goode. "The psychological consequences of money." *Science 314,* 17. 11 2006: 1154–1156.

von Weizsäcker, C. Christian. *Logik der Globalisierung.* Göttingen: Vandenhoek & Ruprecht, 2000.

von Weizsäcker, and Ernst Ulrich. *Grenzen der Privatisierung. Wann ist des Guten zu viel?* Stuttgart: S. Hirzel, 2007.

Wilkinson, Richard, & Kate Pickett. *The Spirit Level. Why Greater Equality Makes Societies Stronger.* New York, Berlin, London, Sydney: Bloomsbury Press, 2011.

———, Amory B. Lovins, and L. Hunter Lovins. *Faktor Vier. Doppelter Wohlstand – halbierter Naturverbrauch.* Munich: Droemer, 1995.

Wateractive. *Wateractive.* 2009. www.wateractive.co.uk/news/courts_urged_to_issue_higher_fines_for_environmental_pollution (Accessed 27. 04 2019).

Worldbank. *China 2020.* Washington, DC: The World Bank, 1997.

Worldbank. *Cost of Pollution in China.* Washington, DC: The World Bank Group, 2007.

Zhuang, Juzhong, Zhihong Liang, Tun Lin, & Franklin De Guzman. *Theory and Practice in the Choice of Social Discount Rate for Cost-Benefit Analysis: A Survey.* ERD Working Paper Series No 94, Manila: Asian Development Bank, 2007.

6

POLITICS

Lack of effective governance for global issues

Contents

Issues of (non-)sustainability are global in nature.[1] For the environmental dimension of sustainability this relates, for instance, to climate change, ocean acidification, loss of biodiversity, the large bio-geo-chemical cycles of phosphorus and nitrogen, but also to stratospheric ozone depletion and the dispersion of pollutants like nuclear waste or POPs. For many of these issues one can define global thresholds, beyond which the risk of abrupt and irreversible change increases significantly. The concept of planetary boundaries (Rockström et al. 2009; Steffen et al. 2015; Steffen et al. 2018) describes the safe operating space for humanity. Many social issues, which can be distinguished but not separated from environmental ones, are also of global relevance. Issues of migration, of global inequalities, international terrorism, organized crime, human trafficking and many more cannot be addressed by national policies alone.

The challenge to sustainability arises from the fact that we have not (yet) established effective mechanisms for dealing with these global issues. There do exist, of course, several organizations which operate globally. Paradigm examples are globally operating MNEs which maximize their profits by capitalizing on the differences in national jurisdiction – often legally. Then there are, of course, civil society organizations like NGOs or initiatives like #fridaysforfuture which are operating globally. They certainly contribute to a common global awareness and discussion of the issues and articulate clear political demands for change towards sustainability, thereby facilitating the establishment of a global civil society. However, most NGOs focus (for good reasons) on rather distinct aspects of sustainability, like combating hunger or poverty, fighting for human rights, protecting rainforests, animal rights, etc. – all critically important for sustainability. Although awareness of the interdependency of the issues has certainly increased, alignment is still needed between the different programs, initiatives and organizations. How can it be avoided that different programs or organizations counteract each other? How can trade-offs be mitigated? Furthermore, it must be noted that NGOs are not democratically legitimized, they mostly run on tight budgets and are therefore in danger of losing their independence, and they occasionally face governance issues. In other words, there is some void of legitimacy at the NGO level – which is not a big issue at the moment but might become one if the credibility of and trust in NGOs were threatened in future.

The strongest mandate for legitimate representation of the public interest is what the nation states agree upon in the best interest of their people, as codified in international agreements and treaties and public international organizations. The latter have considerable improvement potential, for public institutions and organizations on a global level (hereafter: international governmental organizations, IGOs) "provide, at best, only partial solutions, and implementation of even these solutions can be undermined by international competition and recalcitrance" (Walker et al. 2009, 1345).

6.1 Challenges of IGOs and multilateral international treaties

Global issues call for global resolutions, for some kind of governance. There have been significant milestones towards a better governance of these issues at the international level (e.g. Agenda 21, COP 21, 2030 Agenda) but the global public organizations and international treaties face several severe challenges which need to be addressed.

In the following we will list some of the challenges multilateral international agreements and treaties and IGOs are confronted with which impede effective regulation of global issues.

1.   *Agreement on general goals but not on operational targets*

Multilateral agreements or even global treaties are difficult to negotiate, simply due to the diversity of parties involved and their conflicting interests. Sometimes agreement is impossible even after lengthy negotiations. For instance, the WTO's Doha Round broke up inconclusively after several years of intense negotiations. In other cases, the world's nations were able to reach an agreement – but at the expense of rather generic goals and quite some wiggle room to allow for individual interpretations. The Agenda 21, which the world community agreed on in Rio 1992, is such an example. Nobody can really object to the general visions it calls for – combating poverty (I.3), changing consumption patterns (I.4), integrating the environment and development in decision making (I.8), or protecting the atmosphere (II.9) or combating deforestation (II.11) (UNCED 1992a). But its claims are largely generic and appellative – it contains more than 400 appeals introduced by a "should". In a similar way, the Rio declaration, authored at the same event and listing 27 principles, uses phrases which express good intentions but almost revokes them in the same breath:

National authorities should endeavour to promote the internalization of environmental costs and the use of economic instruments, taking into account the approach that the polluter should, in principle, bear the cost of pollution, with due regard to the public interest and *without distorting international trade and investment.*

(UNCED, Rio Declaration 1992b, principle 16; emphasis added)

If any "distortion" of international trade is excluded, how can you then, for instance, do justice to the call for internalizing the external cost of global logistical processes? To be sure, "Rio" was a huge leap forward: for the first time, the world agreed on a sustainability agenda. This was probably the most one could get. However, *global consensus was facilitated by neglecting concrete goals and operational details.*

2. Dependence on voluntary commitments

More than twenty years after UNCED, the Paris Agreement, i.e. the closing communiqué of COP 21, was more concrete in its targets. Of course, compared to Rio the scope was more confined (namely climate only) and agreement on a specific target could be reached. The parties agreed to hold the "increase in the global average temperature to well below 2°C above pre-industrial levels and pursuing efforts to limit the temperature increase to 1.5° C above pre-industrial levels" (UNFCCC 2015, Art. 2.1 (a)). Each party "shall prepare, communicate and maintain successive nationally determined contributions that it intends to achieve" (UNFCCC 2015, Art. 4.2). These contributions were to be "ambitious" (Art. 3). However, the Achilles' heel is that there is no consequence if the NDCs are not ambitious enough. In fact, this is exactly the current situation. There is an ambition gap between the sum of all the NDCs and the global target. The NDCs communicated so far would "imply global warming of about 3°C by 2100" (UNEP 2018). *Agreement on concrete targets was facilitated by a lack of obligation.*

3. Limited sanctioning mechanisms

Mainstream opinion on law enforcement is that regulations require a high compliance rate, which in turn, requires a monopoly of legitimate force (Zürn 2005, 3f.).[2] In the absence of a global monopoly of a legitimate force, one can question that a high compliance rate of international agreements will ever be possible in light of limited sanctioning mechanisms. The WTO does provide sanctioning mechanisms but they do not really work in practice because of lengthy decision processes, power imbalances and the fact that the mediation processes were shaped in favour of the industrialized countries (see Raffer & Singer 2001; Busch & Reinhardt 2004; Kress 2014, 53f.). In the case of conflict between WTO member states, for instance, poor countries often do not even have the capacities for the settlement process (Kress 2014, 53f.).The former General-Secretary of WTO, Pascal Lamy, was well aware of these issues when he conceded: "while the political decolonization took place more than 50 years ago, we have not yet completed economic decolonization" (Lamy 2007).

Therefore, even if consensus on specific and binding targets can be reached, *limited sanctioning mechanisms for non-compliance impede their effectiveness.*

4.   *No general acceptance*

The control of global public institutions is simply not accepted by all parties, in particular not by all major actors. This is true, for instance, for the International Criminal Court in The Hague, which is not supported by the US, Russia, or China – all permanent members of the UN Security Council, and all nuclear powers. Recent trends towards unilateralism impede the acceptance of international treaties even more. Especially the present US government's disrespect of international treaties poses a great risk to the acceptance of global treaties. By cancelling or withdrawing from the Intermediate-Range Nuclear Forces (INF) treaty, the Iran nuclear deal, the Paris Agreement, even calling into question NATO and openly demanding that a global public institution such as the World Bank should "serve American interests and defend American values" (Trump, BBC News 2019), undermines the struggle for establishing a framework of global cooperation.

The control of global public institutions requires that *the powerful subdue themselves under the rule of law* and do not abide by the law of power. This most basic principle of justice and fairness seems to be hardest to realize.

5.   *Existing (national or international) regulation not effectively enforced*

There is no equivalent to national police authorities at the international level. That's why national or international laws are frequently not enforced, as for instance the case of illegal, unreported and unregulated (IUU) fishing demonstrates: "IUU fishing is broadly defined as the use of fishing methods or practices that contravene fisheries laws, regulations or conservation and management measures" (EJF 2018, 5). Illegal fishing is, for instance, "conducted by national or foreign vessels in waters under the jurisdiction of a State, without the permission of that State, or in contravention of its laws and regulations". It "represents up to 26 million tonnes of fish caught annually, valued at USD 10 to USD 23 billion" (FAO 2019). Other examples of the violation of existing regulation is the treatment of e-waste, which is still illegally exported out of the EU despite increasing political pressure to stop this.

This also has to do with the fact that international cooperation between governmental authorities needs to be advanced, despite some cooperation in certain areas (e.g. criminal prosecution).

6. *Regulatory white spaces – control deficits for international actors (e.g. MNEs)*

There are topics on the international stage which are not (or not sufficiently) covered by treaties or IGOs. One such major field is the market, in particular the financial markets and the operations of MNEs. The truly globalized mechanisms of the market are lacking their regulatory counterpart which is a substantial barrier towards better control of global issues. On a national level, jurisdiction has more or less succeeded in setting the boundary conditions for the market in the way that society demands. National jurisdiction could mitigate the problematic effects of capitalism and keep it within limits, although issues of public goods and common-pool resources are certainly not sufficiently resolved. The social welfare state, for instance, which Germany developed after World War II, can be seen as a great success story insofar as it allowed for social welfare, economic development, employee participation and many more benefits. However, national economies compete with one another, which MNEs and the financial markets exploit to full capacity. Trade issues are being regulated by the WTO (however limited), but there is, for instance, no global competition authority. Authorities for competition and monopoly do exist and function at a national level and in some cases at an international but regional level (e.g. the EU), but there is no corresponding institution at the global level.

7. *International public organizations lack legitimacy and democracy*

International public institutions have a legitimization and a democracy problem. Several of these institutions reflect the geopolitical power balance of the past: The Permanent Members of the Security Council represent the post-war world order but not today's geopolitical situation. The UNFCCC, adopted in Rio in 1992, grouped the countries of the world basically in three categories: developing countries, developed countries and countries "undergoing the process of transition to a market economy" (UN 1992). These categories are still being used although some emerging economies, principally China, have undergone very rapid development in the past three decades and have become economic superpowers. They are surely not adequately described as developing countries. This grouping is important, however, since it determines whether a country will be given support or is supposed to support others. The selective Western-based membership clubs like the G7/G8 or the OECD do not include BRICS countries (although the OECD is already closely working with Brazil, China and India (OECD 2018): "The shifting global power configuration challenges each type of multilateral setting" – be they international institutions like (WTO, OECD, G7/G8) or multilateral settings (like UNFCCC) (Jang, McSparren & Rashchupkina 2016, 3).

Another reason for the lacking legitimacy of international public institutions is the frequently expressed allegation of nepotism, lack of democratic principles, bureaucracy and mismanagement. Josef Stiglitz, as a former executive of the

World Bank and most familiar with the Bretton Woods institutions (i.e. the World Bank and the IMF) and their internal processes, sharply criticises them. According to him, it is precisely our systems of governance which call for reform. *"The most fundamental change that is required to make globalization work in the way that it should is a change in governance"* (Stiglitz 2002, 226; original emphasis), entailing changes at the leading global public institutions such as the WTO, the IMF and the World Bank (e.g. a change in voting rights for the IMF and World Bank). Stiglitz argues that we are lacking effective systems for dealing with global issues because "globalization, by increasing the interdependence among the people of the world, has enhanced the need for global collective action and the importance of global public goods" (224).

Daniel Cohen, a leading French economist, shares this critique of the great global public organizations. The WHO, IMF and WTO all have their own statutes and regimes, and they act like government ministries which are left to their own devices, without being controlled by anybody in their actions (Cohen 2006, 193). Cohen's criticism points to yet another issue for IGOs – the lack of alignment of their agendas.

8. *Agendas not aligned*

The agendas of global public organizations are not harmonized and aligned with each other, as Cohen argues (Cohen 2006, 193f.). The WTO is concerned with global trade but does not bother about environmental or health issues on the production side (only on the consumer side). The IMF as the guardian of international monetary transactions is primarily concerned with stability but only marginally cares about the often brutal impact of monetary crises on employment and poverty. The WHO is fully mandated to tackle health problems but cannot legitimately address related issues of social inequality or concrete issues of survival for local people (ibid., 194). Similarly, some standards of the International Labour Organization (ILO) contradict WTO rules without being reconciled so far; conflict potential is also present in WTO's patent protection and health rights (manifested by drug prices which are unaffordable in developing countries) (Kress 2014, 53).

Moreover, there are trade-offs even within one area whose resolution is not well-defined: "Addressing climate change through forest plantations, for example, may replace ecosystems targeted by the U.N. Biodiversity Convention. Similarly, promotion of biofuels can accelerate deforestation and erode the food security of impoverished nations" (Walker et al. 2009, 1345).

These are the main reasons why the current set-up of multilateral international treaties and IGOs challenges progress towards sustainability.

Solution perspectives

What can be done to facilitate a better governance of global sustainability issues? There are several aspects on which consent seems possible, regardless of how the specific issues of global governance are viewed.

1. *Respect (and enforce) the rule of law*

As difficult as international collaboration may be, there is, in my view, no alternative to establishing, strengthening and expanding the rule of law beyond the borders of nation states if we are seeking an effective and just control mechanism for global issues. This entails a plea for international public organizations and treaties – despite their obvious and many shortcomings.

Investigating candidates for the control mechanisms of a new world order, the German political scientist Harald Müller discusses four control systems (Müller 2008). Power, market, and moral can all contribute to establishing a global order, but the decisive mechanism, the *primus inter pares*, is law.

- Power is needed to some degree because it can control divergent interests and enforce their own will against resistance. However, it also provokes the dialectic of power and resistance – because resistance increases the need to exercise power, which in turn increases the need for resources to break resistance, which again increases the need to exercise power because unbroken resistance encourages further deviant behaviour and so on and so forth.
- The market is the most efficient way of allocating scarce resources. But it does require a legal framework, it does produce negative externalities, and it often supports illicit behaviour.
- Ethics works well as a control mechanism as long as there is a morally integrated collective in which all members share certain codes of conduct. However, in today's rapidly changing environments, this morally integrated collective can no longer be assumed. According to Müller, ethics is therefore not suitable as a global control mechanism today.
- Finally, Müller's last candidate is law, which he sees as humanity's greatest invention apart from writing and the wheel. No other instrument is able to coordinate the behaviour and actions of such a wide range of people with minimal direct coercion – while being flexible and changeable at the same time. Governing the world sustainably has to respect the power of law: this means that all actors must acknowledge the rule of law.

Each of these four control mechanisms has its strengths and opportunities, but it is law, as Müller argues, which is the key foundation of all:

> A sustainable world order can function if it leans on law, which controls the other control mechanisms power, market and moral. Our complex and threatened world can only function if those in power subordinate themselves to law, understanding that a situation without rights which only builds on exercising power threatens their own well-understood interests in a most critical way.
>
> *(Müller 2008, 257)*

In his book about the age of sustainable development, Jeffrey Sachs, director of the UN SDSN, argues in a similar way regarding the connection of governance and law:

> Good governance ... means many things. It applies not only to government but also to business. It means that both the public sector (government) and the private sector (business) operate according to the rule of law, with accountability, transparency, responsiveness to the needs of stakeholders, and with the active engagement of the public.
>
> *(Sachs 2015, 42)*

The successful closure of the Uruguay Round within trade politics demonstrates, as Stiglitz argues, that "principles, not just power, can govern trade relations" (Stiglitz 2006, 284). The beneficial effect of law in shaping globalization would be much stronger

> if it were enforced ... America's refusal to do anything about global warming can be considered a major and unwarranted trade subsidy. The enforcement of regulations against such subsidies could be an important instrument both in creating a fairer trading system and in addressing one of today's most important global problems. We have an imperfect system of global governance without global government; and one imperfection is the limitations on our ability to enforce international agreements and stop negative externalities. We must use what instruments we have – including trade sanctions.
>
> *(Stiglitz 2006, 284)*

A precondition for any global order which can hardly be overstated is the respect of the rule of law, especially by governments and public authorities. Defying, violating, or unilaterally terminating international treaties, however, has negative ramifications far beyond the particular contract at stake because it erodes the trust in the meaning of international collaboration in general.

2. *Reform and align existing global public institutions*

As different as they are, major global public institutions like the UN and the WTO, as well as the IMF and World Bank, all share a need for reform. Of course, the degree to which reform is seen as necessary greatly depends on the individual position, and the challenges of the institutions mentioned are quite diverse. Furthermore, one of the fiercest critics of global public institutions, Joseph Stiglitz, concedes that it is actually not the institutions themselves that are to be blamed but the countries that de facto control them (i.e. the USA and some industrialized countries) and their voters (Stiglitz 2006, 277; Stiglitz 2010). This dominant influence of some countries, principally the USA, has long been criticized but rarely expressed so blatantly as when the current US president nominated the new president of the World Bank and stated that the World Bank should serve the USA's interest (Trump, BBC News 2019).

This makes evident that a reform of the main IGOs and the alignment of their agendas is essential. The good news is that there is at least some awareness of the need for reform within these institutions themselves:

- The UN Secretary-General António Guterres proposed a reform of the United Nations, which covers the areas "Development", "Management", and "Peace and Security". According to Guterres, the goal of this reform "is a 21st-century United Nations focused more on people and less on process, more on delivery and less on bureaucracy. The true test of reform will be measured in tangible results in the lives of the people we serve – and the trust of those who support our work" (Guterres 2019).
- Reinhart and Trebesch argue that, despite all criticism, "an international lender of last resort remains indispensable" and the IMF as an institution should be constantly reinventing itself (Reinhart & Trebesch 2016, 23).
- Following the subprime crisis, the UN mandated a commission for "Reforms of the International Monetary and Financial System" which came up with concrete suggestions for a better set-up of the international financial system (UN 2009; Stiglitz 2010).
- IGOs themselves come up with improvement suggestions: the World Bank, for instance, issued a report on the reform of the WTO and concludes that "complementary mechanisms are clearly needed to promote regular dialogue and cooperation on regulatory matters as these are increasingly the source of market segmentation and the focus of concern of firms" (World Bank 2011, 22).

However, drawing on a variety of sources (e.g. Diehl 2001; Cohen 2006; Stiglitz 2006; UN 2009; Reinhart & Trebesch 2016, 23), the following seems to find common agreement among most authors:

a. *Address democratic deficit*

Votes within the IMF and World Bank are largely given on the basis of economic power – but on the economic power of the post-war world of half a century ago (Stiglitz 2006, 281f.). There is the tendency that powerful states "control the interstate system rather than to broaden, and thereby strengthen, participation in global problem solving" (Diehl 2001, 497). Powerful states prefer to decide about the global economy in the IMF, where they have a voting advantage, or the Group of Seven, or OECD, instead of the UN General Assembly and UN Economic and Social Council (ECOSOC) (ibid.). As the President of the 63rd Session of the United Nations General Assembly, Miguel d'Escoto Brockmann stated:

According to democratic principles those who are deeply affected by a policy should have a say in their formulation, and those who are responsible for massive failures and injury should be held accountable. Our present system of global economic governance does not meet either of these fundamental tests of democratic governance.

(UN 2009, 9)

Thomas Bollyky from the Council of Foreign Relations – an influential foreign-policy think tank in the USA – recommends a "voting system that requires leaders to win over a majority of countries, not simply the votes of the primary shareholders" (Masters & Chatzky 2019).

b.   *Better represent developing countries*

Democratizing decision processes goes hand-in-hand with a better representation of developing countries in these processes. There is even a dedicated target within the SDGs, i.e. SDG 10.6: "Ensure enhanced representation and voice for developing countries in decision-making in global international economic and financial institutions in order to deliver more effective, credible, accountable and legitimate institutions" (UN 2017). Such a better representation would not only be a matter of justice and democratic participation. It would also help strengthen the role of the global public institutions. According to the Council on Foreign Relations, for instance, an increase in the voting shares of emerging economies within the IMF would "reinforce the sense of ownership that these countries feel" toward that institution (CFR 2012).

c.   *Increase transparency*

Especially because the respective institutions are not democratically legitimized, transparency is even more important – but often even less present than in other organizations (Stiglitz 2006, 282). We will later discuss the need to increase transparency in a variety of social institutions and suggest it even as one of the principles for sustainable action to increase transparency (see 16.3).

d.   *Better align the agendas of major players (e.g. the WTO and UNEP)*

The mandates, methods and resources of most international organizations represent a portfolio structure of departments, largely similar to most governmental administration offices at the national level: trade (e.g. the WTO), development (e.g. the World Bank, IMF and UNDP), environment (e.g. UNEP), etc. However, as frequently mentioned above, the global sustainability challenges extend across these structures. Even worse, sometimes the best-intentioned actions in one sector can yield detrimental results in other sectors if not advisedly implemented. These conflicting agendas need to be aligned. According to the WTO principle of non-discrimination, products must not be discriminated against because of any circumstances during the production process as long as they are identical in quality. In other words, the very fact that a product is produced more sustainably would not allow it to be preferred or vice versa: unsustainable production (how that is measured is, of course, another difficult question) does not justify discriminating against the product over more sustainable ones (ICTSD 2010; WTO 2019).

Of course one cannot expect trade law to resolve all global issues. However, especially in light of the siloed structure of our administration processes, the deficient executive force on international level, and the power imbalance between the rich and the poor players, I do think that better alignment among the IGOs is needed.

Why does the WTO, for instance, allow the health effects of products to be considered if these affect consumers but not if they occur on the producer side? If carpets harm the consumer's health, their import might well be forbidden under WTO regulations – but should their production by child labour be deemed acceptable? Why can one product produced in a more environmentally friendly way than another not be preferred over a polluting one? Vice versa, why does a country violate WTO regulation if it prohibits the import of products that were produced more unsustainably than others? The WTO is aware of the potential conflicts – and the judgement whether the response is appropriate or not will vary depending on personal position:

> Environmental requirements may affect international trade [*sic!*], especially if they are used to shield domestic producers from international competition, or when they are discriminatory. As countries continue efforts to 'green' their economies, environmental requirements increasingly will become significant determinants of access to foreign markets. The design of the measures, how transparent they are, and issues related to their harmonization or recognition can all give rise to concern.
>
> *(WTO 2011, 10)*

It can also give rise to concern, I would add, that environmental regulation is only considered as a threat to international trade. Many people would argue the opposite.

This is only one example, highlighting the implications of global trade policies. However, there are many more cases in which the agendas of international public organizations need to be aligned and adjusted to the most urgent global challenges.

How can this be done? Different ways are conceivable. Cohen calls for mobilizing the members of the world public to align the agendas and operations of the large public institutions. He suggests that the WTO would adjust its regulations in such a way that it respected recommendations given by the WHO or UNEP (Cohen 2006). Another option would be to strengthen existing institutions as a mediating body. Müller suggests re-evaluating the UN ECOSOC as a mediating institution in case of conflicting interests or trade-offs between different organizations (Müller 2008, 278).

Finally, establishing new institutions, for instance as a mediation authority or arbitration body, is yet another option.

3. Establish new global institutions

There are suggestions for new global institutions which would, alternatively, help align the different agendas of the existing institutions and mediate any trade-offs between them, and address so far uncovered topics in the international regime.

Following the financial crisis of 2008, the UN mandated a commission which was to propose reforms of the international monetary and financial system. This commission recommended, among other ideas, the establishment of a Global Economic Coordinating Council. This council would assess the global economic situation, co-ordinate policies and identify gaps in the global institutional arrangement and propose solutions (Stiglitz 2009; UN 2009).[3] The need for a global competition authority was already mentioned above. With the increasing accumulation of economic power, some kind of regulation authority will be needed to ensure the functioning of a global market.

Cohen calls for two supranational organizations: one which facilitates poor countries' access to international trade, and another independent non-profit one which can establish a global world public (Cohen 2006).

4. Close enforcement gaps and strengthen government networks

Businesses, NGOs, journalists, religious institutions and many other actors are either already operating globally or have at least established forms of collaboration via networks, associations, or similar. As discussed above, there are serious violations of both national and international law in areas like IUU fishing or e-waste management, and these gaps need to be closed. Among other solutions, better collaboration between government offices of different countries is needed. In some areas, national governmental offices are already collaborating quite well, while in others they are gradually catching up. Collaboration of police authorities, for instance, already has a long tradition and is relatively advanced. The idea of international police cooperation had been conceived in 1914; in 1923, the International Criminal Police Commission was founded and named INTERPOL in 1956. Today, almost all states globally are members (INTERPOL 2019). Interestingly, however, INTERPOL is an unincorporated association under French law – it is not controlled by anybody and no government has ever approved it (Kampf 2015).

In other areas there has also been increasing cooperation in recent years to fight organized crime, money laundering, tax evasion, etc. Slaughter sees a great potential in concerted collaboration of government networks and sees these networks as "a key feature of world order in the 21st century, but they are underappreciated, undersupported, and underused to address the central problems of global governance" (Slaughter 2004, 1) They can "promote convergence, compliance with international agreements, and improved cooperation among nations on a wide range of regulatory and judicial issues" (Slaughter 2004, 261).

6.2 Geopolitics and the struggle of establishing a world order

The best international treaties and the most ambitious global sustainability programs will be foiled by geopolitical conflicts. It is therefore essential to understand different concepts for the world order and facilitate the one which is best suited to sustainable development.

The overarching question of this chapter is how global developments can be governed towards sustainability. The answer to this question varies considerably according to the school of thought and worldview. Different proposals have been made for mechanisms which would describe the existing world order. This *empirical* question – how is the world order best understood? – needs to be supplemented by the *normative* one: what are not only promising but also *legitimate* candidates for establishing such a world order? We can only highlight a couple of extreme examples here before we lean towards and argue for the "mainstream" position of global governance.

Scholars in the tradition of political (neo-)realism doubt international organizations and treaties have any value at all. They argue it is the nation states' self-interest which determines whether or not treaties will be accepted and adhered to (Kress 2014, 40). According to political realism, "statesmen think and act in terms of interest defined as power" (Morgenthau 2005, 5).[4] This is essentially debasing international treaties to the status of mere cosmetics. This position not only neglects the achievements global public institutions facilitated – a prime example is the Montreal Protocol of 1987 with its de facto-ban of chlorofluorocarbons (CFCs), which effectively stopped these chemicals further depleting the ozone layer. It also disregards the new influence of multinational enterprises and NGOs on the global arena, which will be discussed below.

Another rather extreme position questions even more generally the usefulness of political dominance over economic forces. Not politics but the free market would be able to resolve the urgent dominant issues. As quoted above (see Chapter 5, note 3), German economist C. Christian von Weizsäcker frankly states: "The global problems are being resolved by surrendering the leading role to the economy over politics" (von Weizsäcker 2000, 166). Weizsäcker demands that markets be de-politicized, only then can true democracy take place. This optimism in the healing forces of free markets, which was expressed almost two decades ago, seems to breathe the neo-liberal paradigm of the 1980s and 1990s.

However, one can argue that it is precisely neo-liberalism which paved the way for the concept of global governance, a concept which became popular after the fall of the Iron Curtain:

> Global governance is a product of neo-liberal paradigm shifts in inter-
> national political and economic relations. The privileging of capital and
> market mechanisms over state authority created governance gaps that have
> encouraged actors from private and civil society sectors to assume authori-
> tative roles previously considered the purview of the State ... Global gov-
> ernance is concerned with issues that have become too complex for
> a single state to address alone.
>
> *(Jang, McSparren & Rashchupkina 2016, 1)*

Jeffrey Sachs argues for a strong role of the government in establishing governance because it is the government which is in charge of building infrastructure
(roads, rail, power transmission, port services, connectivity, water, sewerage),
which is, in turn, necessary for any economy to develop; and the government is
in charge of "human capital development: the health, education, and nutrition
of the population, especially of the children. If the government is not performing, public schools will be miserable" (Sachs 2015, 129).

Chandran Nair also advocates a strong role of the state but he is more sceptical regarding the power of global governance. He bemoans issues which were
touched on above: lack of enforcement or punishment mechanisms, difficulties
in regulating global commons, and Western-style institutions that reflect an outdated geopolitical pattern (Nair 2018).

What actually *is* global governance? This concept is not used equivocally. All
agree that governance is to be distinguished from government. Whereas government relies on rule systems which are rooted in formal and legal procedures,
governance is building on informal rule systems: "The sum of the world's
formal and informal rule systems at all levels of community amount to what can
properly be called global governance. It is a highly disaggregated and only
a minimally coordinated system of governance" (Rosenau 2003, 13).

So global governance is a type of world order which relies on binding norms
and rules (different to an anarchy of nation states) but does not rely on a supra-
national authority as in case of a world state. The table below lists four types of
world orders according to two categories: whether or not binding norms and
rules exist, and whether or not a supra-national authority exists (see Müller
2008; Rittberger, Zangl & Kruck 2013, 254ff.).

According to Biermann and Pattberg, there are three developments at the
core of and specific to global governance:

> first, the emergence of new types of agency and of actors in addition to
> national governments, the traditional core actors in international environmen-
> tal politics; second, the emergence of new mechanisms and institutions of
> global environmental governance that go beyond traditional forms of state-
> led, treaty-based regimes; and third, increasing segmentation and fragmenta-
> tion of the overall governance system across levels and functional spheres.
>
> *(Biermann & Pattberg 2008, 280)*

TABLE 6.1 Typology for types of world order

	No supra-national authority (horizontal coordination)	Supra-national authority (vertical coordination)
No binding norms and rules	Anarchy of nation states	World hegemony
Binding norms and rules	Heterarchical global governance	World state

Source: (Rittberger, Zangl & Kruck 2013, 254)

As an analytical concept, global governance just describes the changing landscape of actors on the global stage, for instance, that NGOs and large multinationals have joined the group of global players, in parallel to the public global institutions: businesses and non-governmental organizations, and also other scientific experts are contributing not only to the global public discourse on sustainability issues but are also being invited to both formal and informal, and private and public, international meetings and conferences. Four general structures within global governance can be identified: IGOs like the WTO or UN, Public-Private Partnerships like the UN Global Compact (UNGC), private governance (e.g. corporations), and tripartite governance mechanisms (EITI, etc.). (Jang, McSparren & Rashchupkina 2016, 1).

The UN ECOSOC, which maintains official relations with NGOs since the inception of the UN, granted NGOs a Consultative Status at the UN Conference in Stockholm 1972. Since 1992 (UNCED), NGOs could contribute even to formal negotiations, and since 2002 (the WSSD), they could even participate in round-table discussions on an eye-to-eye level with government officials (Brühl & Rosert 2014, 356).[5]

Another group of scholars sees global governance not only as a descriptive but also a normative concept. There was some enthusiasm about global governance in the 1990s, as it induced hopes of some kind of a new global order after the fall of the Iron Curtain in 1989 (Messner & Weinlich 2016, 4). The idea of a "world domestic policy", for a gradually emerging global civil society was discussed with some optimism (Messner & Nuscheler 1996). However, looking back to those days one can be disillusioned.

> 25 years after the end of the Cold War, one can no longer detect a sense of global optimism concerning a cooperation-based global governance. The international community does not shape globalization; it meanders and staggers its way through transnational crisis scenarios in security policy, the global economy, global environmental policy, and the international system as a whole.
>
> *(Messner & Weinlich 2016, 4)*[6]

There seems to be no consensus on the question of whether global governance is promising for global sustainability or not. According to James Rosenau, the "prospects for effective governance leading to sustainability are, on balance, quite bleak" (Rosenau 2003, 11). Anthony Giddens bemoans the weakness of international institutions:

> Just at the time when the world needs more effective governance, international institutions look weaker than they have been for some years ... A more multipolar world could, of course, provide a better balance for cooperation, but it could just as easily produce serious divisions and conflicts with no arbiter to resolve them.
>
> *(Giddens 2009, 207)*

Giddens wrote this a decade ago and one might wonder if the situation has not still worsened since then. However, with the 2030 Agenda and the Paris Agreement, two major milestones of international cooperation have meanwhile been agreed. Moreover, there is not really an alternative to some form of global governance – that's why Jang et al. praise it as a necessity for human livelihood:

> Global governance is arguably inevitable for the survival of the human race in present and future generations. Although global governance sometimes appears fragile and ineffective in response to current challenges, the trend of globalization and the demand for global governance approaches have already passed the point of no return.
>
> *(Jang, McSparren & Rashchupkina 2016, 3)*

In sum, from a purely *empirical* point of view, the current global order exhibits elements of most of the types in the table above. Which candidate is seen as most influential depends on the individual perspective, of course. Ritterberger et al. see the heterarchical global governance as de facto already closest to the existing order (Rittberger, Zangl & Kruck 2013, 259). I will also argue from a *normative* point of view, however, that some kind of global governance is best suited to combine the maintenance of independent and diverse nation states with binding norms and rules for a multitude of different actors in the international arena. A heterarchical world order does not rest on top-down control but on horizontal, networked policy coordination and collaboration among states (and their administration offices), international organizations and non-state actors (ibid.).

Solution perspectives

1. *Ensure a strong voice of civil society in global public discourse*

Global governance accounts for the rising importance of new players like NGOs and businesses in the international public discourse and that they gain access to formal public meetings and negotiations. This is a great opportunity, since civil society attains at least some kind of direct access to decision-making processes on an international level. On the other hand, it is not only non-profits who sit at the table in multi-stakeholder panels, but also corporate lobbyists. Corporations are using their considerable influence both on national and international politics, and they have a much greater budget to invest in lobbyism that NGOs. There is a fine line between legitimate influence and corporate lobbying impeding NGOs from being be heard. In early stages of climate politics, an association of coal and oil companies, the "Global Climate Coalition" was trying to influence the discussion despite the obvious conflicting position to "green" NGOs (Brühl & Rosert 2014, 361). It will be important to ensure that NGOs, as voices of civil society and advocates of the common good, continue to be heard in international negotiations.

It is even more important to provide full transparency on all stakeholders involved in public international consultations or negotiations, on institutional affiliations, financial dependencies and potential conflicts of interests.

Politicians will be well advised not to neglect the voice of civil society, lest their legitimacy be threatened because people feel no longer represented by those above.

2. *Compliance with law does not require monopoly of legitimate force*

Political (neo-)realists argue that jurisdiction on an international level does not make much sense since it can be overruled by power anyhow. Real politics seems to verify this position, since not even the possibility of sanctioning mechanisms prevent states from breaching international law, as the US invasion in Iraq or the Russian annexation of Crimea showed (see Kress 2014, 40). Given this argument, a monopoly of legitimate force would be a precondition for law because in order to be effective, law requires a high compliance rate for any given regulation. The latter, however, would require an established monopoly of legitimate force, and a "national identity that determines the consent of those who are targets of a regulation" (Zürn 2005, 5) – in this way, questions of compliance are the Achilles heel of international regulations. However, as Zürn demonstrates with reference to several real cases (e.g. in EU politics), legal compliance is not triggered by coercion alone.[7] Rather, it is the power of the legitimacy of legal norms, the way legal norms work once they are established and the smart management of non-compliance which lead to compliance (Zürn 2005, 5). This linkage between legitimacy and compliance is emphasized by the so-called "managerial school".

Here it becomes evident why a global public discourse is of such eminent importance for proper global governance. The legitimacy of legal norms and the compliance with them, can be assessed by the global public, which has a much more important effect than in previous times. McLuhan already stated that the Vietnam War was decided in the USA's living rooms, not on the battlefields of

Vietnam (McLuhan 1975). This power of the public is presumably much greater today, in times of social media with the potential to capture global attention in terms of minutes and days.

Moreover, I submit that even the very nature of "compliance" is changing. Whereas in the corporate world, compliance used to refer to the regulatory framework of the authorities, today it has a much broader meaning. The Chief Information Officer of a large multinational once told me: "For me compliance means meeting my stakeholder's expectations." This nicely resonates with Zürn's account that enforcement is not so much tied to the means chosen but to the effect that the addressees of a regulation are induced to act in compliance with it.

3. *Work with critical players in "club" format and focus on leverage points*

In an ideal multilateral approach, the entire world community agrees on goals and measures. Unilateralist tendencies and governments should not, however, impede the progress of others. Solutions do not necessarily require that all countries engage to the same degree, as Victor and Jones point out (2018, 1). They argue for an "episodic multilateralism", which works with clubs rather than with multilateral institutions. The idea is that the vast proportion of emissions comes from just a few jurisdictions. So working with these countries could not only be much easier than reaching global agreement (because it reduces the number of participants considerably to the few really important players). It can nevertheless be very effective since the major contributors are included. Furthermore, such an approach allows the inclusion of non-state actors and sub-national governments. In the USA, for instance, the main barrier towards climate mitigation measures is the federal government while there is much going on at the level of cities and states, in business and civil society. Victor and Jones further argue that focus is needed on pivotal technologies and high leverage points. For instance, in many mature economies there has been hardly any progress in GHG emission reduction in the transportation sector because of the dominance of the combustion engine (ibid., 2f.).

This collaboration potentially crosses societal sectors, industries and countries; it includes business, NGOs, the sciences and civil society – and much of that can be done, and is taking place, independent of any public authority.

4. *Strengthen cross-sectoral collaboration*

Thinking and operating in siloes is not a promising approach for issues of sustainability in any case (see Chapter 9). Cross-sectoral collaboration is therefore an important tool for expanding people's horizon, canvassing understanding and strengthening social cohesion – all attributes most welcome for a sustainable society (see 15.4). However, in the context of global governance, it is even more important because it can foster societal goals, establish social norms, and thereby increase the pressure on politics to move into the respective direction.

An important feature of global governance which can be seen already today is the collaboration of actors of different kinds, context-dependant and varying: NGOs, businesses, governmental organizations, scientific experts collaborate in multiple settings. It is precisely this collaboration towards a common objective which allows for the horizontal alignment typical for global governance – having (relatively) binding norms and rule but no central authority.

Giddens argues in a similar direction. Although the Doha Round failed, there has been progress on regional and bilateral trade agreements. Such deals

> can support the overall objectives of the WTO's multilateral trading system rather than, as might seem to be the case on the surface, act to undermine them. Regional agreements have allowed countries to go beyond what has been possible to achieve universally; but these concordats have subsequently paved the way for progress made at the level of the WTO.
>
> *(Giddens 2009, 221)*

Contrary to some pessimistic spirit in the recent global governance discourse, Messner and Weinlich carve out findings from such diverse fields as neuroscience, cognitive psychology, evolutionary anthropology and evolutionary biology, which deal with collective human action to achieve common goals.

> Cooperation, as it is seen in these strands of research, comes about not only as a result of complementary or easily reconciled individual interests, but rather is an original mode of action that people often resort to, even in social dilemma situations.
>
> *(Messner & Weinlich 2016, 9)*

The authors postulate to consider these insights in global governance research.

In a contribution to the Policy Forum of *Science*, Walker et al. argue in a similar direction. The legitimacy of international agreements would "depend on acceptance by numerous and diverse countries and by nongovernmental actors, such as civil society and business". They argue that this common agreement was the basis for the greater success of the Montreal Protocol relative to the Kyoto Protocol (Walker et al. 2009, 1346).

5. *Conclusion: Global governance as orchestration of a multitude of actors on a multitude of levels*

Going back to the typology of world order-types shown above, there is good argument that some form of global governance is what we can see emerging in the global arena. Whether purely analytically and empirically, this water-tight evidence is not our main objective. Unilateralist statements of representatives of major global players might call people to think whether the concept of a hegemonic power or an anarchy of independent states might become a more accurate representation of the world order in the future. We have not studied this on an empirical basis in any detail here, but in my view there are signs for the prevalence of some kind of heterarchical, horizontal coordination.

In his book, *The Politics of Climate Change*, Anthony Giddens commemorates the Bush administration's desire "for a world in which power counts and where the US is pre-eminent in wielding such power". This worldview went along with a contemptuous attitude towards climate change. And yet, the United States,

> the world's greatest military power, was unable to pacify a single medium-sized country, Iraq, in spite of an easy initial military victory. It was not able to fight two wars at the same time, even with the help of allies, and as a consequence the project to bring stability to Afghanistan is meeting with, at best, limited success. The US has the world's largest economy, but the country, acting alone, has very limited capacities to influence the world marketplace – as the financial crisis has shown all too clearly.
>
> *(Giddens 2009, 212)*

These words, written a decade ago, have not lost their validity, although unilateralist voices from similar ideological background have even become louder more recently.

However, even if the empirical signs are rather pointing to a hegemonic or anarchical structure, I would still argue from a normative point of view that the concept of global governance is to be pursued as the most promising option for controlling global issues, since it combines binding rules and norms with independence and flexibility and accounts for a great variety of actors.

The environmental philosopher Konrad Ott comes to a similar conclusion in comparing an institutionalistic (Kantian) view with neo-realism. With regard to "explanation and analysis" institutionalism would at least be astride with neo-realism:

> But, to ethical universalism, it is clearly superior with respect to its principles of peaceful cooperation, respect for human rights, mutual aid, and institution building beyond states. Institutionalizing sustainability beyond states will be a major challenge for decades to come. Discarding neo-realism is a necessary intellectual precondition for facing such challenge.
>
> *(Ott 2014, 909)*

To be sure, the concept of global governance will face challenges. Kress sees a threefold challenge (Kress 2014, 49ff.): First, complexity and fragmentation increase in the global arena due to an increasing number of actors and ever more complex challenges, to which cultural issues arising from different values and the acceleration of technological development (see Chapter 8) contribute; second, asymmetries between actors in economic power, institutional and personal resources (50); thirdly, deficits in legitimacy and democracy.

Nevertheless, there are also promising signs in favour of global governance: "Cross-fertilisation of global governance research and practice outside disciplinary and scholar-practitioner silos has only just begun. There is much to be gained from further interpenetration across disciplinary lines" (Pegram & Acuto 2015, 596).

In sum, I do not really see an alternative to some kind of global governance – both descriptively and prescriptively. It is clear and will hopefully gradually become more evident that humanity can only thrive on this planet in the long run if we manage jointly to move in the direction we want. The spaceship image of Planet Earth is still a powerful symbol for human fate: we will jointly win or jointly despair. I follow Kenneth W. Abbott et al. in seeing global governance rather in a soft mode of orchestration:

> We introduce the concept of orchestration, a mode of governance that is soft and indirect; orchestration stands in contrast to modes of governance that are direct and/or hard, including hierarchy, collaboration and delegation. In orchestration, one actor, the orchestrator, enlists the voluntary assistance of a second actor, the intermediary, to govern a third actor, the target, in line with the orchestrator's goals.
>
> *(Abbott, Genschel & Snidal 2015, 349)*

The role of the orchestrator can be filled by different entities. International governmental organizations are the ones Abbott et al. study, but other types are conceivable.

Of course it will be important that this "wide variety of numerous actors, both individuals and collectivities", are willing to somewhat subordinate "to the interests of their great grandchildren" (Rosenau 2003, 26).

"Although global governance sometimes appears fragile and ineffective in response to current challenges, the trend of globalization and the demand for global governance approaches have already passed the point of no return" (Jang, McSparren & Rashchupkina 2016, 3f.). Just as in a phase transition (e.g. from liquid to gas), in which every particle "knows" what to do once the conditions are given, I envision a phase transition towards a more sustainable world order. This will not be a matter of months or years – rather decades or centuries – but I am confident that it will occur one day. Of course, this is a crude analogy – however, in a similar way a phase transition can occur in a remarkably short time if trigger points are given.

Notes

1 There is, of course, also a lack of sustainability governance on a national level, which we will have to skip here because these issues are mostly execution issues (and not systemic ones) and/or are being addressed in other chapters of this book. For instance, national sustainability governance is impeded by a faulty market system, by lobbyism and corruption, by inequalities and conflicting interests, etc.

2 As we will see below, Zürn argues that law enforcement might not necessarily rely on force monopoly.
3 The commission states:

> Our analysis suggests that not only is there a need for substantial reforms in existing institutions, but that in addition there is also a need to create a new institution, a Global Economic Coordination Council (GECC), supported by an International Panel of Experts. While we understand the concern about the proliferation of international institutions and the hesitancy to create any additional bodies, the need for such a GECC is compelling and spelled out in greater detail below.
>
> *(UN 2009, 87)*

4 "The main signpost that helps political realism to find its way through the landscape of international politics is the concept of interest defined in terms of power ... We think that statesmen think and act in terms of interest defined as power" (Morgenthau 2005, 5). Later Morgenthau continues:

> Political realism is aware of the moral significance of political action ... Realism maintains that universal moral principles cannot be applied to the action of states in their abstract universal formulation but that they must be filtered through the concrete circumstances of time and space.
>
> *(Morgenthau 2005, 14)*

5 As Brühl and Rosert explicate, the growing influence of NGOs is not just a linear trend that can be extrapolated. However, the willingness of states to grant access to negotiation tables follows a cost-benefit analysis. As long as the NGOs bring sufficient resources to the table – in terms of power, knowledge or values – they are accepted (Brühl & Rosert 2014, 356).
6 It should be mentioned that some authors do see a third group of thinkers, which use global governance as a normative concept but in a negative way.

> Third, some writers have adopted the programmatic definition of global governance, yet without its affirmative connotation. We describe this literature here as the critical usage of the global governance concept. For example, some neoconservative writers see global governance as the attempt of the United Nations and other international organizations to limit the freedom of action of powerful states, in particular the United States.
>
> *(Biermann & Pattberg, 2008, 279f.)*

7 Zürn distinguishes between coercion and enforcement. Enforcement is softer than coercion and simply means "compliance generation", but independent of the measure (Zürn 2005, 6, footnote 8).

Bibliography

Abbott, Kenneth W., Philipp Genschel, Duncan Snidal, & Bernhard (eds.), *International Organizations as Orchestrators*. Cambridge: Cambridge University Press, 2015.

Biermann, Frank, & Philipp Pattberg. "Global environmental governance: Taking stock, moving forward." *Annual Review of Environment and Resources 33*, 2008: 277–294.

Brühl, Tanja, & Elvira Rosert. *Die UNO und Global Governance*. Wiesbaden: Springer, 2014.

Busch, Marc L., & Eric Reinhardt. *The WTO Dispute Settlement*. Stockholm: Sida – Swedish International Development Cooperation Agency, 2004.

CFR. "The global finance regime." *Cfr.org*. 23. 01 2012. www.cfr.org/report/global-finance-regime) (Accessed 12. 02 2019).

Cohen, Daniel. *Globalisierung als politische Herausforderung, french original: La mondialisation et ses ennemis*. Hamburg: Europäische Verlagsanstalt, 2006.

Diehl, Paul F. *The Politics of Global Governance. International Organizations in an Interdependent World*. Boulder, CO; London: Lynne Rienner Publishers, 2001.

EJF. *Out of the Shadows. Improving Transparency in Global Fisheries to Stop Illegal, Unreported and Unregulated Fishing*. London: Environmental Justice Foundation, 2018.

FAO. "Illegal, unreported and unregulated (IUU) fishing." *Fao.org*. 2019. www.fao.org/iuu-fishing/background/what-is-iuu-fishing/en/(Accessed 08. 02 2019).

Giddens, Anthony. *The Politics of Climate Change*. Cambridge: Polity Press, 2009.

Guterres, António (UN Secretary General), 27 November 2018, https://reform.un.org/ (accessed 11 September 2019).

ICTSD. *Sustainability Criteria in the EU Renewable Energy Directive: Consistent with WTO Rules? (ICTSD Information Note 2)*. International Centre for Trade and Sustainable Development, 2010.

INTERPOL. "Our history." www.interpol.int. 2019. www.interpol.int/About-INTERPOL/History (Accessed 12. 02 2019).

Jang, Jinseop, Jason McSparren, & Yuliya Rashchupkina. "Global governance: Present and future." *Palgrave Communications 2*, 19. 01 2016: 1–5.

Kampf, Lena. "Der viel zu lange Arm des Gesetzes." *SZ-Magazin 3*, 19. 01 2015.

Kress, Daniela. "Internationale Übereinkommen als künftiges Herzstück einer Global Governance? Auf dem schwierigen Pfad zwischen Anspruch und Wirklichkeit." In *Globalisierungsgestaltung und internationale Übereinkommen*, von Armin Frey, Thomas Jäger, Dirk Messner, Manfred Fischedick, & Thomas Hartmann-Wendels (eds.), 15–58. Wiesbaden: Springer,2014.

Lamy, Pascal. "Lamy says relatively small concessions needed for reaching Doha agreement." *Wto.org*. 02. 07 2007. www.wto.org/english/news_e/sppl_e/sppl64_e.htm (Accessed 07. 02 2019).

Masters, Jonathan, & Andrew Chatzky. "The World Bank Group's role in global development." *Cfr.org*. Herausgeber: Council on Foreign Relations. 09. 04 2019. www.cfr.org/backgrounder/world-bank-groups-role-global-development (Accessed 31. 05 2019).

McLuhan, Marshall. *Montreal Gazette*, 1975 (May 16).

Messner, Dirk, & Franz Nuscheler. *Global Governance: Herausforderungen an die deutsche Politik an der Schwelle zum 21. Jahrhundert*. Bonn: Development and Peace Foundation, 1996.

Messner, Dirk, & Silke Weinlich. "The evolution of human cooperation: Lessons learned for the future of global governance." In *Global Cooperation and the Human Factor in International Relations*, Dirk Messner, & Silke Weinlich (eds.), 3–46. London, New York: Routledge,2016.

Morgenthau, Hans J. *Politics Among Nations. The Struggle for Power and Peace*, 7th ed. New York, Boston, MA McGraw-Hill, 2005.

Müller, Harald. *Wie kann eine neue Weltordnung aussehen? Wege in eine nachhaltige Politik*. Frankfurt a.M: Fischer Taschenbuch Verlag, 2008.

Nair, Chandran. *The Sustainable State. The Future of Government, Economy, and Society*. Oakland, CA: Berrett-Koehler Publishers, 2018.

OECD. "Our global reach." *Oecd.org*. 2018. www.oecd.org/about/membersandpartners/ (Accessed 09. 02 2019).

Ott, Konrad. "Institutionalizing strong sustainability: A Rawlsian perspective." *Sustainability 2014 6*, 2014: 894–912.doi:10.3390/su6020894.

Pegram, Tom, & Michele Acuto. "Introduction: Global governance in the interregnum." *Millennium: Journal of International Studies 43*, 2015: 584–597.

Raffer, Kunibert, & Hans W. Singer. *The Economic North–South Divide: Six Decades of Unequal Development*. Cheltenham, UK; Northampton, MA: Edward Elgar. 2001.

Reinhart, Carmen M., & Christoph Trebesch. "The international monetary fund: 70 years of reinvention." *Journal of Economic Perspectives 30(1)*, 2016:3–28.

Rittberger, Volker, Bernhard Zangl, & Andreas Kruck. *Internationale Organisationen*. Wiesbaden: Springer, 2013.

Rockström, Johan, Will Steffen, Kevin Noone, Asa Person, et al. "A safe operating space for humanity." *Nature 461* 24. 09 2009: 472–475.

Rosenau, James. "Globalization and governance: Bleak prospects for sustainability." *IPG 3*, 2003: 11.29.

Sachs, Jeffrey D. *The Age of Sustainable Development*. New York: Columbia University Press, 2015.

Slaughter, Ann-Marie. *A New World Order*. Princeton, NJ: Princeton University Press, 2004.

Stiglitz, Joseph. *The Stiglitz Report. Reforming the International Monetary and Financial Systems in the Wake of the Global Crisis*, London, New York: The New Press, 2010.

Steffen, Will, Katherine Richardson, Johan Rockström, Sarah E. Cornell, Ingo Fetzer, et al. "Planetary boundaries: Guiding human development on a changing planet." *Science 347*, 13. 02 2015: 736.

———, Johan Rockström, Katherine Richardson, Timothy M. Lenton, Carl Folke, et al. "Trajectories of the Earth system in the anthropocene." *PNAS 115 (33)*, 14. 08 2018: 8252–8259.https://doi.org/10.1073/pnas.1810141115.

Stiglitz, Joseph. *Globalization and Its Discontents*. New York, London: W.W. Norton, 2002.

———. *Making Globalization Work*. New York: W.W. Norton, 2006.

———. *The Guardian*. 27. 03 2009. www.theguardian.com/commentisfree/2009/mar/27/global-recession-reform) (Accessed 12. 02 2019).

Trump, Donald. *BBC News*. 06. 02 2019. www.bbc.com/news/business-47148638 (Accessed 08. 02 2019).

UN. "UNFCCC." In *United Nations Framework Convention on Climate Change*. New York: UN, 1992.

———. *Report of the Commission of Experts of the President of the United Nations General Assembly on Reforms of the International Monetary and Financial System*. New York: United Nations, 2009.

———. *Resolution 71/313*. New York: United Nations, 2017.

UNCED. "Agenda 21." Rio de Janeiro: United Nations,1992a.

———. "Rio declaration." *Report of the United Nations Conference on Environmenta and Development; Annex I: Rio declaration on environment and development*. New York: UN, 12. 08 1992b.

UNEP. *Emissions Gap Report 2018*. Nairobi, Kenya: UNEP, 2018.

UNFCCC. *Paris Agreement*. Paris: UNFCCC, 2015.

Victor, David G., & Bruce D. Jones. *Undiplomatic Action. A Practical Guide to the New Politics and Geopolitics of Climate Change*. Washington, DC: Brookings, 2018.

von Weizsäcker, C. Christian. *Logik der Globalisierung*. Göttingen: Vandenhoek & Ruprecht, 2000.

Walker, Brian, Scott Barrett, Stephen Polasky, Victor Galaz, Carl Folke, et al. "Looming global-scale failures and missing institutions." *Science 325 (5946)*, 11. 09 2009: 1345–1346.doi:10.1126/science.1175325.

Worldbank. *Proposals for WTO Reform: A Synthesis and Assessment.* Washington, DC: The World Bank, 2011.

WTO. *Harnessing Trade for Sustainable Development and a Green Economy.* Geneva: World Trade Organization, 2011.

WTO. "The Doha Round." *Wto.org.* 2019. www.wto.org/english/tratop_e/dda_e/dda_e. htm (Accessed 22. 01 2019).

Zürn, Michael. "Introduction: Law and compliance at different levels." In *Law and Governance in Postnational Europe,* von Michael Zürn, & Christian Joerges, 1–39. Cambridge: Cambridge University Press, 2005 10422.

7

LAW

Insufficient institutions and the challenge of the common good

Contents

The previous chapter discussed the difficulties of establishing, implementing and enforcing regulation on a global level. There are promising examples of international treaties like the Montreal Protocol or the Paris Agreement and agreements like the 2030 Agenda. However, there are severe challenges like ambition gaps and limited obligation, as well as limited enforcement mechanisms. Therefore, a number of authors think that, until further notice, the main responsibility for environmental protection rests with the nation state (Müller 2008; Bosselmann 2017, 176ff.; Nair 2018; SRU 2019, 62), although nation states are experiencing decreasing influence in the rapidly changing global arena (Kress 2014, 22). In case of the European Union as a closely integrated super-national jurisdiction, this responsibility is largely transferred to the super-national level, since, in particular, environmental law is harmonized to a large extent across the EU. The chartered rule of law in constitutional states provides an irreducible and unquestionable basis for all members of society and grants legal certainty and thereby security of expectations. It is therefore the ultimate common frame of reference for the members of a society.

Legal systems are more long-lasting, stable and robust than political moods and decisions, which support the stability of public institutions on the one hand, but are a challenge if new topics or themes enter the public arena on the other. As a political topic, sustainability is certainly not new any more – as a legal topic, however, much more so.

Sustainability needs to gain more influence in the corpus of legislation and administration. Presumably in few countries (if any) do issues of sustainability or the environment receive institutional, budgetary, or political recognition or authority similar to that given to issues of finance, economy, labour, or defence. This does, of course, reflect the weight of this topic in the political discourse and in the public. But there are specific juridical questions related to sustainability which touch on moral and philosophical subject-matters.

7.1 Sustainability concerns not institutionalized

Sustainability concerns need to be institutionalized to ensure their continuous consideration in the state.[1] It is primarily the state's responsibility to ensure this. However, institutionalizing sustainability is impeded by different requirements. Environmental concerns are challenged by structural disadvantages compared to other policy fields. Furthermore, institutionalized sustainability has to be integrative (i.e. cross-departmental) and long-term, considering the demands of future generations. Both requirements are insufficiently accounted for by existing institutions.

The state needs to respond to the specific challenges of sustainability and environmental protection by means of laws and regulations and make sure that the requirements of environmental protection are codified in environmental law (Heselhaus 2018, 16). Sustainability concerns demand an integrated view which includes the consideration of all sectors of society, and they require the consideration of a long-term perspective. Both concerns have so far only been partially realized by institutional measures. The generic explanation for this is that many challenges are relatively new compared to the endurance of the main body of laws and regulations. However, there are specific reasons why such institutionalization is particularly difficult for sustainability concerns.

Already environmental law and environmental politics face several challenges. German environmental law, for instance, has been developed over several decades in light of different environmental challenges, which led to sectoral environmental laws concerned with certain environmental media (air, water, soil) (Umweltbundesamt 2018). These different laws are not consistent in their usage of terminology and regulation approaches, and they evaluate different environmental concerns differently. Enforcement of environmental law is impeded if it is distributed to a great variety of individual laws. An additional challenge in the German context is that environmental law is effective on three different levels: the state level, the federal level, and the EU level. Although a unified environmental code of law (*Umweltgesetzbuch*) was already agreed on in the coalition treaty of the first Merkel government (2005–09), it finally failed due to political controversies among the coalition partners (BMU 2018).

Günther and Krebs argue that the success of environmental politics depends upon the organizational structure at the highest level of the political administrative system (Günther & Krebs 2000, 3) – in other words: it is institutional insufficiency which causes the ineffectiveness of environmental politics. For instance, most environmental issues are caused by sectors which are not under the direction of the ministry for environment. Transportation, trade, agriculture, business and finance are often represented by ministries more influential than environment, although their environmental effect is considerable. The very same issue is currently being discussed in the context of a climate protection law (*Klimaschutzgesetz*). The idea of such a law is its cross-sectoral scope – i.e. each ministry needs to commit to GHG emission reduction goals with regard to its own domain. Furthermore, environmental politics is often confined to bans, constraints, or reductions and has little to offer to other ministries in exchange (Günther & Krebs 2000, 20).

These are structural challenges rooted in the institutional set-up of the government, the competencies of the ministries and its subsequent public authorities. As will be discussed later (see Chapter 9), government departments do often operate in a rather siloed manner. Ministerial bureaucracy functions by some sort of "negative coordination", i.e. each department tries to protect its own initiatives from the interference of others (Scharpf 1993; Günther & Krebs 2000, 17). Collaboration is often viewed with suspicioun, which might also be due to the fact that collaboration leads to vulnerability – whereas reliability, security of expectations and established procedures are common qualities of governmental bureaucracies.

Moreover, how can the demands of future generations be considered in political decisions today? Politicians need to win over their electorate and are rather likely to please them instead of confining them on behalf of the demands of the unborn.

Finally, when public goods or common-pool resources are damaged (e.g. waste illegally dumped in the countryside) it is often not possible to enforce the respective legislation, simply for practical reasons. Omnipresent monitoring of the environment is neither possible nor desirable. In economic terms, this is a problem of "imperfect information", i.e. the regulator's inability to fully observe the damage function and/or the polluter's private abatement cost function (Phaneuf & Requate 2017, 86). Environmental administration need to be sufficiently well-equipped and there has to be political will on the part of all stakeholders involved (SRU 2019, 124).

Solution perspectives[2]

Implement and optimize a national sustainability strategy

Following the first political awareness on sustainability issues in the 1980s, UNCED's Agenda 21 called nation states to walk the talk by adopting national strategies for sustainable development (UNCED, Agenda 21 1992, §8.7). Ten years later, at the 2002 World Summit on Sustainable Development (WSSD) in Johannesburg, 85 states had reported corresponding initiatives – with varying effectiveness (UN 2019). Right before the WSSD, the German Federal Government published their first Sustainability Strategy, which has since been updated regularly, usually every four years, with data updates in between (Bundesregierung 2002, 2012, 2016; Destatis 2019). The following remarks refer to this strategy and corresponding improvement suggestions.

At the outset of the first issue of the Sustainability Strategy, the Federal Government asserted that it had "recognised sustainability as a cross-sectional task and has made it a fundamental principle of its policy. The most important reform projects of this parliamentary term are oriented towards sustainability" (Bundesregierung 2002, 1). Within the framework of the national sustainability strategy the Federal Government formulates "management rules for sustainability" (50). The basic rule reads:

> Each generation must solve its own problems and not burden the next generations with them. It must also make provisions for foreseeable future problems. This applies to the conservation of the natural foundations of life, to economic development, as well as to social cohesion and demographic change.
>
> *(Bundesregierung 2002, 50)*

A great deal of effort was put into this strategy, and to coordinate different institutions in line with the strategy. Goals were set for indicators, data collected and reported, updates published, etc. All valuable efforts – but they could not avoid the fact that several targets were repeatedly not met (e.g. for nitrogen input and GHG emissions), which is most likely due to the fact that different governments had different political priorities. However, in order to bring more continuity to such national strategy, the governance process for federal sustainability management would need to be improved along the following lines (Berg et al. 2012, 89f.):

a) An independent committee should *evaluate the appropriateness of indicators and targets* of the national sustainability strategy. Indicators are always selective – but are the ones chosen appropriate? Furthermore, targets and indicators might need to be adjusted from time to time to reflect changed conditions and visions of desirable futures;

b) *Targets should be linked to specific measures*, otherwise success will remain uncertain. The Federal Government should subject the performance in its sustainability management to political interpretation once a year and report to the Parliament on this.

c) Key actors and *responsibilities should be named* across department boundaries. There needs to be clear political responsibility for the achievement of sustainability goals (which is, of course, a challenge in light of the cross-departmental structure of the issues).

d) *Monitoring needs to be improved.* Clear procedures need to be defined what will happen in case targets are missed.

Assessment of sustainability impact of laws

Another suggestion is to assess the sustainability impact of potential new laws during the legislative process. Something similar is operational regarding the bureaucratic burden of new laws. The National Regulatory Control Council aims to "tangibly reduce bureaucracy and compliance costs" of laws (NKR 2019). The council evaluates every draft of laws and regulations of the Federal

Government and evaluates the impact it will have on bureaucracy. Likewise, the sustainability impact of new laws and regulations is, in principle, already checked by the government during a sustainability test – but it is basically a generic form sheet and not really effective. This could be improved by better documentation and communication of the results, which makes it available for examination by Parliament. Transparency should be improved regarding the test criteria and the results of the tests (Berg et al. 2012, 90).

Constitutional anchoring of sustainability principle

An even more effective measure (but also more fundamental and more difficult to realize) would be to anchor the sustainability principle in the constitution. This could, for instance, require that the sustainability principle "must be taken into account in the definition and conduct of all state policies and measures" (ibid., 91). An independent expert council with constitutional power could supervise the sufficient implementation of this principle in the legislative process and could issue a suspensive veto in case of serious doubts.[3] Such a council would be the advocate of the needs of future generations.

7.2 Limiting of individual liberties for the sake of the common good?

It might be one of the most fundamental and most difficult questions of this book: to what extent may the state limit individual liberties for the sake of the common good? The answer to this will depend on the view of the prime role of the state. Where is the proper balance between seeing the purpose of the state primarily in facilitating the *res publica*, public affairs, the commonwealth, or seeing it in ensuring the *freedom* of its citizens?

7.2.1 Betterment of individual rights compared to public goods

There is an imbalance in the rights of the polluter, the one who suffers from pollution (the sufferer) and the common good of society – this is presumably the case in most liberal constitutional states (SRU 2019, 129f.).[4] If the state wants to limit the basic liberties of the polluter (which are especially protected, like property, professional freedom, or freedom of action), the state needs to substantiate that:

- the polluter causes a concrete harm or risk,
- state intervention would reduce this harm or risk,
- this intervention would be commensurable.

Since the basic individual rights hold such an important status in Western democracies, any constraint of these rights by the state needs to be well justified and is (in principle) subject to examination by law courts. On the other side, the individual contribution of the polluter is difficult to evidence because environmental damages are often

- caused by multiple sources and/or multiple actors,
- diffuse and difficult to measure,
- long-term and therefore subject to some uncertainty.

In other words, the polluter has a much stronger legal position than the sufferer. Furthermore, the state's requirements for protecting individuals against pollution is much more difficult to enforce.

Moreover, as future generations have no institutional support in most jurisdictions, the polluter's position is particularly strong with reference to "unborn sufferers". State intervention into basic rights are therefore particularly challenging if the prime objective is the prevention of a risk or the blocking of environmental damage in the distant future. Interfering with the basic rights of the polluter today requires very concrete state action, while precautionary measures to prevent future harm of a future sufferer remain abstract.

In light of the high value of basic liberal rights in the Western democracies, it is therefore one of the most critical questions whether democracies will be able to limit the liberties of the current generation for the sake of future ones. The widespread opportunism of politics and the tendency to please their own voters can raise substantial doubts about this.

7.2.2 Challenges to the concept of the common good

The state is legitimized by protecting the common good, which is substantiated by the state's goal to provide security to its citizens (Calliess 2001, 101, 149). As the German jurist Christian Calliess argues, security also includes ecological security (Calliess 2001, 149), i.e. the prevention of ecological disasters. Moreover, the state is not only *authorized* to protect natural resources and humanity's living conditions, it is *mandated* to do so by its very goal to ensure security lest the state loses its legitimacy (102). Securing the common good of a society implies the prevention of overutilization of public goods (like clean air). The state needs to ensure that excessive strain on environmental goods is prevented (SRU 2019, 70f.).[5]

However, how is the common good defined in substance? The idea of the common good, understood as "that which benefits society as a whole" (Britannica 2019) has been present in practically all concepts of political thinking since Plato, despite some great variety in these concepts (Dupré 1993, 687). A brief look at its historic genesis will help understanding its challenge today.

A considerable part of the Western tradition did not even separate the individual from the community, which is particularly true for ancient thinking. Plato, for instance, uses the metaphor of the human body to describe the perfect state. Under such perfect conditions, the good or evil experienced by any one of the citizens will lead the whole state to "make this case their own and will either rejoice or sorrow with him" (Plato, 462b).[6] Interestingly, something similar can be stated for a quite different but isochronic tradition, namely ancient Israel, as the Old Testament scholar Gerhard von Rad pointed out: "There was no isolated individuum in such a way on its own that its actions could remain more or less without relation to the community; there were only collectives which knew themselves with all their limbs as living bodies" (von Rad 1992, 399). The individuals' loyalty to the community and the well-being of the community were always associated, as, vice versa, everybody benefited from the well-being of the people (ibid.).

In the scholasticism of the Middle Ages, this thinking was challenged – but rather indirectly and subtly. The scholasticists took up an old argument, namely whether universal concepts (e.g. human being) can be ascribed ontological existence (the *realist* position) or whether they are only names (*nomina*), mental abstractions (the *nominalist* position). It is quite likely that this discussion impacted subsequent Occidental thinking because at the dawn of the Middle Ages, the nominalist position prevailed in most universities (Klima 2017). This turn to the individual facilitated not only the rise of science and technology, but the related conflict with religious authorities also paved the way for the Protestant Reformation.

The individual became the measure of all things, which also had fundamental consequences on the concept of the community. While the "idea of community lost its ontological ultimacy" under the impact of nominalist thought, as Dupré explicates, "a struggle originated between the traditional conception of the community as an end in itself and that of its function to protect the private interests of its members" (Dupré 1993, 687).

> Once the idea of society came to rest upon individualist premises, the common good was inevitably reduced to a collective well-being of its individual members ... Since civil society had no other purpose than to protect the life, liberty, and security of its members, it provided scant support for any normative concept of the common good.
>
> *(Dupré 1993, 687, 696f.)*

This implies a challenge for a material determination of the common good.

Since modern pluralist societies have difficulties in determining the common good in substance, they rely on a formal, procedural understanding of the common good. A precondition for the deliberation of the common good is, for instance, that all members of society can take part in that discourse and can agree on a rational procedure without reference to any particular worldview. However, such formal procedure will not provide any guidance on the normative content of the common good.

> One therefore ends up with a descriptive phenomenon, for example, a public interest constituting any interest that the public has. This is

problematic to the extent that it often equates the public interest with the view of the majority, and this is not always a good thing. John Dewey said that it is the media's job to interest the public in the public interest, and this definition hints at the more normative content of the concept. Procedural approaches, although signalling a very critical aspect of the common good, are not sufficient in themselves .

(Simm 2011, 561).

If society were just a collective of atomistic individuals who only have interests, who insist on their liberties, who closely tie liberty to property but neglect the responsibility of the latter, it would be no wonder if the common good went to rack and ruin. What can be done if the majority of society, which is mandated to determine the common good, prefers to follow a variety of private interests rather than care about that which benefits society as a whole? What can be done if the public has no interest in the public interest? Is this, then, the common good because it is procedurally legitimized? Or do we need to have at least a minimum of a material definition of the common good?

In light of the fact that the "idea of community [has] lost its ontological ultimacy", Dupré calls for at least some common understanding of what is intrinsically good: To accept the

idea of a common good presupposes a minimum agreement on what is intrinsically good, and this itself rests on the presence of a certain amount of *virtue* among the citizens. One of the main predicaments of the modern liberal state is that virtue has been replaced by choice.

(Dupré 1993, 711).

This difference between virtue and choice is taken up later as the difference between principles and interest (see 14.1). If a person does not have any *virtues*, any obligations which are independent of the consequences – in philosophical terms, if the ethics is purely *consequentialist* and does not include elements of *virtue* or *obligation* – there will only remain interests. If there are only interests left, however, the rational choice model is fully adequate. Interests will follow the same instrumental rationality, which dominates both technology and economy – and the consequentialist ethics of utilitarianism.

These are critical issues, and they become even more difficult in light of specific sustainability requirements. What is the proper balance between the common good of the present generation and the common good of future generations? Under the conditions of today, the common interest is a very fragile being and dependent on public discourse.

The open and decisive question is whether liberal societies will manage to develop, reach agreement on and implement a concept of the common good which sufficiently accounts for the natural environment as well as for the needs of future generations, while preserving the individual liberties of their citizens.

Solution perspectives

There is good reason to think that the limitation of individual liberties for the sake of the common good will have to increase in the future. Any such idea is easily distorted or defamed as "eco-dictatorship" – and there are indeed positions which call for a more authoritarian state to resolve environmental problems (see Chapter 1).

However, such accusations should be countered by pointing out that the modern liberal state is built – both historically and systematically – on the idea that the liberty in a community can only be achieved at the cost of subordinating the individual will under the common good (1.). It is the state's prime mandate to protect its citizens – which will increasingly imply ensuring the planetary boundaries are observed. Protecting these boundaries can be a minimum requirement for a substantive definition of the common good (2). This also holds, although to a lesser degree, for some kind of "global common good" (3.) and for future generations (4.)

1. Despite all their differences in detail, the vast majority of positions in the Occidental tradition has seen the purpose of the state in some kind of common good, which provides as a matter of course also a measure for the limitation of the citizens' liberties. Rousseau already discussed the case that a citizen's self-interest would deviate from the "common interest". It would destroy the body politic, the social compact, if someone enjoys the rights of citizenship without being ready to fulfil the corresponding duties (Rousseau 2017, ch. 7). In order to "protect the social compact from being a mere empty formula, therefore, it silently includes the undertaking that anyone who refuses to obey the general will is *to be compelled to do so* by the whole body ... It means nothing less than that each individual *will be forced to be free*" (ibid.; emphasis added). Ultimately coercion would imply freedom because this "is the key to the working of the political machine; it alone legitimises civil commitments which would otherwise be absurd, tyrannical, and liable to frightful abuses" (ibid.). Rousseau's words elicit negative associations if one considers the coercion exercised by totalitarian regimes, of which the twentieth century saw a number. However, I submit the general idea that freedom ultimately depends upon the acceptance of other people's freedom, which requires some recognition of the common good, still holds true. Kant took up Rousseau's idea in his *Metaphysics of Morals* (Kant 1968, §47, A169): the state is constituted by the act of a people which surrender their (outer) freedom in order to get it back as "limbs of a commonwealth". Kant stresses that nobody sacrifices his own freedom but truly leaves the wild, lawless freedom in order to gain freedom in a lawful state because this dependency originates in his own law-making will. In line with this thinking, Rawls built his concept of fairness, which tries to reconcile liberty and equality. The first of his two principles guarantees

everybody the greatest possible set of equal basic liberties as long as they do not infringe the respective rights of others, while the second principle balances inequalities by granting the greatest benefit for the least privileged.In simple words, the deliberate act of subordinating one's own will to the common interest – or rational consent to the principles of fairness – is the precondition for a liberal society, in which individuals' liberties are limited insofar as they limit the liberty of others or violate the common good.This nicely matches Harry Frankfurt's thoughts on the relation of liberty and necessity (Frankfurt 1999). A man without any ideals – i.e. obligations he is not willing to violate – has no limits for action and can do what he wants. However, Frankfurt argues, he is nevertheless not free because his will is not lawful but anarchic. His will is just a pawn of impulses and appetites.

2. Today the common good is endangered by the overutilization of natural resources and the pollution of ecosystems. One of most important goals of the state, providing security to its citizens, is increasingly threatened by environmental issues (SRU 2019). The right to life and physical integrity, for instance, will be increasingly endangered by droughts, floods, extreme weather events, etc. (SRU 2019, 71). In order to identify a minimum condition for the preservation of natural ecosystems, the the German Advisory Council on the Environment (SRU) refers to the concept of planetary boundaries, which identifies a safe operating space for humanity for critical variables. It is the state's duty to ensure that the planetary boundaries will not be violated. It is such protection of the planetary boundaries, I submit, which can justify a minimum material definition of the common good because protecting these boundaries is the condition of possibility for any future procedural determination of the common good.

3. There is, however, some complication by the fact that the common good cannot be confined to national borders. In the philosophical tradition, the common good was related to law – Thomas Aquinas, for instance, said "every law is ordained to the common good" (Acquinas, STh 90.2), and Rousseau called society to "compel" those who would refuse to obey the general will (Rousseau 2017, ch. 7). Although the implementation of international law faces severe challenges (see Chapter 6), the situation might improve with the establishment of a global civil society. Global protests like Fridays For Future indicate that some sort of global civil society is forming. On a national level, civil society can exert enormous influence on policy makers. There is hope that something similar can occur on global level.

4. Finally, does the common good take future generations into account? In a qualified way, I would say so. Under the assumption that having children as a basic human affair serves the common good (at least some), it would be only be consistent that the same function will be attributed to the generation of the children, etc. By iteration this will be valid also for future generations although it becomes weaker as the future horizon is extended.

On top of this, there are several things that will help protect and strengthen the common good, which we discuss in other parts of this book: strengthen social cohesion (see 15.4), strengthen collaboration and mutual understanding (see 15.2), apply Rawls' principle of granting the least privileged greatest support (see 15.1), foster personal principles (see 14.1), internalize the external costs (see 13.5), and increase transparency (see 16.3).

In sum, the limitation of personal liberties for the sake of the common good has functioned remarkably well in the Rhineland capitalism. The challenge today is not so much different, at least in principle, from establishing the social welfare state in many countries of continental Europe; it is not principally different from the social welfare state of Ludwig Erhard. We "just" need to expand the boundaries considered – from nation state to global context, and we need to account for future generations. But the limitation of individual liberties for the sake of the common good should not be argued hesitantly and ashamedly but simply by pointing to the tradition in which our societies have evolved and that it's only natural to extend the idea of the common good beyond the current understanding.

Notes

1 The following is primarily applied to the German context but presumably similar circumstances hold in other jurisdictions.
2 The following suggestions build on the results of a task force on "Sustainable Economic Activity and Growth", which the German Chancellor Merkel mandated within her Dialogue on Germany's Future about future dialogue in 2011 and 2012 (Berg et al. 2012). They are mostly still up-to-date in essence, unfortunately.
3 Suspensive veto means that the council *could not* finally *prevent* what it considers unsustainable but *postpone the process of enacting and call for public discussion.* Such a suspensive veto would conform to democratic principles because the experts would not decide on any legislation, just decelerate the legislatory process; see Berg et al. 2012, 92, for more details.
4 The following argument is taken from (SRU 2019, 129f.)
5 It is often not easy to decide whether the degree of strain can still be tolerated. In concrete practical cases it is often not possible to determine exactly whether or not a certain public good is overutilized or not. Ultimately a public discourse and the subsequent decision of the parliament need to take place (70).
6 "When but a finger of one of us is hurt, the whole frame … feels the hurt and sympathizes all together with the part affected, and we say that the man has a pain in his finger … Then when any one of the citizens experiences any good or evil, the whole State will make his case their own, and will either rejoice or sorrow with him? Very true, he replied; and I agree with you that in the best-ordered State there is the nearest approach to this common feeling which you describe" (Plato, 462b)

Bibliography

Aquinas, Thomas. *Summa Theologica*, ed. by Christian Classics Etheral Library, https://www.ccel.org/ccel/aquinas/summa.pdf (accessed 11 September 2019.

Berg, Christian, Christian Calliess, Christa Liedtke, Georg Meran, Ortwin Renn, Wolfgang Schmalz, & Miranda Schreurs. "Sustainable economic activity and growth". In *Dialogue on Germany's Future*, Bundespresseamt, ed., 86–95, Berlin: Federal Government Publication Office, 2012.

BMU. "Reformprojekt für ein neues Umweltrecht." *Bmu.de*. 20. 09 2018. www. bmu.de/ministerium/chronologie/reformprojekt-fuer-ein-neues-umweltrecht/ (Accessed 17. 06 2019).

Bosselmann, Klaus. *The Principle of Sustainability. Transforming Law and Governance*. London, New York: Routledge, 2017.

Britannica. *Common good*. 2019. www.britannica.com/ (accessed 27. 04 2019).

Bundesregierung. *Perspectives for Germany. Our Strategy for Sustainable Development*. Berlin: Federal Government Publication Office, 2002.

———. *National Sustainable Development Strategy*. Berlin: Federal Government Publication Office, 2012.

———. *German Sustainable Development Strategy*. Berlin: Federal Government Publication Office, 2016.

Calliess, Christian. *Rechtsstaat und Umweltstaat, Zugleich ein Beitrag zur Grundrechtsdogmatik im Rahmen mehrpoliger Verfassungsrechtsverhältnisse*. Tübingen: J.C.B. Mohr (Paul Siebeck), 2001.

Destatis. "Nachhaltigkeitsindikatoren." *Destatis.de*. 2019. www.destatis.de/DE/Themen/ Gesellschaft-Umwelt/Nachhaltigkeitsindikatoren/_inhalt.html (Accessed 2019. 03 13).

Dupré, Louis. "The common good and the open society". *The Review of Politics 55*, 1993: 687–712.

Frankfurt, Harry G. "On the Necessity of Ideals." In *idem, Necessity, Volition, and Love*, 108–116.

Cambridge: Cambridge University Press, 1999.

Günther, Edeltraud, & Maja Krebs. *Aufgaben- und Organisationsstruktur der Umweltpolitik in der Bundesrepublik Deutschland*. Dresden: Dresden University, Business Management, 2000.

Heselhaus, Sebastian. "Verfassungsrechtliche Grundlagen des Umweltschutzes.". In *Grundzüge des Umweltrechts*, Eckard Rehbinder & Alexander Schink (eds.), 4–63. Berlin: Erich Schmidt Verlag, 2018.

Kant, Immanuel, *Metaphysik der Sitten*, ed. by Wilhelm Weischedel, Volume 8, Frankfurt a. M.: Suhrkamp 1968

Klima, Gyula. "The medieval problem of universals." *The Stanford Encyclopedia of Philosophy*, 2017.

Kress, Daniela. „Internationale Übereinkommen als künftiges Herzstück einer Global Governance? Auf dem schwierigen Pfad zwischen Anspruch und Wirklichkeit.". In *Globalisierungsgestaltung und internationale Übereinkommen*, Armin Frey, Thomas Jäger, Dirk Messner, Manfred Fischedick, & Thomas Hartmann-Wendels (eds.), 15–58. Wiesbaden: Springer, 2014.

Müller, Harald. *Wie kann eine neue Weltordnung aussehen? Wege in eine nachhaltige Politik*. Frankfurt a.M: Fischer Taschenbuch Verlag, 2008.

Nair, Chandran. *The Sustainable State. The Future of Government, Economy, and Society*. Oakland, CA: Berrett-Koehler Publishers, 2018.

NKR. *National Regulatory Control Council*. Berlin: Federal Government of Germany. 2019.

Phaneuf, Daniel J., & Till Requate. *A Course in Environmental Economics*. Cambridge: Cambridge University Press, 2017.

Plato. *Republic*. Übersetzung: Benjamin Jowett. The Internet Classics Archive by Daniel C. Stevenson, Web Atomics, 1994–2000.

Rousseau, Jean-Jacques. *The Social Contract*. Jonathan Bennett, 2017 (1762).

Scharpf, Fritz W. "Positive und negative Koordination in Verhandlungssystemen.". In *Policy-Analyse – Kritik und Neuorientierung (Politische Vierteljahresschrift, Sonderheft 24)*, A. Héritier, (ed.), 57–83, Opladen: Westdeutscher Verlag, 1993.

Simm, Kadri. "The concepts of common good and public interest: From Plato to biobanking". *Cambridge Quarterly of Healthcare Ethics 20*, 2011: 554–562).

SRU. *Demokratisch regieren in ökologischen Grenzen – Zur Legitimation von Umweltpolitik.* Berlin: German Advisory Council on the Environment, 2019.

Umweltbundesamt. "Umweltgesetzbuch." *Umweltbundesamt.de.* 19. 12 2018. www.umwelt bundesamt.de/themen/nachhaltigkeit-strategien-internationales/umweltrecht/bessere-umweltrechtsetzung/umweltgesetzbuch#textpart-1 (accessed 17. 06 2019).

UN. "National Sustainable Development Strategies (NSDS)." https://sustainabledevelop ment.un.org/. 2019. https://sustainabledevelopment.un.org/topics/nationalsustainable developmentstrategies (accessed 28. 04 2019).

UNCED. "Agenda 21." Rio de Janeiro: United Nations, 1992.

von Rad, Gerhard. *Theologie des Alten Testaments.* Munich, 1992 (1960).

8

TECHNOLOGY

Mismatch between impact and governance

Technology has not really been the focus of philosophical reflection in Occidental philosophy, which is rooted in the disregard for mechanical skills in ancient Greek thought.[1] This changed in the twentieth century, when scholars began to reflect on the enormously increased power of technology, for which nuclear technology is the paradigm. The bombs on Hiroshima and Nagasaki at the end of World War II and the arms race during the Cold War caused increasing concern and stimulated reflection on the role of technology in society. Scholars like Albert Einstein, Max Born, Joseph Rotblat and Bertrand Russell pointed to the danger of nuclear weapons and stood up against the arms race. Philosophers like Günter Anders and Hans Jonas discussed the "metaphysical" character of technology, a result of its unprecedented power and its potential to annihilate humanity, for which 1945 was a watershed (Anders 1980, 20). In his seminal book, *Das Prinzip Verantwortung* (*The Imperative of Responsibility*), Jonas offered a substantial account of an "Ethics for the Technological Age", in which he argued that today's new kind of technology with its vast expansion of destructive potential calls for a new ethics, in which the "obligation to the future" (*Pflicht zur Zukunft*) plays an eminent role (Jonas 1984).[2] Jonas demanded that the uncertainty of future projections in ethical reflection be considered. The impact of technology is so huge and potentially so destructive, its future path yet unpredictable that the "negative projection" should always "precede the positive one" (Jonas 1984, 70ff.). Today's widespread peaceful usage of nuclear technology documents that Jonas' call has not been heard – its "negative projection" has meanwhile been realized twice: in Chernobyl and Fukushima. The aftermaths of these nuclear accidents will be felt by many generations to come.

In 2017, the German government passed a law which defines the criteria for the search for a final storage facility of nuclear waste. On that occasion, the former Federal Environmental Minister, Barbara Hendricks, pointed to the fact that the aftermath of nuclear technology will affect 30,000 generations, although the nuclear power plant's operating period was just 60 years. The final waste storage facility will have to safeguard the waste for one million years (Bundesregierung 2017). Never before has humankind had to take provisions for time spans in the order of magnitude of their own existence as a species! No civilization has lasted longer than a few thousand years, the entire history of human culture is counted in tens of thousands of years. The very idea of constructing storage facilities which would endure a million years is bizarre.

All environmental problems are brought about by technology in one way or another. Apart from the harm caused by accidents (e.g. Chernobyl, Fukushima, the Bhopal chemical disaster, Deep Water Horizon, Exxon Valdez, etc.), the destructive impacts occur mostly as "side effects" of well-intended technological usage: chemicals like pesticides or CFCs were *intentionally* designed to be as they are (e.g. biocidal or chemically inert and stable), but their harmful effects had been disregarded or underestimated. Rachel Carson, a US-American biologist who was one of the earliest initiators of the environmental movement, described the devastating effect of pesticides already occurring in the early 1960s:

> For the first time in the history of the world, every human being is now subjected to contact with dangerous chemicals, from the moment of conception until death. In the less than two decades of their use, the synthetic pesticides have been so thoroughly distributed throughout the animate and inanimate world that they occur virtually everywhere.
>
> *(Carson 1962, 24)*

Compared to the harmful effects of these special chemicals, the presumably even greater challenge stems from the massive global spread of just "normal", everyday technologies. POPs are found in penguins in Antarctica and microplastics found in human bodies. The industrial metabolism of humanity consumes such vast amounts of renewable and non-renewable resources, interferes with the large bio-geo-chemical material fluxes, and produces and emits into the environment such great amounts of GHGs, plastics, POPs, heavy metals, pesticides, fungicides, fertilizers and nuclear waste that pristine natural habitats become utterly impossible.

The societal challenge is, of course, to manage and control these negative effects of technology. The unplanned socio-cultural aftermath of technology challenges humanity, as Jürgen Habermas noted in the 1960s. This is "a challenge of technology which cannot be met by technological means". (Habermas 1969, 118). Habermas' call for a politically effectual discussion which would "relate the societal potential of technological knowledge and capability to our practical knowledge and desire in a rationally binding manner" has lost nothing of its timeliness half a century later (ibid., 107). Where are such places for a rational discourse on technology? To be sure, well-established academic disciplines like technology assessment, and philosophy and ethics of technology meanwhile explore the societal consequences of technologies, assess their impacts and deliberate ethical implications (see e.g. Grunwald 2000; Hubig, Huning & Ropohl 2000; Grunwald 2002; Ropohl 2009; Grunwald 2018) – but is there a public discourse on the question which technology we want? Rarely. Is it even possible to control the direction of technological development? There are serious doubts about that.

Ulrich Beck, whose book *Risikogesellschaft* (*Risk Society*) (Beck 1986) was a milestone in the ethics and sociology of technology, was sceptical that the course of technological development can be influenced substantially. Political institutions are mandated to manage developments which they can hardly influence. The dynamics of technological change and its inherent laws threaten political institutions because these institutions become trustees (*Sachverwalter*) of developments, which they cannot shape and have neither planned but are still somehow responsible for (Beck 1986, 305). At the same time, decisions in business and science come along with political implications for which the actors are by no means legitimized:

> The decisions which alter society have no place where they could emerge. They become voiceless and anonymized. In the economy they are embedded in investment decisions which relegate their society-changing potential capability to "unseen side-effect". In their self-conception and their institutional embeddedness, the empirical-analytical sciences, which think ahead the innovation, remain truncated from the technological consequences and the consequences of the consequences. The unknowability of the consequences, their non-accountability is the development program of science.
>
> *(Beck 1986, 306)*

Politicians depend upon experts for understanding technological developments, but these experts might have their own agenda and are not legitimized democratically. Technological "progress" becomes inevitable, its non-decidability challenges democratic legitimization. The unseen side-effects of technology take control of the regime in Western democracies (Beck 1986, 306).

Moreover, it is not only that our public institutions have not yet managed to effectively control technological developments; one can argue that our *ethical awareness* is also not suited to reflect on these new opportunities. The philosopher Vittorio Hösle sees an increasing chasm between accelerated scientific-technological development and stagnating, if not regressing ethical awareness (Hösle 1997, 23). In 1997, Hösle called for a double expertise in both science and ethics to bridge that gulf between the two cultures. His call for new study programs has meanwhile been realized in many universities worldwide – but this does not seem to have changed our moral reflection on technology or our ability to manage the direction of technological "progress".

Harry Frankfurt, US-American philosopher, also addresses the widening gap between the possible and the legitimate and discusses the repercussion of increased technological possibilities on ethical reasoning. The "proliferation of possibility engendered by increasing technological and managerial sophistication" corresponds, according to Frankfurt, to

> a steady and notable weakening of the ethical and social constraints on legitimate choices and courses of action. The expansion of freedom has affected not only what can be done but what is permissible as well. This combination of endlessly more masterful technical control and increasingly uncritical permissiveness has generated a tendency whose limit would be a culture in which everything is possible and anything goes.
>
> *(Frankfurt 1999, 108)*

In line with this argument, it is not only the mismatch between today's impact of technology and the ineffectiveness of controlling its development which is worrying, it is also the fact that the expanded possibilities are decreasingly countered by ethical deliberation. What *can* be done *will* be done, and what will be done is largely driven by economic interests. Steve Jobs' dictum according to which "sometimes people don't know what they want until you show it to them" (The Times 2011) accentuates that technological development is largely driven by commercial interests. Products are being developed for which the demand needs to be created in the first place.

The gap between moral reflection, societal discussion and political governance on the one hand, and the autonomous laws of technological development in combination with the "market pull" on the other, is an unresolved issue of today's market-based societies. Habermas' question about the rational discourse by which we control the relationship between technological progress and social living environment remains open. In fact, it is all the more worrying because of the extreme speed with which changes occur.

Solution perspectives

1. *Rising awareness of the problematic side-effects of technologies*

The good news is that global awareness about environmental pollution and the harmful (side) effects of our technologies is increasing. The focus of public discussion does not necessarily match the issues which the scientific account would prioritize – plastic, climate, biodiversity is presumably the order of priority in the public eye, while the scientific account would reverse the order (Steffen et al. 2015) – but it is nevertheless important that people pay attention. The public discourse, which is currently more centred around the visible symptoms, will at some point hopefully broach the issue of the underlying causes, too. This will entail our production and consumption patterns and thereby also the question which technologies we actually need and want.

2. *International treaties*

Moreover, considering some especially harmful technologies, substances, or controversial practices, there are encouraging examples which illustrate how international treaties could limit the harmful effects of technologies or substances. The de facto-banning of CFCs is an often-quoted case is this context. CFCs had been used as refrigerants and aerosol propellants in the 1960s and 1970s. Their molecules are very long-lived and accumulate in the atmosphere. Their potentially ozone-depleting feature was discussed since the early 1970s (Douglass, Newman & Solomon 2014). In 1985 it was detected that the ozone layer over Antarctica had dropped by 30%. Only two years later, in 1987, the international community adopted the Montreal Protocol on Substances that Deplete the Ozone Layer, which was successively improved in subsequent years and reversed the depletion of the ozone layer. Although the full recovery to ozone concentration levels of 1980 will still take until 2050 or even 2070, the trend is positive and the "hole" is closing (Douglass, Paul & Solomon 2014, 48). To be sure, there were several conditions favourable for such rapid political agreement and the effectiveness of the respective policy measures: there was *solid scientific evidence*, a clear, *well-described and significant threat to human health*, combined with a *straightforward policy measure* which could be implemented quite *easily and effectively*. Other contracts in the international regime restrict or forbid the usage or spread of certain technologies or techniques like the Treaty on the Non-Proliferation of Nuclear Weapons from 1968.

3. *Public discourse and stakeholder consensus*

In line with the discussion on global governance above, a broad agreement among critical stakeholders might be as effective as an international treaty. Adèle Langlois argues along these lines using the example of the international discussion on reproductive human cloning. State consent was difficult to achieve but a broad consensus among various stakeholders was given. Langlois suggests that it is better to have "thick stakeholder consensus" than "thin state consent", which was the "hallmark of the old hierarchical approach to governance". Treaties might be formally recognized by international law but could be based on "back-room deals between undemocratic states", in which case they would have no guarantee of legitimacy (Langlois 2017, 5). Drawing on Pegram and Acuto (2015), Langlois sees this as a promising example of a "pluralist concept of global governance" that would be "focusing less exclusively on intergovernmental politics" (ibid., 1).

4. *Technology Assessment*

The examples discussed so far relate to contexts in which either morally reprehensible *practices* or harmful *effects* of a technology should be constrained. What is to be done, however, if the effects of a *new* technology need to be assessed? In the "classical concept" of technology assessment, new technologies would be evaluated, i.e. their likely effects on society and natural environment would be explored by an interdisciplinary group of technology experts; the results would be made available to the public, and the pros and cons of the different policy options would be publicly discussed. Political regulation would finally be enacted for corresponding policies (Grunwald 2002, 124). This "classical" approach to technology assessment is compromised by a number of difficulties:

- *Lack of information.* New technologies are often developed by corporations and information about them is limited at an early stage. However, once a given technology is better known, it has often already been rolled out into the markets and its assessment might be without consequences.
- *Speed of development.* There is an amazing and still increasing speed of new technological developments. Speed equals competitive advantage, which is essential in a globally competitive market. However, investigating and evaluating the societal consequences of any new technology takes time – not only because of the intricacies of the subject matter but also because the methods of knowledge generation, for instance by Delphi methods which build on expert judgement, are time consuming.
- *Intricacy and complexity.* New technologies are ever more difficult to under-stand and require sophisticated technical skills, not just for selected individ-uals but for entire teams, since no technology is developed in isolation any more.
- *Recruiting.* The smartest brains often work for companies because of strong financial incentives, making it more difficult for research institutions or gov-ernment agencies to recruit similarly skilled experts.
- *To assess the impact of a given technology, its social effects need to be considered.* However, it is difficult to anticipate social response since it depends on a variety of factors (e.g. market demand, future developments, inventions, needs, resource prices, etc.).
- *Value judgements.* Technology assessment is not just a technical skill, it involves value judgements which ultimately require the involvement of society. Expert judgement cannot replace public discourse.

Technology assessment was first institutionalized in the USA. In 1972, the US Congress enacted an Office of Technology Assessment.[3] This forward-looking institution was the first of its kind globally and enjoyed an excellent reputation. Nevertheless, it was shut down in 1995, when the Republicans, who gained majority in both houses in 1994, intended to push back governmental influence (Grunwald 2002, 104). Other countries are still operating similar institutions (e.g. the UK, Germany, Denmark), and numerous valuable studies have been produced (see e.g. TAB 2019), although the net effect on the course of techno-logical development is rather limited.

5. *Precautionary measures*

One political instrument to halt the development and application of potentially harmful technologies is the precautionary principle, which will be discussed later (see 13.6). This is a principle of informed prudence, it requires precautionary measures to be taken when potential future harms are anticipated early. Calliess argues that the precautionary principle can justify a reversal of the burden of proof. The state is obliged to act especially *in the face of concrete threats* to environmental goods. The precautionary principle is a statutory requirement for environmental policies in the European Union (e.g. EU 2012, Art. 191). As such, it is a guiding principle in European environmental law (Calliess 2018, 101).

Notes

1 The main ideas of the following paragraph are taken from Berg (2010, 9f.).
2 The book appeared in English as *The Imperative of Responsibility. In Search of an Ethics for the Technological Age.*
3 The respective law says: "(a) As technology continues to change and expand rapidly, its applications are – (1) large and growing in scale; and (2) increasingly extensive, pervasive, and critical in their impact, beneficial and adverse, on the natural and social environment. (b) Therefore, it is essential that, to the fullest extent possible, the consequences of technological applications be anticipated, understood, and considered in determination of public policy on existing and emerging national problems" (US Congress 1972).

Bibliography

Anders, Günther. *Die Antiquiertheit des Menschen. Über die Zerstörung des Lebens im Zeitalter der dritten industriellen Revolution, Band 2.* Munich: Beck, 1980.

Beck, Ulrich. *Risikogesellschaft. Auf dem Weg in eine andere Moderne.* Frankfurt: Suhrkamp, 1986.

Berg, Christian. *Handlung als überindividuelles Konzept? Eine Untersuchung zur Rede vom 'technischen Handeln'.* Münster, Hamburg. Berlin, London: LIT, 2010.

Bundesregierung. "Endlagergesetz in Kraft getreten." *Bundesregierung.* 16. 05 2017. www.bundesregierung.de/breg-de/aktuelles/endlagergesetz-in-kraft-getreten-394898 (Zugriff am 28. 04 2019).

Calliess, Christian. "EU-Umweltrecht." In *Grundzüge des Umweltrechts*, von Eckard Rehbinder & Alexander Schink (eds.), 65–144. Berlin: Erich Schmidt Verlag, 2018.

Carson, Rachel. *Silent Spring.* New York: Crest Book, Houghton Mifflin, 1962.

Douglass, Anne R., Paul A. Newman, & Susan Solomon. "The Antarctic ozone whole. An update." *Physics Today*, 07 2014: 42–48.

European Union (EU), Consolidated Version of the Treaty on the Functioning of the European Union (C 326/47, 26.10.2012), Brussels: European Union, 2012. https://eur-lex.europa.eu/resource.html?uri=cellar:2bf140bf-a3f8-4ab2-b506-fd71826e 6da6.0023.02/DOC_2&format=PDF (accessed 20 September 2019).

Frankfurt, Harry G. "On the necessity of ideals." In *Necessity, Volition, and Love*, Harry G. Frankfurt (ed.), 108–116. Cambridge: Cambridge University Press, 1999.

Grunwald, Armin. *Technik für die Gesellschaft von morgen. Möglichkeiten und Grenzen gesellschaftlicher Technikgestaltung.* Frankfurt, New York: Campus, 2000.

———. *Technikfolgenabschätzung – eine Einführung*. Berlin: edition sigma, 2002.

———. *Technology assessment in practice and theory*. Abingdon: Routledge, 2018.

Habermas, Jürgen. "Technischer Fortschritt und Soziale Lebenswelt." In *Technik und Wissenschaft als 'Ideologie'*, Jürgen Habermas (ed.), 194-118. Frankfurt: Suhrkamp, 1969.

Hösle, Vittorio. *Die Krise der Gegenwart und die Verantwortung der Philosophie. Transzendentalpragmatik, Letztbegründung, Ethik*. Munich: C.H. Beck, 1997.

Hubig, Christoph, Alois Huning, & Günter Ropohl. *Nachdenken über Technik*. Berlin: edition sigma, 2000.

Jonas, Hans. *Das Prinzip Verantwortung*. Frankfurt: Insel (also: Suhrkamp, stw 1085), 1984.

Langlois, Adèle. "The global governance of human cloning: The case of UNESCO." *Palgrave Communications 3*, 2017: 1–8. 17019.

Pegram, Tom & Michele Acuto. "Introduction: Global Governance in the Interregnum." *Millennium: Journal of International Studies 43*, 2, 2015: 584–597.

Ropohl, Günter. *Allgemeine Technologie. Eine Systemtheorie der Technik*. Karlsruhe: Universitätsverlag Karlsruhe, 2009.

Steffen, Will, Katherine Richardson, Johan Rockström, Sarah E. Cornell, & Ingo Fetzer. "Planetary boundaries: Guiding human development on a changing planet." *Science 347*, 13. 02 2015: 736.

TAB. *Office of Technology Assessment at the German Bundestag (TAB)*. Berlin: Office of Technology Assessment at the German Bundestag (TAB). 02. 05 2019. www.tab-beim-bundestag.de/en/index.html (accessed 01. 06 2019).

The Times. "Sometimes people don't know what they want until you show it to them." *The Times*, 08 2011.

US Congress. "Technology Assessment Act of 1972 (PUBLIC LAW 92-484-OCT. 13, 1972)." *Technology Assessment Act of 1972*. Washington, D.C.: US Congress, 1972.3176

9

STRUCTURAL SILOS

Fragmentation of knowledge, administration and responsibility

Contents

9.1 Fragmentation of knowledge

"The greatest improvement in the productive power of labour, and the greater part of the skill, dexterity, and judgement with which it is anywhere directed, or applied, seem to have been the effects of the division of labour" (Smith 2007, 8 (Glasgow ed., p. 13)) This is how Adam Smith begins his *Inquiry into the Nature and Causes of the Wealth of Nations*, in which he dedicates the first three chapters to the remarkable effects, the principles and the limitations of the division of labour. Smith was certainly not the first one who deliberated division of labour. Already Plato knew about the great importance of the division of labour; he considered it foundational to the formation of the state (Plato, Republic Book II, XI). While Smith was praising the division of labour as an engine for productive power, Karl Marx already knew about the other side of that coin, "craft-idiocy": "What characterizes the division of labour inside modern society is that it engenders specialized functions, specialists, and with them craft-idiocy" (Marx 1955; see Elias 1996, 48, footnote 15).[1]

The issues arising from specialism, "craft-idiocy", compartmentalized knowledge, siloed thinking and departmental structures have much increased since Marx's times and pose a continuous challenge to sustainability endeavours because the problems of modern industrial societies do not do us the favour of defining themselves as problems for disciplinary specialists, as the philosopher of science Jürgen Mittelstraß formulated it (Mittelstraß 1992, 99).

The flipside of specialization and their implications for the academic world has long been discussed – but rarely resolved. August Comte identified specialization as the weakest part of the scientific system already in the first half of the 19th century and urged us to ensure that the human intellect does not lose its way in a pile of particulars (Elias 1996, 49).[2] In the 1970s, Erich Jantsch called for an "Inter- and Transdisciplinary University",[3] which would have to "contribute to the development of a common policy for society at large" (Jantsch 1970, 427). Helmut Schelsky argued that for every step of specialization in an academic field, a countermeasure of integration would need to be carried out. (see Schelsky 1971, 208, quoted from Mittelstraß 1992, 96).

Botkin et al. pointed to the need for integrated knowledge in the light of global challenges in their 1979 Club of Rome report:

> Nowhere is the impact of over-specialization so keenly felt as in the context of global issues. It is simply not possible to analyze and formulate policies for global issues from any exclusive disciplinary perspective. The economic approach, the legal approach, the social or political approach are each, by themselves, insufficient for dealing with problems that require an integrated and holistic understanding. Such specialization virtually guarantees irrelevance.
>
> *(Botkin, Elmandjra & Malitza 1979, 70)*

In the early 1990s, Mittelstraß illustrated the mismatch between the shape of academic disciplines and the shape of real-world challenges: by referencing "three keywords: environment, energy, technology impact", he sees an "asymmetry between the development of problems and the development of disciplines, and this is growing at the same rate at which disciplinary development is characterized by increasing specialization" (Mittelstraß 1992, 99).

In 2010, a group of scientists communicated a correspondence in the journal *Nature* in which they urgently called for more funding for transdisciplinary research:

> Europe's future hinges on funding transdisciplinary scientific collaboration. But career paths, peer recognition, publication channels and the public funding of science are still mostly geared to maintain and reinforce disciplinarity. We do not properly understand the effects of technology on the evolution of the systems on which we all depend. To take on global challenges such as climate change, growing urbanization and loss of biodiversity, we need to build a new science community that will explore common themes in natural, artificial and social systems.
>
> *(Vasbinder et al. 2010)*

In more recent years, we could notice that "[t]ransdisciplinary research is surely gaining momentum" (Brandt et al. 2013, 6), even a "proliferation of contributions about transdisciplinarity" can be percieved, which has been adopted in the natural as well as the social sciences and the humanities:

> However, conceptual and institutional barriers for transdisciplinary inquiry are still common whereas incentives remain rare. This is not only due to

> the scepticism of decision makers in academic institutions, in conventional funding agencies and in policy decision making but also to the formal education and personal motives of scientific researchers in academic institutions.
>
> *(Lawrence 2015, 1)*

Many contributions to the transdisciplinary discourse lack sound conceptual frameworks. *Come on!*, the 2017 report to the Club of Rome also criticizes the fragmentation of knowledge and increasing specialization. The authors not only discuss the fragmentation on the theoretical level, they also observe discrepancies between theory and practice (von Weizsäcker & Wijkman 2017, 173).

Moreover, the fragmentation of knowledge is also a *question of mindset* and of *culture*, including corporate culture. In his book *Harmony*, HRH The Prince of Wales calls for "A New Way of Looking at the World". He bemoans the fragmented view of the world, which would even

> extend to the way people are expected to behave ... I have lost count of the number of people I have spoken with who tell me quietly of how, even though privately they may feel deeply anxious inside themselves about the consequences of this whole mechanistic approach, when at work they are expected to lock those feelings away.
>
> *(HRH The Prince of Wales, Juniper & Skelly 2010, 21, 22)*

Fragmentation of knowledge, one could summarize this issue, relates to a fragmentation of worldview, which impacts our mindsets, behavioural patterns, and culture.

9.2 Fragmentation of administration

These challenges of specialization in the academic world are mirrored by similar issues in other sectors of society. The disciplinary focus in science corresponds to the departmental structure of government administration and public authorities. The ministers have the mandate for their respective domain, which is partly even secured by the constitution. In Germany, for instance, article 65 of the Basic Law states that "each Federal Minister shall conduct the affairs of his department independently and on his own responsibility" (Bundestag 2014). To this extent, it is clear that the governments' operating mode is essentially oriented towards departments, which delimit their assignment of duties and which severely impedes environmental integration and effective environmental policies (SRU 2019, 125) The ministers do not only represent their own field and department but as politicians they are also competing against each other and have an interest in getting their own agenda realized. Therefore, ministerial bureaucracy is often driven by some sort of "negative coordination", i.e. each department tries to protect its own initiatives from the interference of others

(Günther & Krebs 2000, 17). Apparently such "negative coordination" does not only have a structural or institutional element but also a cultural one. In a "silo culture", in which bureaucrats feel obliged to defend their own departmental goals and achievements, collaboration is suspicious. Given the inertness of bureaucracies discussed above (see 4.1), one can well imagine how much effort it will take to change this mode of operation.

However, this illustrates that a sustainability transition obviously must be supported by a new kind of culture, a new kind of collaboration, a new kind of leadership (Künkel 2019). By definition, people feel comfortable in their comfort zone, in which they know both content and processes. Leaving the comfort zone makes people vulnerable – something which many feel increasingly challenging in times of rougher competition and rapid changes in work environments, value systems, lifestyles, etc.

Physicist and economist Robert Ayres calls "the traditional governmental division of responsibility into a large number of independent bureaucratic fiefdoms ... dangerously faulty" (Ayres 1994, 36). The perspective of industrial metabolism[4] would be essentially "holistic" because it jointly considers "whole range of interactions between energy, materials, and the environment" – but

> the departmental structure of environmental protection policy has systematically ignored this fundamental reality by imposing regulations on emissions *by medium* ... Typically, one legislative act mandates a bureaucracy that formulates and enforces a set of regulations dealing with emissions by "point sources" only to the air. Another act creates a bureaucracy that deals only with waterborne emissions, again by "point sources." And so forth.
>
> *(Ayres 1994, 35)*

In addition to this, the departments which are particularly important for the future of society and environment, e.g. those for education, family, or environment are typically not the most powerful ones. The environmental management economists Günther and Krebs have argued that in the case of Germany, the environmental department has often only minor bargaining power within the government Cabinet; it has a small budget and its political goals are often impacted by policy measures from other departments like transportation, economics, construction, or finance (Günther & Krebs 2000).

9.3 Fragmentation of responsibility

The specialization in the sciences and the departmentalization in bureaucracy imply yet another challenge: the fragmentation, the "dilution" of responsibility (Lenk 1993, 125f.). The individual is embedded in ever longer chains of interdependency, which defy individual control (Elias 1996, 159). The individual is only a small cog in the machine, with the effect that the individual contribution is getting ever smaller. This is a challenge for motivating responsible behaviour as well as for attributing accountability. We are living in times of "organized irresponsibility" (Beck 1988) because we have set up our systems in such ways that far too often there is no one to be held accountable for accidents, failures, or collapses of socio-technological systems.

This is not just an accidental and ephemeral defect of science and technology today. Rather, according to the sociologist Beck, the structure of science and technology pose systemic challenges:

> The sciences as they are constituted – in their overspecialized division of labour, in their understanding of methods and theories, in their heteronomous ivory tower mentality are in no position to respond adequately to the risks to civilization because they are pre-eminently involved in the inception and growth of precisely these risks.
>
> *(Beck 1986, 78)*

What Beck bemoans about the constitution of the academic world, the sciences, also applies to our societies at large. The legal and administrative processes in highly differentiated societies have reached a point at which responsibility seems to evaporate. Time and again, accidents occur due to human failure but no single person is convicted. The 2010 tragedy of the Love Parade in Duisburg, Germany, is one such example. During a festival, a mass panic broke out, in which 21 people died and more than 650 were wounded because the crowd was funnelled through a single disproportionately small egress. There were numerous indications about the unsuitability of the terrain, many people were involved in the planning and approval process – but almost ten years after the tragedy it looks as if the ensuing lawsuit will end without a verdict. To date, nobody could be held accountable enough to justify a verdict (Deutsche Welle 2019).

Solution perspectives

Institutional support for integrated thinking and acting

The continuous differentiation which characterizes the history of modernity cannot be turned back. But we need to establish control mechanisms addressing the related downsides. Siloed thinking – which prevails in the sciences, in government agencies, and even within large corporations – needs to be banished as a dead end. As Lichtenberg (1984) already knew, understanding only one field implies that even this field is not fully understood – how much more valid is this in today's interconnected world than it was two hundred years ago? Issues of sustainability require a new kind of interdisciplinary collaboration, probably more than any other topic, since its demand is so comprehensive, the tasks are so different in kind, that literally all academic disciplines can and should contribute to this discourse. Most academic institutions, and reward and incentive systems, as well as quality management systems, are still tuned towards disciplinary research. Departments in governmental administration, in turn, need to adjust their structures, processes and collaboration mechanisms to address pressing issues effectively – issues which can be ever less resolved by one department alone.

Effective institutions need to be established which counteract the more or less self-organized process of differentiation and specialization. Institutions for quality management, recognition and reward systems, career paths, even most of the successful role models for young professionals – all these are tuned to focus on higher specialization. There is too little institutional support for activities which run counter to mainstream progress. The sociologist Norbert Elias calls for strengthening the effectiveness of institutional control of the most critical societal roles. In his view, it is a critical task for highly differentiated societies to increase the effectiveness of institutional control over all the interacting and collaborating societal positions which are indispensable (Elias 1996, 159).

- There is a need for greater institutional support for collaboration across disciplines and sectors (see 15.4). Corresponding incentives need to be established for career paths, public funding, etc. Fellowship programs, as offered by large corporations, offer employees the chance to work in another team for some time, which greatly facilitates information exchange, corporate identity and employee motivation. Similar programs are conceivable for the exchange between public-sector and private-sector organizations, between NGOs and MNEs, between theorists and practitioners.
- A promising sign is the emergence of nexus thinking in academia, indicating a "transition of scientific thought and policies towards integrative thinking to address global change and challenges". With the water-energy-food nexus, the "integration debate" acquires an important new emphasis: "This debate moves now from focusing on the interdisciplinarity and sector-based integration of resource use issues to highlighting the intersectoral interdependence of decisions and the interactionality of impacts" (Al-Saidi & Elagib 2017, 1137).
- As we discussed above in the context of trade-offs (see 3.4), there are various voices calling for an integrative and inclusive formulation of SDG policies: "Policy options developed by sectorial and technical specialists must also be subjected to assessments of total system effects outside the bounds of their silos" (Obersteiner et al. 2016, 5).

Integrated education

Of course there are great differences in countries' educational systems. For many countries, however, I submit that Erich Jantsch's vision for universities, which he expressed half a century ago, would still be valid:

The university has to become a basic unit in a decentralized, pluralistic process of shaping the national – and, beyond that, a future global – science policy. It has to contribute to the development of a common policy for society at large, participate in the competitive process of formulating strategies, but be fully responsible for its own tactics which include the support of basic science and the development of technological skills ... The task of turning the university from a passive servant of various elements of society and of individual and even egoistic ambitions of the members of its community into an active institution in the process of planning for society implies profound change in purpose, thought, institutional and individual behavior.

(Jantsch 1970, 427, 428)

Concluding their review of transdisciplinary research in sustainability science, Brandt et al. remind us that societal transitions require "commitment from all societal actor groups including scientists, policy makers and civil society" (Brandt et al. 2013, 8). The authors call scientists to participate in the realization of a sustainable future:

They should seize the initiative to act together with real-world practitioners and take the responsibility to tackle real-world problems with objective and reproducible methods. This engagement requires that both scientific institutions and societal actors need to acknowledge and promote such transdisciplinary research approaches. Current deficiencies in communication and political will result in scientific and governance structures that adapt too slowly to the rapid changes in socio-ecological systems. If such transformative and collaborative research endeavours are not fostered, we run the risk that the potential of sustainability science will never be fully realized and urgent sustainability problems remain unsolved.

(Brandt et al. 2013, 8)

In light of an ever-increasing specialization in our educational programs – German universities alone offer 19,559 different study programs (HRK 2018)! – we might need to consider some mandatory courses for integrated, cross-disciplinary education for all students. On top of this, special programs for generalists need to be fostered and incentivized.

Furthermore, in elementary and grammar schools, we need to put much greater emphasis on the value of integrated thinking and integrated knowledge (see 15.6, 16.1), the cultivation of soft skills, of cooperation, of listening, etc.

Responsibility and liability of corporations

Today's dilution of responsibility is partly a consequence of specialization and division of labour, but it also stems from the fact that we have allowed important new actors on the global scene to escape their responsibility – corporations, especially the large multinationals. Large corporations are dominant players today, but the balance between chances and risks needs to be reconsidered, for the investors and for the corporation as such.

Stockholders can often capitalize wins but do not take responsibility for losses beyond their own investment. The Japanese operator of the Fukushima power plant would have been bankrupt a few weeks after the maximum accident in March 2011 if it had not been supported by the government. Eight years after the disaster, the estimates for the duration of the clean-up and deconstruction range between 30 and 200 years and the costs are estimated to range between US$ 200 billion and US$ 640 billion (Denyer 2019). If nuclear power generation were a truly private business – covering full opportunities and full risks, insurance costs would be unpayable. But now profits are privatized and losses socialized. Furthermore, managers' personal liability is often limited and does not reflect the potential or real harm they do to society.

The role and responsibility of corporations also needs to be revisited. MNEs are dominant players in today's world and to a large degree share the constituting aspects which also characterize the action of individuals: causal effectiveness, intentionality, and some kind of a reflecting organ ("consciousness" represented by supervisory boards) (Berg 2010). This might also imply attributing moral standards to corporations – which is de facto already happening anyhow.

Cultural shift

Siloed thinking cannot only be blamed on specialization (which is a societal necessity), it is also a result of a certain culture because openness for new people and curiosity for new concepts often requires leaving one's own comfort zone, against which most of us have a personal bias. Personally I am often irritated by the fact that apparently intelligent people seem to be incapable or unwilling to transfer their system-internal logic to the context of outsiders, to put themselves in the position of the dialogue partner. I have experienced such situations in all sorts of areas, from governmental agencies via academic disciplines to different organizations within large corporations. There are probably several reasons for this – be it simply unintended or wilful ignorance, routine, or anxiety – and it might also reflect the fact that most people are much better at talking than listening. I am absolutely convinced, however, that it will enhance societal progress if the critical stakeholders have a better understanding of the different needs, demands and interests of respective other fields. We are all more likely to judge or condemn things which we do not know – and mutual understanding is a precondition for cross-sectoral collaboration (see 15.2).

In her book, *Stewarding Sustainability Transformations*, Künkel develops "aliveness principles", one of which matches well the cultural shift we discuss here: The principle "Mutually-enhancing Wholeness" would mean "[t]apping into the human capability to engage with a bigger picture, the larger story, and the greater system". This would foster a "culture of contribution to the larger context" and staying tuned to new trends and developments (Künkel 2019, Table 8.2). This runs against a prevailing individualism and calls for systemic thinking (see Chapter 16). Corresponding change will take time and will require a mind shift. It will also require that leadership becomes more diverse (see 16.2) and more collaborative (see 15.4). Such a cultural shift thus relates to several of the action principles for sustainability which will be discussed in the second part of the book.

Notes

1 The division of labour does not only distribute the work among members of a society, it also allows for automation, as Marx already discussed in dispute with Adam Smith (see Marx 1887, Chapter 14). Although Marx was fiercely criticizing the division of labour, he also needs to explain how productivity gains and economies of scale could be reached in a world without it. In fact, the economic success of the real socialist societies seems to be correlated to the degree to which this economic principle was adhered to. This might be one of the most fundamental challenges of the economic system (see Mises 1922, 151).
2 The solution suggested by Comte, a philosophy of science, is meanwhile well established but rarely resolves the issue – mainly because all real-world issues involve value judgements which cannot be delegated to experts but require social discourse and deliberation.
3 Half a century after Jantsch published this article, one can realize today how visionary his account was: "It is obvious that the traditional concepts of 'value-free' science and 'neutral' technology will become completely dissolved in the unified approach, as the university proceeds to inter- and transdisciplinarity. On the other hand, the normative and psychosocial disciplines also, such as law and sociology, will lose their abstract disciplinary identity and concepts and become aspects of social systems design. Through a transdisciplinary approach, the university will maintain its flexibility also for future situations in which there may be less emphasis on scientific/technical aspects of social systems design, and more on human and psychosocial development. Some people expect such a shift in emphasis to become significant before the end of the century" (Jantsch 1970, 427).
4 Ayres coined the phrase "industrial metabolism", which he defines as "the whole integrated collection of physical processes that convert raw materials and energy, plus labor, into finished products and wastes in a (more or less) steady-state condition" (Ayres 1994, 23).

Bibliography

Al-Saidi, Mohammad, & Nadir Ahmed Elagib. "Towards understanding the integrative approach of the water, energy and food nexus." *Science of the Total Environment 574*, 2017: 1131–1139.

Ayres, Robert U. "Industrial metabolism: Theory and policy." In *The Greening of Industrial Ecosystems*, von National Academy of Engineering (ed.), 23–37. Washington, DC: The National Academies Press, 1994.

Beck, Ulrich. *Gegengifte. Die organisierte Unverantwortlichkeit*, Frankfurt: Suhrkamp, 1988.

Beck, Ulrich. *Risikogesellschaft. Auf dem Weg in eine andere Moderne*. Frankfurt: Suhrkamp, 1986.

Berg, Christian. *Handlung als überindividuelles Konzept? Eine Untersuchung zur Rede vom 'technischen Handeln'*. Münster, Hamburg, Berlin, London: LIT, 2010.

Botkin, James W., Mahdi Elmandjra, & Mircea Malitza. *No Limits to Learning. Bridging the Human Gap*. Oxford, New York, Toronto, Sydney, Paris, Frankfurt: Pergamon Press, 1979.

Brandt, Patric, Anna Ernst, Fabienne Gralla, Christopher Luederitz, Daniel J. Lang et al. "A review of transdisciplinary research in sustainability science." *Ecological Economics 92*, 2013: 1–15.

Bundestag. „Grundgesetz der Bundesrepublik Deutschland." *Basic Law for the Federal Republic of Germany, translated by Ch. Tomuschat and D.P. Currie*. Deutscher Bundestag, 2014.

Denyer, Simon. "Eight years after Fukushima's meltdown, the land is recovering, but public trust is not." *Washington Post*. Feb. 20, 2019.

Deutsche Welle. "Lawyers: Love Parade trial likely to be terminated." *Deutsche Welle*. Jan. 16, 2019. www.dw.com/en/lawyers-love-parade-trial-likely-to-be-terminated/a-47111199 (accessed 01. 06 2019).

Elias, Norbert. *Was ist Soziologie?* Weinheim, Munich: Juventa, 1996.

Günther, Edeltraud, und Maja Krebs. *Aufgaben- und Organisationsstruktur der Umweltpolitik in der Bundesrepublik Deutschland*. Dresden: Dresden University, Business Management, 2000.

HRK. *Statistische Daten zu Studienangeboten an Hochschulen in Deutschland*. Berlin: Hochschulrektorenkonferenz, 2018.

HRH The Prince of Wales, Tony, Juniper, & Ian Skelly. *Harmony. A New Way of Looking at the World*. London, New York: HarperCollins, 2010.

Jantsch, Erich. "Inter- and transdisciplinary university: A systems approach to education and innovation." *Policy Sciences 1*, 1970: 403–428.

Künkel, Petra. *Stewarding Sustainability Transformation. An Emerging Theory and Practice of SDG Implementation*. Springer Nature, 2019.

Lawrence, Roderick J. "Advances in transdisciplinarity: Epistemologies, methodologies." *Futures 65*, 2015: 1–9.

Lenk, Hans. "Über Verantwortungsbegriffe und das Verantwortungsproblem in der Technik." In: Technik und Ethik, Hans Lenk and Günter (eds.), Stuttgart: Ropohl, 1993, pp.112–112148.

Lichtenberg, Georg Christoph. *Sudelbücher, insel taschenbuch 792*, Frankfurt a.M.: Insel, 1984.

Marx, Karl. *Capital. A Critique of Political Economy* (first English Edition of 1887). Moscow: Progress Publishers, 1887.

———. *The Poverty of Philosophy. Answer to the Philosophy of Poverty by M. Proudhon*. Progress Publishers, 1955 (1847).

Mises, Ludwig. *Die Gemeinwirtschaft. Untersuchungen über den Sozialismus*. Jena: Gustav Fischer, 1922.

Mittelstraß, Jürgen. *Leonardo-Welt. Über Wissenschaft, Forschung und Verantwortung*. Frankfurt: Suhrkamp (stw 1042), 1992.

Obersteiner, Michael, Brian Walsh, Stefan Frank, & Peter Havlík. "Assessing the land resource–food price nexus of the Sustainable Development Goals." *Science Advances*. Sept. 2016.

Plato. *Republic*. Übersetzung: Benjamin Jowett. The Internet Classics Archive by Daniel C. Stevenson, Web Atomics, 1994–2000.

Schelsky, Helmut. *Einamkeit und Freiheit. Idee und Gestalt der deutschen Universität und ihrer Reformen*. Düsseldorf: Bertelsmann-Universitätsverlag, 1971.

Smith, Adam. *An Inquiry into the Nature and Causes of the Wealth of Nations* (ed. by S. M. Soares). MetaLibri Digital Library, 2007 (1784).

SRU. *Demokratisch regieren in ökologischen Grenzen – Zur Legitimation von Umweltpolitik.* Berlin: German Advisory Council on the Environment, 2019.

Vasbinder, Jan W., Bertil Andersson, W. Brian Arthur, Maarten Boasson, Rob de Boer, et al., "Transdisciplinary EU science institute needs funds urgently – Correspondence." *Nature 463*, 18.022010: 876.

von Weizsäcker, Ernst Ulrich, & Anders Wijkman. *Wir sind dran. Was wir ändern müssen, wenn wir bleiben wollen.* Gütersloh: Gütersloher Verlagshaus, 2017. https://www.springer.com/gp/book/9781493974184 (accessed 21. 09 2019).

Extrinsic barriers 2 – Zeitgeist-dependent barriers

It is much easier to assess a certain situation from the outside than from within. Outsiders can identify weaknesses more easily because they are more detached and see the difference in the surroundings, which makes them good reviewers. Insiders have difficulties in identifying their own paradigms. This also holds for the time period in which people live. By watching images or videos from the past, one can see the specialties of that period: customs, fashion, preferences, moods, etc. Less obvious than fashion and habits but nonetheless a formative force for people's worldviews, their values and their perception of reality is the zeitgeist of an era. "Zeitgeist" is a German word introduced by the great eighteenth-century scholar Johann Gottfried Herder.[1] It tries to capture the peculiarity of an epoch, its specific character (Hiery 2001). Merriam-Webster depicts Zeitgeist as the "general intellectual, moral, and cultural climate of an era" (Merriam-Webster 2019). By its very nature, the zeitgeist is not easy to locate or describe because the historian would actually need to leave behind the zeitgeist of their own epoch – which would require them to know it (Hiery 2001).

In the following, I will nevertheless propose two barriers to sustainability as "zeitgeisty" (i.e. zeitgeist-dependent): today's focus on the short term, and the prevailing consumerism in many parts of the world. To be sure, both of these barriers are closely related to the market framework; in fact, they have their source in the market domain. Increasingly global competition, ever-shorter cycles of production and consumption, or the demands of quarterly reports all contribute to short-term orientation. Industrial mass production, globally distributed value chains and the ubiquity of cheap commodities foster consumerism. However, while the origin might lie in the economic domain, both short-termism and consumerism have become so dominant in influencing the entire intellectual, moral and cultural climate of our times that I suggest we consider them as zeitgeisty.[2]

10

SHORT-TERM ORIENTATION AND ACCELERATION[3]

Enabled by technological development and fostered by market mechanisms, the timescales of many processes have dramatically decreased in recent decades. The focus on the short term by its very nature contradicts a sustainable development because it does not allow for any longer-term considerations, let alone future generations.

Today, life is inconceivable without smartphones, which did not exist twenty years ago. The rapid speed of technological progress today hides the fact that contemporaries over the last two hundred years felt similarly regarding the changes in their time. It was particularly the transportation technologies and those of information and communication which enabled and boosted global markets, global cultural exchange, global travel and tourism (see Berg 2005, 2008). The railway shortened travel times by orders of magnitude. Around 1850, the German railway reduced travel time between Cologne and Berlin from one week to 14 hours: "Time was more fast moving: Mail needed only days instead of weeks for the same distance" (Weber 1997, 172). The steamship had the same effect at sea. While travelling from Liverpool to New York took more than a month in a sailing ship around 1850, this was shortened to just ten days in the following decades (ibid., 156, 158). One contemporary is quoted excitedly: "Time and space are suspended; … ten days from shore to shore across the vast water desert" (ibid., 158). For the first time in history, telegraphy rendered possible instantaneous global communication (the first fixed telegraphic connection between London and New York was established in 1866) (ibid., 218), which boosted globalized markets because supplies and demands could be matched in real time.

These developments further accelerated in the twentieth century with the automobile, the airplane, radio, television, fax machines, and then – most importantly, of course – the internet with all its related applications. Modern IT is enabled by integrated circuits (ICs), which have seen an amazing increase in calculating power due to Moore's law.[4] Given the critical importance of modern IT for business competitiveness, the speed of IC improvement is transferred to business, accelerating all phases of value creation in all industries and infecting society with the delusion that speed is always gain.

Businesses operate in quarterly increments, high-frequency traders in nanoseconds. Time is money: in some industries, long-term planning means 3–5 years, while in IT this is ages. Being the first in the market is often more important than being the best. Incentive and reward systems are (still) short-term oriented, although some countermeasures have been implemented more recently.

Politicians think in terms of legislative periods, but due to long campaigns often little more than half of the legislation period is used to draw up effective legislation. Furthermore, opinion polls constantly feel the pulse of the people and influence decision makers.

Employees are challenged by rapid changes. Automation kills jobs. New skills are required. The short-termness of business and the demand for flexibility leads to a *Corrosion of Character* (Sennett 1998). Those who are not able to adapt their pace to changing environments are rapidly left behind. Acceleration and short-termism operate contagiously and discriminate against people. Academic programs have been streamlined and matched, not only to facilitate international exchange, but also to reduce the effective program duration.

On the consumer level, speed is often treated as quality. New technologies, new business concepts infiltrate all areas of life. Long-term planning, building up stock for the future and for hard times, which was a matter of course throughout most of human history, have become obsolete in times of "overnight delivery", "24/7 availability" and "instant access". Speed might not even be required, but it is offered because it is possible.

In the media, respected journalism becomes more difficult because social media spread news within hours – even more so if the stories are fake. Quality requires deliberation, but deliberation is impeded by great rush.

On a societal level, lasting, long-term solutions will be inferior if short-term interests prevail. The promise of short-term return beats long-term success. Short-term thinking threatens sustainability almost by definition: Sustainable solutions to almost any problem are difficult and intricate – and one cannot expect to find such solutions quickly.

Short-termism is closely related to efficiency, which is the great paradigm of instrumental rationality and the dominant driver of the process of rationalization, as Max Weber described it.[5] It is the call for efficiency which lies at the heart not only of capitalism, but also of technology and of utilitarian ethics: the mandate is to achieve the most efficient allocation of scarce resources (economy), the most efficient use of resources or the maximum effect in technology, and the greatest good for the greatest number in utilitarianism. Without any doubt, this efficiency paradigm with its instrumental rationality has been a major driver of technological and societal progress in the past two hundred years. Yet it becomes a serious threat to human culture if it proliferates into realms of life which cannot be adequately captured by instrumental rationality – as discussed above in the section on the pervasiveness of economic thinking. Hösle considers this as the greatest fallacy of modern political and intellectual history: the delusion to think that all substantial questions can be transformed into instrumentally rational ones (Hösle 1994, 66).

It is precisely this logic which lies at heart of modern societies. The technological development and the infectious logic of the efficiency of markets mutually reinforce each other and preclude any reasonable discourse on essential, fundamental questions by enforcing a logic of maximization to circumstances and entities which cannot be maximized. It does not make sense to ask about the best colour or the nicest piece of music.

The philosopher Georg Picht, in a text on technology and tradition, criticizes the permanent call for expansion: "There is an optimum only if one knows measure. If the demand for permanent expansion, however, lets every measure as such appear as negation, the chasing of the unreachable and therefore constantly expanding maximum replaces the potential optimum" (Picht 1959, 10).

In the affluence of the consumerist societies, time becomes the most valued resource (see, e.g. Paech 2012, 127). Within the logic just described, it is therefore only natural that acceleration, speed and the focus on the short term determine our societies.

Solution perspectives

What can be done? It is already a step forward to realize the nature and origin of the present short-termism. The incredible pace of today's life is made visible by looking at historical records of people who were overwhelmed by accelerations at which we can only smile. This puts our own situation into perspective. At the same time, it is important to understand that it is a certain kind of rationality which lies at the heart of this acceleration: The instrumental rationality of maximization drives both technological progress and market efficiency, which mutually reinforce each other.

Instrumental rationality does have its legitimate role, but raises substantial questions it cannot answer. It is therefore important to resist the pervasiveness of this kind of rationality and maintain spaces in which efficiency is not the measure (see 5.2).

Apart from a revision of the market framework which is required anyhow (see 5.1), there need to be incentive structures for long-term orientation. Starting from compensation and reward systems, via systems of consumption and production, systems of political responsibility, of public procurement to the financing of pension systems – there is a need for long-term incentive structures in all domains. This has partly begun already but there is still a long way to go.

Wolfgang Sachs, researcher at the German Wuppertal Institute, called for deceleration already in the early 1990s (Sachs 1993). Sachs recommended four Ds (in the German original, four Es) to reach greater sufficiency: decelerating, de-cluttering (less accumulation), decentralizing and decommercializing. It is noteworthy that three of the four (decelerating, de-cluttering (less accumulation) and decommercializing) are directly related to market principles.

There are trends today which resist the ever-accelerating pace of change. There seems to be a new awareness of what is lost in speed, a new appreciation of slow movements, of slow food, slow travel, etc., as depicted by Sten Nadolny in his book *The Discovery of Slowness* (Nadolny 2005). Will slow become the new trend? Is it possible to actively influence the zeitgeist?

Ever-more accelerated processes have made us much more efficient – but left us with a void of meaning which instrumental rationality cannot fill. We are so involved with our products; we identify ourselves with products – with having, not being. Pressure is constantly increasing and, as if in a hamster wheel, we chase short-term effects which look promising and grant moments of security but no lasting satisfaction. Short-term orientation is therefore closely related to consumerism, which is the next zeitgeisty barrier.

Notes

1 As Hiery explicates, Herder just created the word but not the concept, which was actually taken from the Latin *"genius seculi"* (Hiery 2001).
2 It should be mentioned that zeitgeist does not necessarily impact sustainability negatively. It could well be that the urgency and importance of sustainable development will become so evident at some point that there will be a zeitgeist which might describe this move towards environmentalism.
3 Why subsume short-term orientation under zeitgeist-dependent barriers? One can well argue that humans' short-term orientation is deeply rooted in our physical condition with its stimulus-reaction scheme, which gives preference to the immediate versus the long term. For good evolutionary reasons, humans have learned (as have many animals) to respond to rapid movements. By grouping this barrier into the zeitgeist category, I want to emphasize that there is an element of short-termism which goes beyond this biological foundation. Furthermore, subsuming the zeitgeist-dependent barriers under the extrinsic ones implies that they are not necessarily tied to the concept of sustainability. This does not mean, however, that they could easily be altered. We cannot discuss here whether and to what extent it is possible to influence the zeitgeist of an era. I tend to think that some resistance is possible, but it would not be possible to simply change it.
4 Named after Gordon Moore, who discovered a regular increase of the capacity of integrated circuits. Inititially, in 1965 he considered the capacity would double every 18 months; later this was adjusted to two years (Moore 1965). Although Moore formulated his principle half a century ago, it is still remarkably valid today (Roser 2019). Similar characteristics apply to storage capacities and bandwidth. Moore himself did not call it Moore's law, obviously.
5 "Instrumentally rational (*zweckrational*)", as Max Weber defined it, is a type of action which is "determined by expectations as to the behavior of objects in the environment and of other human beings; these expectations are used as 'conditions' or 'means' for the attainment of the actor's own rationally pursued and calculated ends" (Weber 1978, 24).

Bibliography

Berg, Christian. *Vernetzung als Syndrom. Risiken und Chancen von Vernetzungsprozessen für eine nachhaltige Entwicklunge.* Frankfurt: Campus, 2005.

———. "Global networks. Notes on their history and their effects." In *Futurology – The Challenges of the XXI Century*, A. Kuklinski, & K. Pawlowski (eds.), 199–209. Novy Sacz: MiasteczkoMultimedialne, 2008.

Hiery, Hermann Josef. "Zur Einleitung: Die deutschen Historiker und der Zeitgeist." In *Der Zeitgeist und die Historie*, Hermann Josef Hiery (ed.), 1–6. Dettelbach: J. H. Röll Verlag, 2001.

Hösle, Vittorio. *Philosophie der ökologischen Krise.* Munich: Beck, 1994.

Merriam-Webster. www.merriam-webster.com/dictionary/. 2019. www.merriam-web ster.com/dictionary/zeitgeist (accessed 27. 02 2019).

Moore, Gordon E. "Cramming more components onto integrated circuits." *Electronics 38 (8)*, 19. 04 1965: 114–117.

Nadolny, Sten. *The Discovery of Slowness*, Philadelphia, PA: Paul Dry Books, 2005.

Paech, Nico. *Befreiung vom Überfluss. Auf dem Weg in die Postwachstumsökonomie.* Munich: oekom, 2012.

Picht, Georg. *Technik und Überlieferung. Die Überlieferung der Technik, die Autonomie der Vernunft und die Freiheit des Menschen.* Hamburg: Furche, 1959.

Roser, Max. *Ourworldindata.org.* 2019. https://ourworldindata.org/ (accessed 28. 02 2019).

Sachs, Wolfgang. "Die vier E's : Merkposten für einen maß-vollen Wirtschaftsstil." *Politische Ökologie 11 (33)*, 1993: 69–72.

Sennett, Richard. *The Corrosion of Character.* New York: W. W. Norton, 1998.

Weber, Max. *Economy and Society. An Outline of Interpretetive Sociology. Vol. 1*, ed. by Guenther Roth & Claus Wittich. Berkeley, Los Angeles, London: University of California Press, 1978.

Weber, Wolfhard. "Verkürzung von Zeit und Raum. Techniken ohne Balance zwischen 1840 und 1880." In *Netzwerke Stahl und Strom (1840–1914)*, Wolfgang König and Wolfhard Weber (eds.), 9–261. Berlin: Propyläen, 1997.

11

CONSUMERISM

There can be no doubt that the excessive levels of consumption by the global elites are not sustainable. A small share of the global population is consuming most of the earth's resources, is responsible for most of the waste generated, and most of the GHGs emitted. The richest half-billion people are emitting 50% of the world's CO_2 emissions (Pearce 2009).

However, consumption has become much more than just satisfying existing needs – it has become an end in itself. A variety of authors from quite different disciplines have critically discussed consumerism. They have highlighted different aspects of it, ranging from questions of the market system, to those of personal identity and of community and society.

Günther Anders, philosopher and man of letters, elaborates on consumption as a constituent factor of industrial society, as a necessity to keep up the industrial process. He argues that our industrial system "depends on the manufacture of products which themselves serve as means of production to produce other products, which etc., etc., – until a final machine ejects *final products*, which are no longer means of production but means of consumption: products which, *by their very essence or final purpose of use are to be consumed, like bread or grenades*" (Anders 1980, 15; original emphasis).[1] This is then, according to Anders, the "second industrial revolution", in which the acts of consumption become the means of production (*Produktionsmittel*), "a truly humiliating circumstance because our role as humans is now confined to the consumption of products (for which we even have to pay) and thereby providing for the perpetuation of the production" (ibid., 16). Today we would no longer need to pray "Give us today our daily bread", but being honest we would need to pray "Give us today our daily hunger" to ensure the manufacture of bread. Even worse, actually it would be the products that pray today: "Give us today our daily eaters" (ibid.).

The logic of the industrial society would define every-*thing* and every *being* as a resource within the production process: "Being a resource is *criterium existendi*, being is resource-being – this is the foundational metaphysical thesis of industrialism" (ibid., 33).

Therefore, consumption is, according to this view, rooted in this "foundational metaphysical thesis of industrialism"; it is tied to the logic of the industrial society.

One can challenge Anders' view by pointing out that this phenomenon is, to some extent, prevalent in any society which is based on the division of labour, since anyone who offers a product or service depends upon the demand for it. This is the case even in a command economy, the only difference being that somebody then guarantees the demand. However, in my opinion, Anders rightly points to industrial society's dependence on the creation of ever-new needs and wants. Steve Jobs' above-mentioned dictum that people do not know what they want until someone shows it to them, validates this view. Needs for consumption are being developed to keep the machine going.

However, apart from this market-systemic aspect, today's consumption goes beyond the economic: it has become a defining moment of human identity, as several authors bemoan.

The Polish-British sociologist and philosopher Zygmunt Bauman emphasizes that consumption has become the answer to the anxieties of "institutional erosion" and "enforced individualization" (Bauman 2001). In an increasingly insecure and changing world, consumption provides a shelter of security and predictability, as Bauman illustrates with reference to the holiday expectations of tourists from the global North (Bauman 2001, 26). Tourists want wildernesses experiences – but these "ought to have exits well mapped and signed" (ibid.). The

> powers and the weaknesses, the glory and the blight of the consumer society – a society in which life is consuming through the continuous success of discontinuous consumer concerns (and is itself consumed in its course) – are rooted in the same condition, the anxieties born of and perpetuated by institutional erosion coupled with enforced individualization.
>
> *(28)*

Long before Bauman, psychoanalyst and sociologist Erich Fromm held a similar view. In his early book, *The Fear of Freedom*, Fromm argues that the fate of humans is to be made to contribute to the success of the economic system, to become a cogwheel in a huge economic machine (Fromm 1942, 95). According to Fromm, consumption is a form of having, as opposed to being. To be sure, consumption would lessen anxiety because that which one consumes cannot be taken away any more. On the other hand, once consumed, the consumed can no longer satisfy and therefore the next act of consuming needs to take place. Modern consumers could say: I am what I have and what I consume (Fromm 1976). In line with this argument, one could say: if consumption constitutes identity, then identity becomes ephemeral and depends on the insatiable desire for consumption. Fromm observes that

> while the selfish person is always anxiously concerned with himself, he is never satisfied, is always restless, always driven by the fear of not getting enough, of missing something, of being deprived of something. He is filled with burning envy of anyone who might have more. If we observe still closer, especially the unconscious dynamics, we find that this type of person is basically not fond of himself, but deeply dislikes himself. The puzzle in this seeming contradiction is easy to solve. Selfishness is rooted in this very lack of fondness for oneself. The person who is not fond of himself, who does not approve of himself, is in constant anxiety concerning his own self. He has not the inner security which can exist only on the basis of genuine fondness and affirmation. He must be concerned about himself, greedy to get everything for himself, since basically he lacks security and satisfaction.
>
> *(Fromm 1942, 100)*

Consumption, one could continue this argument, addresses the selfish person in his or her lack of self-love. In the absence of inner security, greedy self-interest tries to constitute identity – consumption due to a lack of self-esteem. Consumerism would therefore reflect a crisis of identity, a crisis of meaning, as the psychoanalyst Viktor Frankl diagnoses. Frankl sees a crisis of meaning as being at the root of our insatiable quest for consumption. The welfare state could "satisfy practically all human needs. Indeed, some needs are actually only created by the consumer society. Yet one need is missing: that is the human need for meaning" (Frankl 2013, 46).

The conclusions of contemporary lateral thinkers in economics like Tim Jackson or Nico Paech resonate remarkably well with this finding. Nico Paech, a German economist, criticizes that consumption has become a self-sustaining activity which diverts us from the essential:

> [We] dissipate our energy in a stimulus-satiated bubble of consumption, which consumes our scarcest resource, our time. If we dropped off our affluence ballast we would have the chance to concentrate on the essential – instead of suffering dizziness attacks in the hamster wheel of buyable self-actualization.
>
> *(Paech 2012, 11)*

The British economist Tim Jackson criticizes that consumption has defined our sense of identity, our expressions of love, and even our search for meaning and purpose in the language of goods (Jackson 2011, 175ff.).

Jackson also points out that consumerism affects the community, the third critical aspect of consumerism mentioned above. Jackson is scathing about the devastating social effects of consumerism, which "promotes unproductive status competition and has damaging psychological and social impacts on people's lives" (83). Jackson's critique is backed and substantiated by the work of two British epidemiologists, Richard Wilkinson and Kate Pickett. In their book, *The Spirit Level*, they deliberate about consumerism's social dimension. Consumerism would by no means be an expression of a fundamental human material self-interest and possessiveness. Quite the contrary:

> Our almost neurotic need to shop and consume is instead a reflection of how deeply social we are. Living in unequal and individualistic societies, we use possessions to show ourselves in a good light, to make a positive impression, and to avoid appearing incompetent or inadequate in the eyes of others. Consumerism shows how powerfully we are affected by each other. Once we have enough of the basic necessities for comfort, possessions matter less and less in themselves, and are used more and more for what they say about their owners. Ideally, our impressions of each other would depend on face-to-face interactions in the course of community life, rather than on outward appearances in the absence of real knowledge of each other ... The weakening of community life and the growth of consumerism are related.
>
> *(Wilkinson & Pickett 2011, 230)*

In sum, it is time to challenge the prevailing consumerism, not least because of a growing global middle class which will boost consumption in the next decades beyond any historic measure. We cannot rely on efficiency, since any decoupling of resource consumption from economic growth has so far only been relative. Of course, the consumption level for many people at the base of the pyramid will have to increase. It will therefore be even more important that the affluent countries soon find ways to tackle their consumerism.

On a *systemic* level, excessive consumption has become a constituent *driver* of the market system which needs to be altered to constrain the exploitation of resources as well as curbing waste and pollution (see Chapter 5). On a *personal and societal* level, however, excessive consumption is rather a *symptom* – a symptom for the quest for meaning, identity and security in an ever-more individualized society. It remains an empty promise that fugitive consumption experiences would provide self-actualization and endurance. Excessive consumption rather reveals an unrestrained, dissolute character, for it is the self-imposed constraints of will that determine the shape, the gestalt of humans as persons – which points to the importance of personal principles (see 14.1).

Solution perspective

'Consumer' as a swear word?

Addressing consumerism will require measures on different levels and in several ways.

- *Adapt the market system to alleviate the damage of consumerism*
  At the level of the market system, we need to facilitate less damaging, less harmful ways of consumption as a kind of contingency measure. External social and environmental costs need to be fully reflected by market prices as discussed above (see 5.1). New, less resource-intensive business models need to be promoted (e.g. featuring access instead of possessions) (see 13.2). Production systems need to be adapted accordingly and products need to be designed for longevity, durability, upgradeability and recyclability ("design for disassembly"). These measures are more treatment of symptoms than cure, but they can help reduce the environmental and social burden related to consumerism.

- *Facilitate transition to a post-materialist society*
  The empirical sociologist Ronald Inglehardt investigates the development of people's values over time. For this purpose, he established a series of "World Value Surveys", in which he studied the development of people's values over time in close to fifty countries, in each of which he interviewed a thousand citizens (Esmer & Pettersson 2007; World Value Surveys 2019). Inglehardt found a tendency towards post-materialist values in affluent societies, as Delhey describes: There is "an evolutionary pattern in that unprecedented levels of affluence in today's post-industrial societies have fundamentally re-organized the way citizens achieve happiness ... Largely driven by rising standards of living and the widespread sense that existential security can be taken for granted, a sea change in value priorities has been taking place, away from materialist scarcity values towards postmaterialist self-expression values. It is perfectly possible then that existential security has changed individuals' recipes for happiness accordingly, from materialist to post-materialist happiness" (Delhey 2010, 66). Employing a multi-level design, Delhey substantiated "a quite consistent pattern towards postmaterialist happiness as we move from poor to rich societies. This pattern seems to be driven by both a devalorization of material concerns and a valorization of post-materialist concerns, although the evidence suggests that the former trend is stronger and more linear than the latter" (65). Although we may not have passed peak consumption in terms of material throughput, one can see the values and attitudes changing in affluent countries. It might need some more time until this is manifested in the material flows, but there are good chances that it will occur.

- *Reduce inequalities because they boost consumption*
  One measure to reduce the incentive for consumption is reducing inequalities – as Wilkinson and Pickett have documented: "Greater equality gives us the crucial key to reducing the cultural pressure to consume" (2011, 226). The authors view status competition as the main driver of consumption: "The problem is that second-class goods make us look like second-class people. By comparison with the rich and famous, the rest of us appear second-rate and inferior, and the bigger the differences, the more noticeable and important they become. As inequality increases status

competition, we have to struggle harder to keep up" (227). Reducing inequalities within societies therefore not only strengthens social cohesion (see 15.4) but also lessens the pressure to consume.

- *Question the paradigm of consumption: "Consumer" as a swear word?*

At a deeper level, we need to challenge the paradigm of consumption. What does consuming ultimately mean? It means taking away something which is then no longer available for use. This paradigm reflects the take-make-waste approach of the last hundred years, which cannot work in the long term, and causes great problems in the short term. The consumerist paradigm must change. Maybe "consumer" will become a swear word one day – because a consumer takes something from somebody or somewhere, ultimately from one of humanity's common natural resources, and consumes it, which means, according to the Merriam-Webster dictionary: destroys, squanders, uses up, devours or engrosses it – and only in its fifth and final meaning offers the meaning 'utilizes' (Merriam-Webster 2019).[2] Except for the final meaning, consumption does not work in the long run on a finite planet and will have to be succeeded by less invasive, less aggressive concepts like utilizing, using, benefiting from, gaining, etc. In her book, *The Great Mindshift*, Maja Göpel asks, "why should we continue to build societal development paths around the idea that constant accumulation is always beneficial to people or society as a whole?" (Göpel 2016, 77). We need to work towards a mindshift, towards a new perception of consumption.

- *Sufficiency and more frugal lifestyles*

It might not be a recommendation for the masses (yet), but more sufficient and frugal lifestyles are certainly important countermeasures against consumerism. Sufficiency will be suggested as one of the principles for sustainable action (see 13.2, and Schneidewind & Zahrnt 2014). As quoted above, Wilkinson and Pickett argued that reduced levels of consumption would not mean sacrificing improvements in quality of life as measured by health, happiness, friendship and community life (see 5.1). Likewise, Wolfgang Sachs recommended decommercializing and decluttering (or less accumulating) as a means towards sufficiency. In Paech's account, the liberating effect of less consumption is stressed – reducing "affluence ballast" liberates us to focus on the essential, and "[u]sing fewer things more intensively and competently, ignoring certain options for precisely this reason means less stress and therefore more happiness" (Paech 2012, 11).

- *Address selfishness and the void of meaning*

If we follow Viktor Frankl and Erich Fromm, consumerism has to do with self-acceptance and a sense of meaning. If excessive consumption reveals selfishness, which is "rooted in this very lack of fondness for oneself" as Fromm argues, questions of consumerism can only be addressed if the broader questions of identity and meaning are addressed. Consumerism can only be tackled by speaking to humans' self-love and self-esteem.

Notes

1 In the German original the last part reads "die durch ihr *Gebrauchtwerden verbraucht werden* sollen, wie *Brote oder Granaten*".
2 The first four (or five) defining phrases for the transitive usage of the verb all have a pretty negative connotation: "1. to do away with completely: DESTROY (Fire consumed several buildings), 2a: to spend wastefully: SQUANDER (consumed his inheritance on luxuries), b: USE UP (Writing consumed much of his time), 3a: to eat or drink especially in great quantity (consumed several bags of pretzels), b: to enjoy avidly: DEVOUR (… mysteries, which she consumes for fun … – Eden Ross Lipson); 4: to engage fully: ENGROSS (consumed with curiosity)" (Merriam-Webster 2019). Only the fifth and final usage defines consume as "to utilize as a customer (consume goods and services)", ascribing to it a more neutral relation to goods and services.

Bibliography

Anders, Günther. *Die Antiquiertheit des Menschen. Über die Zerstörung des Lebens im Zeitalter der dritten industriellen Revolution, Band 2*, Munich: Beck, 1980.

Bauman, Zygmunt. "Consuming life." *Journal for Consumer Culture 1*, 2001: 9–29.

Delhey, Jan. "From materialist to post-materialist happiness? National affluence and determinants of life satisfaction in cross-national perspective." *Social Indicators Research* 97, 2010: 65–84.

Esmer, Yilmaz, & Thorleif Pettersson (eds.), *Measuring and Mapping Cultures: 25 Years of Comparative Value Surveys*. Leiden: Brill, 2007.

Frankl, Viktor E. "Das Leiden am Sinnlosen Leben." In *Der Mensch vor der Frage nach dem Sinn*, Viktor Frankl (ed.), 44–49. Munich: Piper, 2013.

Fromm, Erich. *The Fear of Freedom*. London: Kegan Paul, 1942.

Fromm, Erich. *To Have Or to Be? 'A New Blueprint for Mankind'*. New York: Harper & Row, 1976.

Göpel, Maja. *The Great Mindshift. How a New Economic Paradigm and Sustainability Transformations Go Hand in Hand*. Springer Nature, 2016.

Jackson, Tim. *Prosperity without Growth. Economics for a Finite Planet*, London: Earthscan, 2011.

Merriam-Webster. www.merriam-webster.com/dictionary/. 2019. www.merriam-webster.com/dictionary/consume (accessed 27. 02 2019).

Paech, Nico. *Befreiung vom Überfluss. Auf dem Weg in die Postwachstumsökonomie*, Munich: oekom, 2012.

Pearce, Fred. *Consumption Dwarfs Population as Main Environmental Threat*. New Haven, CT: Yale School of Forestry & Environmental Studies. 13. 04. 2009.

Schneidewind, Uwe, & Angelika Zahrnt. *The Politics of Sufficiency. Making It Easier to Live the Good Life*, Munich: oekom, 2014.

Wilkinson, Richard, & Kate Pickett. *The Spirit Level. Why Greater Equality Makes Societies Stronger*, New York, Berlin, London, Sydney: Bloomsbury Press, 2011.

World Value Surveys 2019, www.worldvaluessurvey.org/wvs.jsp (accessed 11 September 2019).

PART 2

Action principles

The first part of the book looked at sustainability barriers from a systemic ("top-down") per-spective – what are the barriers rooted in nature, in the human condition, in society as well institutional and zeitgeisty ones? Solution perspectives were suggested for each of them. The second part of the book will look from the bottom up, i.e. from the actor's view: how do actors need to behave in order to facilitate more sustainable societies. This is not just an addendum to keep people busy in the absence of any capability to overcome the barriers – very few decision makers are actually capable of triggering the needed systemic change. On the contrary, who should change a system in the absence of any "steersman", if not the actors? It is the actors which have the pivotal, in fact, the only role in change – actors of all kinds and on all levels, from individuals to corporations, from NGOs to government authorities.

PART 3

Action principles

12

WHY ACTION PRINCIPLES?

Contents

12.1 A change in perspective – take the actor's view

The first part of the book looked at sustainability barriers from a top-down, systemic view. However, systemic change does not appear from nowhere, nor is there anybody planning systemic change from scratch. Rather, "[f]rom complex system theory we learn that the direction of the system is determined by the multitudinous actions of the individual agents who constitute the system. Points of intervention are the opportunities to influence the direction of the system" (Brown 2008, 149). Changing the course of "the system" therefore requires aligning the multiple actors and motivating and guiding them towards the same direction. This is the purpose of suggesting principles for sustainable action.

As a central thesis of this book is that the barriers to sustainability cannot be addressed in isolation, it is essential that multiple actors do the right things on multiple levels, since there is no director or any control mechanism to lead the way.

This approach differs from that of the 2030 Agenda, which specifies concrete sustainability objectives, i.e. 17 distinct goals and 169 targets. To be sure, it is essential to have concrete goals and targets. These targets address the systems level, they specify goals on the national or even global level, and in many cases it is the state which is the main addressee.

Focusing on the result has the benefit of output orientation. In today's prevailing focus on the effect, it is the results which count. However, what the targets do *not* answer is the question of how actors of different kinds can contribute to their achievement. In concrete situations which require action, it is often not possible to anticipate the long-term effect of that action. This is a general problem of consequentialism, which values the morality of an action in terms of its consequences – like the utilitarian maxim to strive for the greatest good for the greatest number. However, one can never be sure about the long-term consequences of an action.

From the individual actors' point of view (regardless of whether the actor be an individual person, a corporation, a government agency, or other), the SDGs rarely help in concrete situations if the individual cannot see whether or not a certain course of action contributes to the achievement of a given SDG. The SDGs do not provide general rules which can be applied in many different contexts. This is something, I propose, that principles can do.

12.2 Why principles for sustainable action?

A principle is "a comprehensive and fundamental law, doctrine, or assumption". It can also be "a rule or code of conduct" or a "habitual devotion to right principles" (Merriam-Webster 2019). Immanuel Kant formulated a fundamental rule of conduct with his categorical imperative. Kant wanted to formulate a universal law which would have the same validity as a natural law. Kant's first formulation follows the universalizability principle. "Act only in accordance with that maxim through which you can at the same time will that it become a universal law" (Kant 2002, G4:421). The right moral behaviour is only that kind of behaviour of which you can will that its maxim can function as a universal law. The universality of this principle is a great benefit because it ensures general applicability. However, the more universal a principle is, the more difficult it is to apply to concrete situations.

Two hundred years after Kant, the philosopher Hans Jonas formulated another principle, which copies Kant's structure but adapts its formulation to today's challenges by pointing to the new kinds of threats caused by human use of technology and the ecological consequences: "Act so that the effects of your action are compatible with the permanence of genuine human life" (Jonas 1984).

It cannot be discussed here whether Jonas' principle actually adds something to Kant's categorical imperative. Kant could argue: if an action is not compatible with such permanence, you could not will that it becomes a natural law. Notwithstanding, Jonas' principle emphasizes that human behaviour threatens such permanent existence, which is surely a valid concern. However, to operationalize sustainability, to overcome the barriers discussed, more concrete principles are needed which can help agents in day-to-day decisions. Such principles for sustainable action would not be categorical imperatives in Kant's sense but rather "hypothetical" ones, since a hypothetical imperative is a "practical necessity of a possible action as a means to attain something else which one wills (or which it is possible that one might will)" (Kant 2002, Ak 4:414). A principle for sustainable action would be a practical necessity of a possible action as a means to attain more sustainability.

There is a trade-off between universal validity on the one side and applicability to concrete situations on the other. The general validity of Kant's categorical imperative or Rawls' fairness principle comes at the cost that they are difficult to apply to day-to-day situations, especially regarding the question of which of two options is the more sustainable. On the other side of the spectrum, very detailed and concrete suggestions, like "purchase locally produced goods", are often not generally valid. In fact, they might stimulate behaviour which counteracts sustainability. Therefore, the intent of the following principles is to be sufficiently concrete to guide in specific contexts of action but still sufficiently generic to apply to a wide range of situations at the same time.

The hope is that the focus on action principles will have two critical benefits: it helps reduce complexity, and it makes change tangible and operational. Nobody can elicit system change alone – not even the most influential politicians. However, not unlike the case of global governance discussed above, change can emerge by aligning the multiple actions of multiple players – even without a central coordinating mechanism.

There are several conceivable objections to such an approach.

- *Consistency?*
  Presumably the most severe issue with the SDGs is the trade-offs among them. It is not clear whether the SDGs are actually consistent, whether it is possible to achieve them all together. This is, as I see it, an inherent problem of the concept of sustainability and the nature of trade-offs, and will certainly remain as an issue if one looks at action principles.I cannot claim consistency for the principles suggested here. Time and research will reveal shortcomings, and the need for complements and revisions. Future research can hopefully distil even more accurate and comprehensive principles. However, I consider it important to start the discussion on such principles because it is the multitude of actors which demand guidance which can hopefully be addressed by such principles.
- *Practicality?*
  A practical objection could be that an agent would be most unlikely to take out a list of action principles and deliberate the most sustainable option in each and every situation. However, many (if not all) action principles discussed below reflect values and convictions which are already present in several important cultural and moral traditions – but they have so far just not been applied to questions of sustainability. Take, for instance, the "polluter pays" principle. As will be shown later (see 13.5), this principle can be traced back to the earliest legal texts conveyed, in the Code of Hammurabi. In other words, the "polluter pays" principle today does not demand anything bizarre; it only requires us to adopt one of the oldest legal principles to systematically include and operationalize environmental and social harm. By occasionally referencing some of these traditions, it will hopefully become easier to concur with the principles suggested.
- *Substitute for systemic change?*
  Action principles cannot, of course, substitute for the systemic change which is needed for a transition towards sustainability. However, that systemic change does not appear from nowhere; it requires pressure from actors on all kinds of

levels. The principles suggested below do not only apply to individuals, or even more narrowly, address individuals as moral agents. They target different kinds of actors on different levels. The transformation towards sustainability "calls for collective action by myriad actors on scales from local to global" (Künkel 2019, 6). For example, "seek mutual understanding, trust and multiple wins" functions, in my view, from the very small individual level to geopolitical contexts (see 15.2). Systemic change will occur – this is an unprovable hope underlying this argument – once a sufficient number of agents moves in the right direction.

- *Overstraining the individual?*

  Does the emphasis on principles not put too much of a strain on the individual agent – and too little on the systemic level? I would reply that this is a frequent issue of change processes, that change leaders have to bear a special burden. However, once such systemic changes will have taken place, the burden for individual actors will be lessened, since properly arranged systems unburden the agent. For instance, as soon as negative ecological externalities are properly represented in the market system, consumers would no longer need to consider "food miles", carbon footprints, or child labour in their purchasing decisions – they would simply purchase on the basis of preference and price.

12.3 Types of principles

The principles will be grouped according to the domain *which the principle primarily addresses*. The "polluter pays" principle, for instance, refers (primarily) to the natural environment; "celebrate frugality" is a personal principle since it addresses the individual. Of course, the interconnectedness of the issues implies that secondary, higher-order effects also apply to other domains, for instance nature-related principles will also have social effects, etc. However, the ordering will follow the primary effects. In any case, as with the taxonomy suggested for the barriers, this one also does not mean to reflect any substantive categories, but rather heuristic ones.

The logic of the domains of the first part will be followed, i.e. starting with principles relating to nature followed by personal principles relating to the human condition, those related to society and finally those with a systems-perspective.

Not all principles are relevant for all kinds of actors. Some principles mainly address policy makers (e.g. "polluter pays" principle), others, especially the personal principles, of course, speak mainly to individuals as moral agents, whereas still others can be used by all kinds of actors (e.g. increase transparency).

Bibliography

Brown, Casey. "Emergent Sustainability: The Concept of Sustainable Development in a Complex World." In *Globalization and Environmental Challenges*, eds Hans Günter Brauch, Navnita Chadha Behera, Béchir Chourou, Pál Dunay, John Grin et al., 141–149. Berlin, Heidelberg, New York: Springer, 2008.

Jonas, Hans. *Das Prinzip Verantwortung.* Frankfurt: Insel (also: Suhrkamp, stw 1085), 1984.

Kant, Immanuel. *Groundwork for the Metaphysics of Morals (ed. by* Allan W. Wood*).* New Haven, CT, London: Yale University Press, 2002 (1785).

Künkel, Petra. *Stewarding Sustainability Transformation. An Emerging Theory and Practice of SDG Implementation.* Springer Nature, 2019.

Merriam-Webster. "Principle." 2019. www.merriam-webster.com/dictionary/principle (accessed 27. 02 2019).

13

NATURE-RELATED PRINCIPLES

Contents

13.1 Decarbonize

As the Keeling Curve impressively shows, all of our efforts for climate change mitigation have not led to a reversal of the trend of rising carbon concentration in the atmosphere (see Figure 1.1). The Paris Agreement was an important milestone – but has so far not been effective and suffers from a number of issues: First, some countries have already withdrawn from it. Second, there is an "emission gap", i.e. the sum of the emissions comprising the NDCs led to an increase of 3°C instead of the 2°C (or 1.5°C) agreed upon in Paris. The current pledges and targets would lead us to a 3°C increase; the actual current policies would even imply a rise of 3.3°C (ClimateAction Tracker 2018). Third, not even the unambitious NDCs are being adhered to throughout – the German government, for instance, has already decommitted from its reduction goal for 2020.

Therefore, fighting the climate crisis first and foremost means decarbonizing civilization, which is vastly dependent on carbon. Carbon dioxide is the major anthropogenic GHG, accounting for 76% of total anthropogenic GHG emissions in 2010. The energy sector contributes 35% of GHG emissions, agriculture, forestry and land use changes 24%, industry 21%, transport 14% and the building sector 6.4% (IPCC 2014, 46).

In light of this, current energy production is damagingly disproportioned. Humanity cannot afford to burn all existing fossil reserves: "No more than one-third of proven reserves of fossil fuels can be consumed prior to 2050 if the world is to achieve the 2°C goal, unless carbon capture and storage (CCS) technology is widely deployed" (IEA 2012, 3). To be sure, some countries have announced phasing out coal for energy production, but the time horizon is far too distant in the future – 2025 in the UK, for instance, (The Guardian 2018), even 2038 in Germany (Spiegel Online 2019).

There is a need to phase out coal soon (as the source with the highest GHG footprint) and put a general price tag on GHG emissions. The cheap price of fossil fuels, in particular cheap coal, might be the biggest challenge for decarbonization. However, the overall price of coal is certainly strongly dependent on the long-term operating costs, which will increasingly have to factor in the carbon cost, too. Tracking the economic and financial risks of coal power at the asset level throughout the world, considering 95% of operating capacity and 90% of capacity under construction, the British think tank Carbon Tracker calculates that 42% of coal capacity operating today could be losing money (Carbon Tracker 2018). This number is expected to increase in the future: "From 2019 onwards, we expect a combination of renewable energy costs, air pollution regulation and carbon pricing to result in further cost pressures and make 72% of the fleet cashflow negative by 2040" (ibid., 5).

One of the first sectors to be decarbonized is, of course, the energy sector, due to its 35% contribution to the global GHG gases. Energy production is still more than 80% based on fossil fuels (renewable energy already contributes 24% of world electricity production, but the overall share of renewables is just 14%) (European Commission 2018, 16.13). The good news is that the technological solutions for renewable energy production already exist and have become price competitive even in light of fossil subsidies. There are still technological challenges to be resolved, to be sure, like scalable and cheap energy-storage technologies, grid stability, or matching demand and supply. However, the major obstacles are not technological but rather administrative and societal.

Huge investments in infrastructure and new technologies are needed and the inertness of the current carbon-based subsystems is enormous. The German energy transition ("*Energiewende*"), which is particularly challenging because Germany will phase out both nuclear energy (by 2022) and coal (by 2038), demonstrates some of these difficulties. Those negatively affected by such a transition (e.g. coal workers) will have to be sufficiently compensated by the public (e.g. investing in new job opportunities) to secure societal peace and stability.

Decarbonizing societies will therefore require the state to play a predominant role. It is the state which has highest obligation to facilitate this transition, simply because it requires a substantial coordination effort which nobody else could provide. A strong state role, however, will require social cohesion, political stability and trust in society, which evidences that the climate crisis is so closely related to crises in other areas.

In a similar way, the transportation sector calls for such coordination by the state since it implies huge investments in infrastructure and technologies, and investors need security of expectations that the new paths chosen will be continued. Different modes of transport (e.g. trains, road traffic, air traffic, shipping) need to be coordinated, as well as their respective energy supplies, the propulsion technologies, etc.

Not only in energy production and transportation but also in systems of production and consumption, of living and working: the necessary decarbonization requires courageous political action.

13.2 Reduce environmental impact by efficiency, sufficiency and consistency

Humanity's impact on the global ecosystems is unsustainable, and production and consumption patterns exert enormous pressure on the environment. The well-known concepts of "ecological rucksack", "ecological footprint", "planetary boundaries" are all attempts to illustrate that we are constantly reducing nature's capital. Humanity's environmental impact needs to be reduced, not only relatively but in absolute numbers, by combining measures for eco-efficiency improvements, sufficiency and consistency.

As early as 1998, a German parliamentary committee proposed four rules for the sustainable management of material flows:

1. Renewable resources shall only be exploited to the degree which does not exceed their rate of regeneration.
2. Non-renewable resources shall only be used to the extent that a physically and functionally comparable equivalent is created in the form of renewable resources or with higher productivity.
3. Emission of substances into the environment shall be limited by the carrying capacity of the environmental media.
4. The timescales of anthropogenic changes shall be balanced with the timescales of the reaction capacity of the natural processes (Deutscher Bundestag 1998, Section 3.2.3).[1]

Twenty years later, one needs to realize that no nation, let alone humanity in general, can claim to meet these requirements. We exploit renewables beyond their regeneration rate (e.g. fishing grounds), we have not established functionally comparable equivalents in many areas in which we depend on non-renewables (e.g. the transportation sector), we have introduced substances like nitrogen or phosphorus into the environment in excess of the carrying capacity of environmental media, and all this on timescales much shorter than natural ones. Worse still, the global material flows have even increased since the beginning of the twenty-first century. Krausmann et al. investigated the material flows through the global economy in the period of 1900–2015 and observed a growth in global material extraction by a factor of 12 to 89 Gt/yr over the whole period. Despite the global financial crisis of 2008, global material extraction increased by 53% between 2002 and 2015 (Krausmann et al. 2018, 131). The authors call for urgent action to reduce material flows "in industrialized countries, as these countries directly and indirectly still appropriate the largest and a disproportionally high share of key materials extracted globally" (Krausmann et al. 2018, 139).[2]

What needs to be done? From a cybernetic perspective, one can say that the industrial metabolism needs to be adjusted. Reduce the amount of (re-)sources, of sinks (e.g. pollution) and make the operation of the system consistent, compatible with natural cycles. In order to address the first two issues, the material fluxes interacting with the environment need to be reduced,[3] which can be achieved by measures of efficiency and sufficiency. The third issue relates to the consistency with the natural mass and energy fluxes, which are largely addressed by a circular economy. All three concepts – efficiency, sufficiency, consistency – have been discussed for decades. None of them can achieve the needed reduction of the environmental impact alone. It is *their combination which is needed*.

13.2.1 Efficiency and sufficiency

Efficiency means increasing the return or the benefit from a given effort, often by use of technology. Increases in efficiency are of natural interest to market participants since they relate to cost-reduction. Applying the efficiency concept to natural resources ("eco-efficiency") is therefore probably the most frequently given suggestion for reducing the environmental impact – politicians and industry alike praise the concept. Increasing value-creation without increasing the consumption of natural resources and energy is the idea of concepts like "Factor 4" (von Weizsäcker, Lovins & Lovins 1995), "Factor 5" (von Weizsäcker, Hargroves & Smith 2010), or Factor 10 (Schmidt-Bleek 2000). Factor 4 would mean, for instance, doubling welfare while halving resource consumption. In other words, economic growth would be de-coupled from resource use (resource de-coupling) and environmental impact (impact de-coupling) (UNEP 2011). Unfortunately, such a de-coupling of economic growth and resource consumption or environmental impact has only been relative so far, not absolute. To date, efficiency gains have mostly been (over-)compensated by a change in behaviour and/or increased convenience – which is called the "rebound effect". Originally discussed in energy economics, the rebound effect is meanwhile also applied to other areas of environmental economics (Santarius 2012). The average car today consumes only moderately less fuel than fifty years ago – but is much more convenient, heavier, has more power, air conditioning, etc. Due to this gain in convenience, people drive longer distances. In Germany, for example, the number of kilometres travelled increased by more than 30% between 1991 and 2016, the energy consumed by passenger traffic in 2014 was almost the same as in 1991 (*Umweltbundesamt* 2018).

Therefore, *efficiency* will *not* be enough to reduce material consumption. It needs to be complemented by measures of sufficiency, which means reducing material consumption by a change in behaviour. The concept of sufficiency (sometimes called "eco-sufficiency") was introduced by Herman Daly (Daly 1996), founding father of the steady-state economy. As Daly explains, "the correlation between absolute income and happiness extends only up to some threshold of 'sufficiency'; beyond that point only relative position influences self-evaluated happiness" (Daly 2007, 23). Daly calls for a reduction of consumption in candid words:

> instead of vaguely calling for 'changed consumption patterns' we need to specify 'reduced consumption levels' of resources and environmental

services. Once the level of resource throughput is reduced to a sustainable level, the pattern of consumption will automatically adapt, thanks to the market.

(Daly 1996, 17)

Sufficiency is far less discussed than efficiency, as it is often related to relinquishing consumption – and this does not sell very well. The prevailing perception is that consumers are hardly likely to appreciate reducing or downsizing their consumption (Boulanger 2010, 5). No politician is likely to win elections by preaching reduction, frugality and sufficiency. However, advocating for sufficiency can build on the following arguments:

a) *Sufficiency as an act of solidarity with the world's poor*

The prevalence of the rebound effect requires measures other than efficiency to reduce the environmental impact. Especially the rich countries need to be aware that the global distribution of the environmental impact mirrors huge global inequalities. No rational person can want that every citizen of the globe consumes as much as the people in the global North, because this would imply a rapid collapse of our ecosystems. On the other hand, there can be no doubt that the poor countries will need to increase their resource consumption for some time – although this should certainly be restricted as much as possible. The implication for the rich countries can in my view only be that they start with reducing their footprint, not only relatively (compared to GDP), but also in absolute terms. This also includes that everybody reconsiders his or her own consumption pattern. Qualifying a tremendously wasteful consumption pattern in light of billions of people at the base of the pyramid is quite simply demanded by Rawls' fairness principle.

b) *Institutional backing of sufficient behaviour*

More sufficient lifestyles need not, in fact should not, rely on personal motivation and conviction alone. They can also be nudged by political measures. A city tax for cars, for instance, can nudge people to use the Underground. Why are products not configured in a such a way that by default the less consumptive option is chosen (e.g. office printers). Policies of nudging ("libertarian paternalism") could help trigger behavioural changes (WBGU 2011, 78), be they regulatory policies, taxes, subsidies, labels, product declarations, or others (see Schneidewind & Zahrnt 2014). Hirschnitz-Garbers et al. argue that change will be effectively triggered by a policy mix, which includes information, target values, prices, and availability and affordability of more sustainable alternatives (Hirschnitz-Garbers et al. 2016, 28). The authors are sceptical of mandatory measures (e.g. mandatory vegetarian canteen days), since they fear resistance.: "The more drastic a policy instrument is (perceived to be), i.e. adversely affecting addressees, the lower its feasibility … This applies in particular to top-down, government imposed regulatory instruments" (Hirschnitz-Garbers et al.

2016, 28). Kaiser et al. confirmed the efficacy of nudging and found that voluntary measure (e.g. opt-out defaults) are much more effective than compulsory ones (Kaiser, Arnold & Otto 2014, 204).

c) *Sufficiency facilitating enlightened happiness*

Resisting the impulse of expansion and growth is difficult in a consumerist environment but, as will be argued below (see 14.4), simplicity and frugality not only can reduce environmental impact, they can also enhance personal life. Celebrating reduction and frugality can lead the way to an experience of liberation. Nico Paech calls for a "liberation from abundance" (Paech 2012) and suggests the category of "enlightened happiness" (*aufgeklärtes Glück*): This would be inseparably bound to the awareness of practicing a happiness giving art of life within a responsible scope of action:

> How much self-deception is needed to be happy with things for which I can know that I could never take responsibility – measured against my awareness of global well-being? Is happiness which is not honest, because it requires the suppression or extent of inconsistencies, not finally absurd? Enlightened happiness would hence not only presuppose enjoyment but thereby also being on good terms with oneself.
>
> *(Paech 2012, 149)*

It is a combination of efficiency and sufficiency which can help reduce our material fluxes. However, it is essential that the discussion of these two does not remain at the product level – it must be applied to the system level. Neither a more efficient product nor its frugal usage will significantly reduce the environmental impact if the entire system is not considered! Stuchtey et al. have shown nicely how much improvement would be possible if one takes a systems perspective. If one considers the time cars spend in the parking lot or in traffic jams, the average load factor of 1.5 persons, and the efficiency of the combustion engine, the total fuel efficiency in transportation is around 2% (Stuchtey, Enkvist & Zumwinkel 2016, 16ff.). The authors continue by considering similar calculation for the usage of streets. However, we do not even need not to stop here. The next level could be to ask: what is the *reason* for this transport? Is it actually *needed* to get from A to B? What if there is a different way of getting the same service delivered without transporting so much material? It is such considerations which go beyond the dualism of efficiency and sufficiency.

13.2.2 Consistency

While efficiency and sufficiency address the *amount* of resources implied by consumption, consistency looks at civilizations' *mass fluxes* and their compatibility, their consistency with the natural cycles. The current mass fluxes are not compatible with natural cycles because the amount and/or the rate of infiltration is too high and/or because the substances cannot (easily) be degraded by natural processes (see Ayres 1994; Rockström et al. 2009; Schaub & Turek 2016, 83ff.; Steffen et al. 2018). We emit anthropogenic substances and compounds to the environment which are either not degradable by natural processes at all (e.g. POPs, nuclear waste, plastic, etc.), or for which natural degradation takes much longer than the rate of infiltration. The industrial metabolism therefore needs to be more *consistent* with natural cycles (also referred to as "eco-effectiveness"). The idea that there should be no waste because everything is being reintroduced into either the natural or the technological cycle is a key feature of the circular economy, the cradle-to-cradle concept (Braungart & McDonough 2002; McDonough & Braungart 2013), or the "Blue Economy" of Gunter Pauli (see Pauli 1998, 2010). "Waste is food" is the paradigm.

The combination of efficiency, sufficiency and compatibility can facilitate the transition towards a service economy, can create jobs, and promote local value creation. A service economy is "inherently more labour-intensive than the mass production manufacturing and 'throw-away' economy. Treating products as capital goods will create more jobs because repair, renovation, disassembly and remanufacturing are inherently more labour-intensive than original equipment manufacturing" (Ayres 2008, 292).

13.3 Be "net-positive" – build up environmental and societal capital!

One flaw in our market system is that it externalizes costs – to the environment, to others and to the future. This ultimately stems from our focus on the financial, the economic side of things. As has been mentioned throughout this book, this focus needs to be challenged for a number of reasons. In the context of action principles this suggests a principle of being "net-positive", of building up natural and social capital.

Corporations are mostly judged in terms of their (expected) profitability and – to a lesser degree – on their capacity to create jobs. Investors consider share prices and dividends, while politicians and unions focus on job creation or preservation and perhaps labour standards. But what about pollution, what about social standards in the supply chain, what about environmental management, what about waste management – and above all: what about the sustainability of the product portfolio? Why praise profit and jobs when suppliers are badly treated, ecosystems destroyed and taxes avoided? Can we as a society afford to leave this space to NGOs and "enlightened" consumers? There is, in my view, a great need to develop more comprehensive ways of assessing the success of business. Promising first steps have been made in this direction.

In 2011, sportswear manufacturer PUMA was the first company globally to issue an environmental profit and loss (EP&L) account. PUMA calculated the environmental damages which their own as well as their suppliers' operations had incurred, i.e. "the estimated cost to society of PUMA's environmental impacts" (PUMA SE 2011, 15). This very far-sighted project, which calculated the damage from water use, GHG emissions, other air pollution, land use and waste for their own operations, as well as up to tier four of their supply chain, is not only taking seriously the demands of the polluter-pays principle (see below, 13.5). It also implies that this negative ecological impact should be reduced in the long run. Regardless of how PUMA followed up on this,[4] the very fact that they calculated and published their ecological damage validates the point that corporations should not only be measured by their financial success alone. Competing with their peers for the best workforce, the best branding and for long-term investors, companies might reduce their environmental impact to zero – and then potentially even build up natural capital. This will likely also require regulatory changes (e.g. tax reforms, incentives, etc.) to build up capital of different types.

There is a global initiative which moves in a similar direction: the International Integrated Reporting Council (IIRC), which aims for a revision of corporate reporting. IIRC is a global coalition of a variety of stakeholders (e.g. regulators, investors, companies, NGOs) who share the view that "communication about value creation should be the next step in the evolution of corporate reporting" (IIRC 2013). "The primary purpose of an integrated report is to explain to providers of financial capital how an organization creates value over time", it benefits all stakeholders and considers several forms of capital: "financial, manufactured, intellectual, human, social and relationship, and natural" (IIRC 2013, 4.11)

The Economy for the Common Good calls for an even more fundamental change (Felber 2018; Ecogood.org. 2019). It advocates a "more ethical economic model, in which the well-being of people and the environment become the ultimate goal of business". Organizations (not just corporations but also municipalities, NGOs, etc.) are invited to publish "Common Good Balance Sheets", which report on performance regarding "human dignity", "solidarity and justice", "ecological sustainability" and "transparency and participation" for each of the critical stakeholder groups: suppliers, owners, employees, customers and social environment (Ecogood.org. 2019).

Why should we only work on reducing the negative and not facilitate the positive? Why should we only aim at not reducing natural capital if we can also increase it?

The call to building up natural and societal capital is very close to the concept of "net-positive" activities which Stuchtey et al. suggested. Starting from the premise that "less bad" is not enough, they call to establish

> a basic societal norm of all economic activities that their total net impact, including their impact on manufactured capital, natural capital and human capital, is actually positive – in other words, that the economic activity in total does more good than harm – over the lifetime of the asset (and including after-life uses).
>
> *(Stuchtey, Enkvist & Zumwinkel 2016, 24)*

How can environmental capital be built up? One important way of doing so is the restoration of large-scale damaged ecosystems (see Commonland 2019). During the last forty years, nearly one-third of the world's arable land has been lost due to erosion and continues to be lost at a rate of more than 10 million hectare per year (UNCCD 2014, 7). Land restoration has multiple benefits for biodiversity, water cycles, job creation, CDR, nutrition, etc. Combined with soft infrastructure development and energetic utilization of high solar radiation in arid regions, it is likely that such "Desert2Eden" projects can address at least half of the SDGs (Berg 2015).

13.4 Prefer local, seasonal, plant-based and labour-intensive

This principle addresses both individuals and organizations as consumers. As prices do not (yet) reflect the true eco-social costs of products, conscious consumption can support the sustainability transition. One guideline for such consumption can be to prefer local, seasonal, plant-based and labour-intensive products. This will help reduce our environmental footprint.

In an ideal world, products would be priced in such a way that the price would reflect all related lifecycle costs incurred by society and environment. These negative externalities are a critical market failure as discussed earlier (see 5.1). For the time being, however, these externalities must be minimized. One way of doing this is consuming responsibly and consciously. Consumption is, as it were, the interface between personal behaviour and the larger world "out there" with its unsustainable modes of production and logistics. Responsible consumption can impact production because consumers' demand impacts the supply side. The more affluent societies become, the more choices people (or organizations) have for consumption; the more we know about production methods, fair wages and environmental pollution, the more consumers consider their own role and their own responsibility for their consumption behaviour. Again, this is only a provisional approach until the set-up of the market framework becomes more sustainable.

However, sustainable consumption faces several difficulties:

- Being a moral imperative, the call for sustainable consumption will always be heard by a limited number of people, so the overall impact will remain limited.
- Some people might feel obligated to "save the world" with every apple they buy. This is not only a psychological challenge because it has the potential to frustrate those people whom you should actually encourage and motivate —no one can constantly deliberate the complexity of the world out there, reduce it to the point of sale and respond in a morally sound way.
- Calls for sustainable consumption can backfire – on an individual level, because people might become cynical or flee into excessive consumption if they are overwhelmed by the task; on a societal level, if patterns of

sustainable consumption become politically loaded and provoke strong reactions (e.g. some media considering this as being paternalistic).
- It is a social issue – since more sustainable products are frequently more expensive.
- It is also very difficult because it takes a lot of effort to figure out what is the more sustainable alternative. So there is need for clear advice on what to purchase.

To be sure, any principle for such advice should be easy to follow. On the other hand, it is difficult to find a rule for sustainable consumption which is simple and true at the same time, because one can hardly encapsulate the complexity of sustainability requirements in one practical rule. Several such rules have been proposed. Here is just a small selection:

- Source locally because locally sourced products require less logistics (which relates to GHG emissions, pollution and resource consumption), stimulate local exchange and cohesion, and foster local know-how and value-creation. The concept of food miles tries to raise awareness about the distances our food and consumer goods travel. It is not easy, though, to calculate food miles properly because the mode of transport (e.g. container ship or air freight) heavily influences the impact and precludes comparisons (Theis & Tomkin 2012, 455ff.).
- Consume food seasonally because it reduces cost and energy for warehouses, keeps products fresh and avoids their preservation.
- Favour plant-based diets over meat diets because they imply a whole series of benefits: much less energy consumption, far fewer GHG emissions, smaller water footprint, fewer fertilizers, better animal health, less erosion and degradation.
- Choose renewable rather than fossil energy – it is a truism that this reduces GHGs.
- Prefer service intensive to resource intensive because this has a great potential for reducing resource consumption. By changing their business model from selling products to leasing them, producers would have an interest in longevity, durability, good performance, easy maintenance, upgradability and recyclability – and they would stop planning built-in obsolescence. According to Ayres, "long-term sustainability requires that future economic growth must rely much more on services and be less dependent on exergy inputs than in the past. This means it must be correspondingly more dependent on labor and capital." Ayres see this as good news "since labor is now in surplus supply almost everywhere whereas natural resources will inevitably become scarcer" (Ayres 1996, 19).

Ceteris paribus, all other things kept equal, locally sourced is better than globally sourced, plant-based is better than animal-based, seasonal is better than out-of-seasonal, renewable is better than non-renewable and labour-intensive is better than resource-intensive. However, we often *cannot assume* that all other things are kept equal. For instance, production conditions vary in different locations. This is the reason why there are circumstances in which some of these rules can actually be misleading:

- "Buy local" is certainly good advice in many cases, but it might not always the best thing to do. The carbon footprint of a domestic apple increases considerably if it is consumed "at the wrong time". An apple harvested in October and sold in April will have to be kept in an air-conditioned warehouse for half a year, which could imply that an apple sourced from a nearshore origin might be more carbon friendly than a domestic one. In addition, lamb produced in the UK has a higher carbon footprint than lamb imported from New Zealand, because the energy input during the breeding phase is lower in New Zealand and the energy needed for logistics is relatively small compared to that. So buying local needs to be combined with *buying seasonal* (Ledgard et al. 2011).
- "Favour a vegetarian diet over a meat diet" is another such example which might not always be more environmentally friendly, for it strongly depends on the alternative. Diets with reduced meat consumption do not necessarily have a lower carbon footprint than average diets – at least if the substitute for meat are dairy products and tropical fruits (Tom, Fischbeck & Hendrickson 2016).
- Even if you follow several of the rules above, some practices might slip through the cracks: bio-fuel is plant-based, renewable, might even be locally produced and seasonally consumed and is still a serious ecological challenge because it is produced by industrial farming with related monocultures, fertilizers and pesticides, etc.

What follows from these cases? On the one hand, one cannot expect to rely on one simple rule to reduce one's environmental impact. On the other hand, consumers should not be irritated by too many categories or – even worse – by revising previously given advice. Therefore, I suggest that these rules are communicated with the caveat that no single one might always be true but the more attributes considered, the more certain one can be that the corresponding action will have the least environmental impact. Similarly to password security, there is a trade-off between practicability and security. The more security features added (length, character types, etc.) the better. The more attributes suggested in this principle, the more certain one can be that the resulting action minimizes environmental impact.

Finally, it should be mentioned that the rules above have a certain bias towards the carbon footprint and take it as a proxy for environmental impact, which is, of course, not always sensible. Future insights might require a qualification of the rules given here. Furthermore, these rules do not, of course, resolve trade-offs between different environmental goals. Offshore wind-parks might affect the marine ecosystem, concentrated solar power will threaten birds and insects, the production of solar panels and batteries requires substantial amounts of energy and other resources, respectively.

In sum, sustainable consumption is a critical component on the way towards the more sustainable production and consumption patterns that SDG 12 calls for. The rules given above can help consumers on all levels, be they individuals, corporations or organizations in their procurement, and certainly public procurement offices, to identify the more sustainable alternative of different consumption options.

13.5 Polluter-pays principle

The polluter-pays principle is critical for internalizing external costs, as demanded by the deficiencies of the market framework. Furthermore, it can also help address the consumption-oriented zeitgeist-related barriers.

"Clean up your own mess" is one of the things not only Robert Fulghum learned in kindergarten (Fulghum 2003, 2). It is a basic principle of fairness which we not only teach our children but which can be traced back to the earliest codification of law: a person responsible for causing damage or harm is made accountable and has to repair it. This can well be claimed to be one of the oldest principles of human law and custom. The earliest known written legal code, the Mesopotamian Code of Hammurabi, composed about 1780 BC by Hammurabi, the ruler of Babylon, already mentions measures of reparation several times:

> If any one be too lazy to keep his dam in proper condition, and does not so keep it; if then the dam break and all the fields be flooded, then shall he in whose dam the break occurred be sold for money, and the money shall replace the corn which he has caused to be ruined.
>
> *(Code of Hammurabi, §53)*

The timeliness of this 3,800-year-old legislation is shocking – as I write these lines in January 2019, the Vale Dam in Minas Gerais, Brazil, has just collapsed and killed up to 300 people. Three judges have already frozen US$ 3 billion of the mining company's funds for compensating the victims and for the clean-up; the total compensation cost are estimated to be up to US$ 7 billion (Bloomberg 2019). But what is the sum of US$ 3 billion in light of so many fatalities? And what is this number in light of US$ 4.6 billion profit in 2017 alone?

The Code of Hammurabi foresaw drastic penalties for unfaithful herdsmen: if a herdsman caring for an entrusted herd of sheep cheats and tried to "make false returns of the natural increase" he would have to pay the owner ten times the loss (Code of Hammurabi, §265). I think it is humiliating to see how little we value a human life and great environmental harm compared to the standards which our ancient ancestors from Mesopotamia had almost four millennia ago. One may wonder who can be called civilized.

In modern times, the OECD was the first to postulate the polluter-pays principle in 1972, as mentioned above (OECD 1972, 1992, 44). This 1972 mention already nicely describes the nature of the principle:

> Environmental resources are in general limited and their use in production and consumption activities may lead to their deterioration. When the cost of this deterioration is not adequately taken into account in the price system, the market fails to reflect the scarcity of such resources both at the national and international levels. Public measures are thus necessary to reduce pollution and to reach a better allocation of resources by ensuring that prices of goods depending on the quality and/or quantity of environmental resources reflect more closely their relative scarcity and that economic agents concerned react accordingly
>
> *(OECD 1972, Annex, A. a) 2)*

According to the OECD, the application of this principle covers measures of pollution prevention and control, and "as a general rule", member countries "should not assist the polluters in bearing the costs of pollution control whether by means of subsidies, tax advantages or other measures" (OECD 1974, III. 1).

Meanwhile, the polluter-pays principle has become a widely accepted political goal and is encoded in many legal texts worldwide: the Rio Declaration demands it (UNCED, Rio Declaration 1992, Principle 16); the EU demands that "the polluter should pay" as a principle of environmental law, although its substance is not defined (EU 2012), etc. Several developing countries (e.g. India, Malaysia, Taiwan, Ecuador, Chile, Costa Rica, Kenya) have varied this principle in such a way that the government takes liability for cases in which the polluter cannot be identified to ensure victims' compensation (Luppi, Parisi & Rajagopalan 2012).[5]

Despite the long record of authorities claiming the validity of this principle throughout the world, its execution is inadequate in practice. Despite the claim that the polluter should not even be aided in bearing the costs of pollution control (as the 1974 OECD texts framed), the opposite is taking place. The polluter is not only supported in control mechanisms but actually *subsidized in pollution*. There are numerous examples for this – but the most obvious and most harmful domain is the amount of global subsidies for fossil fuels. On a global scale, post-tax subsidies of fossil fuels range in the order of more than 6% of GDP (IMF 2015, 19). A large portion of this, particularly in the advanced economies (which is basically the OECD countries), is attributed to "externalities" (21). Although the major countries – the Group of 20 – agreed in 2009 to phase out these subsidies in the mid-term (G20 2009), hardly any progress is visible.

At the same time, phasing out such subsidies would imply several beneficial effects. Focusing on energy subsidies, the IMF concludes that eliminating post-tax subsidies could "raise government revenue by \$2.9 trillion (3.6% of global

GDP), cut global CO_2 emissions by more than 20%, and cut premature air pollution deaths by more than half" (IMF 2015, 6). According to the IMF, this action would raise global economic welfare by US$ 1.8 trillion or 2.2% of global GDP (IMF 2015, 6).

It is apparently very difficult to abolish subsidies because governments would need to take away something from voters which they consider to be their acquired right. There can be no doubt, however, that reducing our environmental impact cannot work without first ending the incentivization of environmental damage. Policy leaders are called to seek measures to phase out such subsidies. They will first need "to identify the political forces that created energy subsidies in the first place and then to redirect or inoculate those forces" as a World Bank study concludes (World Bank 2017, 33).

It will be important to focus research on questions like:

a) What is the overall rationale of a subsidy (recipient, purpose, spending, etc.)? Who are the recipients and what is their need?
b) Can this need be addressed in a different way without subsidies?
c) If not, can a more environmentally friendly subsidy replace the existing one?

Having identified different scenarios for policy recommendations, a public discourse is needed which considers the pros and cons of every scenario, and then political decisions need to be taken. The polluter-pays principle would obviously need to be implemented by politicians. However, other societal actors can attribute it to their domain. Corporations are starting to consider their pollution and beginning to take responsibility for it, as the abovementioned case of PUMA illustrated (see 13.3). On a consumer level, the polluter-pays principle is reflected by all kinds of compensation services, which raise funds from consumers of environmentally harmful activities (e.g. flying) and invest these funds in projects of restoration and compensation.

Despite the broad consensus on the need for a polluter pays-principle, there are also critical voices. Georgescu-Roegen notes that it would allow the rich to pollute the environment at their discretion, and he calls for strict regulation instead (Georgescu-Roegen 1986, 15). Tekayak's criticism adopts a similar direction: the polluter-pays principle would not really alter corporations' operations if they can simply get away with monetary compensation (Tekayak 2016, 64). I do not think, however, that the enforcement and consequent execution of this principle would not change corporations' operations. The PUMA case demonstrates that the polluter-pays logic is understood in the corporate world. Moreover, Georgescu-Roegen is certainly right that it would be unfair if the rich can simply afford to compensate their wrongdoing by the purchase of indulgences. However, in a situation in which the rich do not even pay for their pollution at all (or are only gradually beginning to do so, i.e. COP21), such a principle can already be a big leap forward.

Nevertheless, as important as the enforcement of the polluter-pays principle would be at the moment, in the long run it will be essential to prevent pollution in the first place. This is actually already codified in EU law, which calls for fighting pollution at its source: "environmental damage should as a priority be rectified at source" (EU 2012, Art. 191).

13.6 Precautionary principle

The precautionary principle demands taking precautionary measures if there is good reason to believe that a development can imply substantial risk, even though the evidence for that risk cannot be fully validated scientifically.

The examples discussed so far relate to contexts in which either morally reprehensible *practices* or harmful *effects* of a technology should be constrained. What needs to be done, however, if the effects of a new technology cannot be assessed, at least not with certainty? New technologies are being developed and rolled out so rapidly that their social and environmental implications can only be vaguely anticipated. Political regulation is often struggling to keep up. In such cases, the precautionary principle needs to be applied. This is a principle of informed prudence; it requires precautionary measures when potential future harms are anticipated early. The precautionary principle has two aspects: precaution of risks and precaution of resources (Calliess 2001, 245ff.) The precautionary principle mandates the state to act, especially in the face of concrete threats to environmental goods. This is particularly true for the cumulative and synergetic effects of environmental impairments (Heselhaus 2018, 32).

The precautionary principle, dating back to the German *Vorsorgeprinzip* in the 1970s, justifies taking measures even in the absence of conclusive evidence (see Gilbert, Van Leeuwen & Hakkinen 2009; Persson 2016, 138). It is a basis for the state to take action in cases of insufficient scientific understanding when outcomes are irreversible and/or widespread (deFur & Kaszuba 2002, 155).

If there is good reason to think that a development can potentially threaten human health or the natural environment, the precautionary principle calls for the state to intervene. This kind of state intervention becomes necessary given today's rapid pace of technological and scientific developments, which makes their effects and side effects often difficult to assess in advance despite potentially considerable risks. As Christian Calliess explains:

> In a situation of this kind, in which it is not possible to attribute responsibility to an identifiable individual, thus causing the failure of private liability law as well as traditional laws to avert imminent danger, safety and security expectations are directed once again to state institutions from whom precautionary measures to protect against damage can be justifiably expected.
>
> *(Calliess 2013, 20)*

The precautionary principle therefore allows state institutions to intervene in the process of development or roll-out of technologies "in the event of an abstract concern and not only in the event of concrete danger (hazard) for which there is concrete evidence" (Calliess 2013, 20). As Calliess argues, the precautionary principle "pre-shifts" the admissible moment for governmental intervention (Calliess 2009). In other cases of averting a threat, reasonably solid scientific evidence of the expectable danger is demanded before the state can legitimately interfere. By its very meaning and purpose, the precautionary principle implies a reversal of evidence (Calliess 2009, 126).

The precautionary principle therefore frames what Hans Jonas called for in his seminal work *Das Prinzip Verantwortung* (*The Imperative of Responsibility*, German edn 1979; Jonas 1984). Jonas insists that the uncertainty about future projections – rooted among others things in the complexity of causal chains combined with the efficacy of modern technology (66) – should imply a new moral prescription: the doom prophecy (*Unheilsprophezeihung*) shall have primacy over the salvation prophecy (70), i.e. in case of doubt about the future implications of a given technology, the pessimistic projection should outweigh the optimistic one.

Erik Persson investigates several studies on the precautionary principle and elaborates their commonalities. He stresses that the principle does not discuss any measures that need to be taken; rather, it describes situations in which special caution is needed (Persson 2016, 135). In line with Persson, we argue that the presence of one of the following circumstances justifies the application of the precautionary principle:

a) When there is a *trade-off between two conflicting goals which both represent critical values*. In such a case, traditional decision methods like cost-benefit analysis are not appropriate. Persson illustrates this with the case that a serious threat to human health could be prevented by use of chemicals which seriously threaten the environment. A decision in favour of one or the other resulting, for instance, from the perspective of a cost-benefit analysis would not do justice to the value of each of the goods involved (Persson 2016, 137).

b) When there are potentially *irreversible and severe consequences*, like a "severe loss that is not just irreversible but also irreplaceable, this puts it in a special category that calls for extra precaution" (Persson 2016, 138).

c) When there is *urgency*. You cannot deliberate about the pros and cons of introducing genetically modified crops into the environment if this is already being done – by then, at the latest, it has lost its point.

d) When it is "*more important to avoid false negatives than false positives*" as Persson calls it: "Scientists do not like to be wrong. In the world of science, making a claim that turns out to be wrong is, in general, worse than abstaining from making a claim that later turns out to be true. This means that scientists tend to be biased to err in favour of false negatives over false positives" (Persson 2016, 139). From the point of view of societal risk management, however, it is rather important to articulate the potential threat even though its probability cannot be calculated.

The precautionary principle has meanwhile found its way into important international politics and legal texts. The Agenda 21 is full of references that "a precautionary approach should be applied" (art. 20.32), that "appropriate use" shall be made "of the concept of the precautionary approach" (art. 22.5.c), or simply that "[t]he precautionary approach is important" (art. 35.5, see UNCED 1992). As with the polluter-pays principle, the precautionary principle is already codified in some regional legal texts; it is a statutory requirement for environmental policies in the European Union (e.g. EU 2012, Art. 191). As such, it is a guiding principle in European environmental law (Calliess 2018, 101).

However, there are striking differences in the perception of this principle in different political contexts. In the USA, for instance, where people tend to have a more sceptical view of state intervention than in Europe, and where business plays an even larger role than in the EU, it does not come as a surprise that "US experience with the precautionary principle is quite different from Europe," as two US-based authors record (deFur & Kaszuba 2002, 156). "US business and corporate interests have argued vociferously against the precautionary principle … largely on the basis of cost. The application of the precautionary principle to genetically modified organisms seems to have drawn the most vitriolic attacks" (deFur & Kaszuba 2002, 156). Such strong resistance from business points to an important challenge of the precautionary principle. As long as it is implemented only regionally, in the current market set-up, the precautionary principle functions de facto as a competitive disadvantage, as a break to innovation. Once the market framework does more justice to environmental concerns and extends the producers' liability, the situation might change and the principle can even boost innovation because it introduces an additional scarcity into the market which guides technological development and incentivizes sustainable innovation.

13.7 Appreciate and celebrate the beauty of nature

With an ever-increasing share of the population living and working in cities in mostly synthetic environments, people tend to be more and more alienated from natural environments. We can only sustain nature, however, if we love it; and we will love it the more we understand it. We need to regain a sense of awe and appreciation for the beauties of nature and their finely balanced ecosystems. This last nature-related principle already points to the next chapter the personal principles.

"In the end we will conserve only what we love, we will love only what we understand, and we will understand only what we are taught" is a dictum of the Senegalese forestry engineer Baba Dioum (Valenti & Tavana 2005, 308). People lose their sense of nature, they have no immediate contact with it, they know only little about nature's wonders. Environmental psychologists speak of an "extinction of experience", as more and more people are disconnected from nature (Soga et al. 2016). Soga et al. conducted a questionnaire among undergraduate university students in Tokyo and determined a positive relationship between their current and childhood frequency of contact with nature and their emotional connectedness to nature, as well as their perceptions of neighbourhood nature (149). "Our results suggest that, given the rapid decrease in children's daily contact with nature, public appreciation of the value of the natural world is likely gradually also to decrease. This can be a major obstacle to reversing global environmental challenges" (143). The authors conclude that especially children "should therefore be encouraged to experience neighbourhood natural environments and their associated biodiversity. To do so, more strategic, well-designed urban planning and broader policy changes, such as social marketing campaigns, educational and outreach programs, will both be necessary" (149). How can one motivate people to protect what they hardly now? How can you argue about unsustainable production and consumption patterns if people do not realize what is at stake?

Evans et al. see it as similarly important that children have an experiential relationship to nature from early childhood on. Having performed a longitudinal analysis over a 12-year period among children from age 6 to young adults of 18, they found three factors positively influencing the young adults' environmental behaviour: the time the kids spent outside, their mothers' environmental attitude, and her level of education (Evans, Otto & Kaiser 2018).

Another study, by Otto and Pensini, looked at the importance of nature-based environmental education. Their starting point is that the promotion of environmental knowledge would generally be viewed as a fundamental component of environmental education and a necessary prerequisite to ecological behaviour. The problem is, however, that environmental knowledge has little effect on actual behaviour. Interestingly, *connectedness to nature is much more important than environmental knowledge* (Otto & Pensini 2017, 88). Therefore, it is essential that children have an experiential relation to nature – feel it, smell it, touch it and "be connected" to it – and this will be a strong motivational force for knowledge acquisition and environmental concern.

Kaiser et al. investigated the relationship between a greater appreciation of nature and ecological behaviour (Kaiser et al. 2014a). Astonishingly, as the authors write, the positive motivational basis of the appreciation of nature for pro-environmental behaviour is not widely discussed, although "[s]ome of the environmental problems now confronting many societies appear to stem from a lack of individual engagement in ecological or pro-environmental behavior" (270). Using longitudinal survey data, they found that "appreciation of nature, measured in various ways, has repeatedly been found to correlate moderately to strongly with ecological behavior" (270). In other words: "It is possible to protect the environment by encouraging appreciation of nature" (269).

As an interim result we can document: the appreciation of nature correlates with pro-environmental behaviour; it is important that especially children have the chance to make their own experiences in nature, and that this is even more important than knowledge. The mother's attitude towards nature also matters.

Although direct interaction with nature therefore has a strong motivational force for pro-environmental behaviour, learning should not be underestimated. Otto and Kaiser looked at the well-corroborated relationship between age and ecological behaviour: "we found that learning rather than maturation explained the relation between age and self-reported ecological behavior. The more exposed people are to information that deals with environmental-conservation-relevant topics, the more pronounced their ecological engagement" (Otto & Kaiser 2014, 331). The authors admit, though, that their study is one of the few that supports the efficacy of learning "in promoting the ecological performance of individuals" (331).

Celebrating and appreciating nature is an important sustainability principle for yet another reason. Contrary to several other principles, this is a *positive* one, i.e. one that has a strong motivational component. The problem with the concepts of eco-efficiency, sufficiency, the precautionary principle, or the polluter-pays principle is that they are all about avoiding, preventing, reducing, denouncing. Celebrating nature is a thoroughly positive principle. There is ample evidence of the importance of positive emotions for personal development and social con-nection. "[P]ositive emotions are vehicles for individual growth and social con-nection: By building people's personal and social resources, positive emotions transform people for the better, giving them better lives in the future"((Fredrick-son 2001); cf. (Fredrickson et al. 2008)). The challenge is therefore to develop a new sense of awareness, a new sense of connectedness, a new appreciation of nature and its beauties. In my view, it is a sense of awe and wonder which lies at heart of all true human activities – definitely at the heart of science, of phil-osophy, of religion, of music, arts – and the respect for the other. We need to regain this sense of wonder for the miracles of creation.

Notes

1 Brown (2008) is critical of such principles because nobody could really determine whether they are met or not:

> we face the challenge of decision-making under great uncertainty. This makes the implementation of many seemingly wise and straightforward concepts of sustain-ability difficult and impossible … A look at some typical principles of sustainability makes this clear. Some examples are … waste emissions not exceeding the natural assimilative capacity, harvest rates not exceed in the rate of biodiversity preserva-tion. These are certainly good ideas, but attempt to enforce them is prohibitive. Simply estimate the rates is problematic, enforcement and monitoring across the global is beyond feasible.
>
> *(Brown 2008, 143)*

In my view, this points to the need of both concrete goals for the systemic level (as, for instance, the SDGs represent) and action principles for concrete actions.

2 Bleischwitz et al. looked at the material consumption of selected materials and documented a certain saturation effect: "Our results confirm the occurrence of a saturation effect for most materials considered." There is strong evidence for such an effect for the apparent per capita consumption of steel, copper and cement in the four industrialized countries investigated (Bleischwitz et al. 2018, 86).

3 Of course, a more detailed account would need to distinguish the kinds of material we are exchanging, because their damaging effect certainly varies. For the current purpose is is enough, however, to generally call for a reduction since we are exceeding sustainable limits in almost every respect: be it fish catch or nitrogen input, metal ore extraction or phosphorus consumption, microplastics or nano-materials, and certainly fossil fuels and carbon emission – none of these can be sustained in the long run if we continue with our current patterns.

4 To my knowledge, no further EP&L account has been communicated.

5 Some economists object that the polluter-pays principle does not always allow for an efficient reduction of environmental harm and therefore consider the polluter-pays principle as outdated and limited (Schmidtchen et al. 2008). They argue for the cheapest-cost-avoider principle, which applies a cost-benefit analysis and decides which policy is most cost effective for the specific context. If people are littering the environment with their trash, it might be cheaper to get someone to clean this up than to search for the causer and charge him. However, in my view, this proves that questions of pollution should not be decided on economic grounds alone. Paying for indulgences may be efficient but not necessarily moral. In my view, it is important to adhere to the polluter-pays principle as a moral concept – not least because it is embodied in countless legal texts – since it makes a normative claim about the attribution of responsibility. However, it might need to be complemented by more efficient allocation mechanisms in certain contexts.

Bibliography

Ayres, Robert U. "Industrial metabolism: Theory and policy." In *The Greening of Industrial Ecosystems*, von National Academy of Engineering (ed.), 23–37. Washington, DC: The National Academies Press, 1994.

———. *Eco-Thermodynamics: Economics and Second Law*. Fontainebleau: INSEAD, 1996.

———. "Sustainability economics: Where do we stand?" *Ecological Economics 67 (2)*, 2008: 281–310.

Berg, Christian. "Desert2Eden – Integrated restoration and development of arid regions to address the Sustainable Development Goals (SDGs)." *GESolutions. Global Economic Symposion 2015*. Kiel: Kiel Institute for the World Economy, 2015: 89–93.

Bleischwitz, Raimund, Victor Nechifora, Matthew Winning, Beija Huang & Yong Geng. "Extrapolation or saturation – Revisiting growth patterns, development stages and decoupling." *Global Environmental Change 48* 2018: 86–96.

Bloomberg. "Vale's Dam Breach in Brazil leaves dozens dead; stock plunges." *Bloomberg.com*. 28. 01 2019. www.bloomberg.com/news/articles/2019-01-28/after-dam-disaster-vale-s-board-considers-suspending-dividends (accessed 19. 02 2019).

Boulanger, Paul-Marie. "Three strategies for sustainable consumption." *S.A.P.I.E.N.S.– Surveys and Perspectives Integrating Environment and Society 3 (2)*,2010: 1–10.

Braungart, Michael & William McDonough. *Cradle to Cradle: Remaking the Way We Make Things*. New York: North Point Press, 2002.

Brown, Casey. "Emergent sustainability: The concept of sustainable development in a complex world." In *Globalization and Environmental Challenges*, Hans Günter Brauch, Navnita Chadha Behera, Béchir Chourou, Pál Dunay, John Grin, et al. 141–149. Berlin, Heidelberg, New York: Springer, 2008.

Calliess, Christian. *Rechtsstaat und Umweltstaat, Zugleich ein Beitrag zur Grundrechtsdogmatik im Rahmen mehrpoliger Verfassungsrechtsverhältnisse*. Tübingen: J.C.B. Mohr (Paul Siebeck), 2001.

———. „Das Innovationspotenzial des Vorsorgeprinzips unter besonderer Berücksichtigung des integrierten Umweltschutzes." In *Innovationsverantwortung. Innovation Und Recht III*, von Martin Eifert & Wolfgang Hoffmann-Riem (eds.), 119–145. Berlin: Duncker & Humblot, 2009.

———. „Rechtsstaat und Vorsorgestaat." *Annual Review of Law and Ethics* 21,2013: 3–22.

———. „EU-Umweltrecht." In *Grundzüge des Umweltrechts*, Eckard Rehbinder & Alexander Schink (eds.), 65–144. Berlin: Erich Schmidt Verlag, 2018.

Carbon Tracker. *Powering Down Coal. Navigating the Economic and Financial Risks in the Last Years of Coal Power*. London: Carbon Tracker, 2018.

ClimateActionTracker. *climateactiontracker.org*. 03. 12 2018. https://climateactiontracker.org/media/images/CAT-Thermometer-2018.12-3Bars.original.png (accessed 11. 05 2019).

Code of Hammurabi. *Sacred-texts.com*. 2019. www.sacred-texts.com/ane/ham/ham05.htm (accessed 15. 02 2019).

Commonland. *Commonland*. 2019. www.commonland.com/en (accessed 29. 04 2019).

Daly, Herman E. *Beyond Growth*. Boston, MA: Beacon Press, 1996.

———. *Ecological Economics and Sustainable Development, Selected Essays of Herman Daly*. Chaltenham, UK; Northampton, MA: Edward Elgar, 2007.

deFur, Peter L. & Michelle Kaszuba. "Implementing the precautionary principle." *The Science of the Total Environment 288* 2002: 155–165.

Deutscher Bundestag. *Die Industriegesellschaft gestalten. Perspektiven für einen nachhaltigen Umgang mit Stoff- und Materialströmen*. Bonn: Economia Verlag, 1998.

Ecogood.org. „Herausgeber: Internationaler Verein zur Förderung der Gemeinwohl-Ökonomie e.V." 2019. www.ecogood.org/en/(accessed 29. 04 2019).

EU. *Consolidated Version of the Treaty on the Functioning of the European Union*. Brussels: European Union, 2012.

European Commission. *EU Energy in Figures*. Luxembourg: European Commission, 2018.

Evans, Gary W., Siegmar Otto & Florian G. Kaiser. "Childhood origins of young adult environmental behavior." *Psychological Science 29*, 15. 02 2018: 679–687.

Felber, Christian. *Gemeinwohlökonomie*. München: Piper, 2018.

Fredrickson, Barbara L. "The role of positive emotions in positive psychology: The broaden-and-build theory of positive emotions." *The American Psychologist 56 (3)*, 03 2001: 218–226.

Fredrickson, Barbara L., Michael A. Cohn, Kimberley A. Coffey, Jolynn Pek & Sandra M. Finkel. "Open hearts build lives: Positive emotions, induced through loving-kindness meditation, build consequential personal resources." *Journal of Personality and Social Psychology 95 (5)*11 2008: 1045–1062,.doi:10.1037/a0013262

Fromm, Erich. *The Fear of Freedom*. London: Kegan Paul, 1942.

Fulghum, Robert. *All I Really Need to Know I Learned in Kindergarten*. New York: Balantine Books, 2003 (1986).

G20. "Leaders' statement – The Pittsburgh summit." *Oecd.org*. 2009. www.oecd.org/g20/summits/pittsburgh/G20-Pittsburgh-Leaders-Declaration.pdf (accessed 15. 07 2019).

Georgescu-Roegen, Nicholas. "The entropy law and the economic process in retrospect." *Eastern Economic Journal 12*, 01. 03 1986: 3–25.

Gilbert, Steven G., Kees Van Leeuwen and Pertti Hakkinen. "Precautionary principle." In *Information Resources in Toxicology*, Philip Wexler (ed.), 387–393. New York: Academic Press, 2009.

Heselhaus, Sebastian. „Verfassungsrechtliche Grundlagen des Umweltschutzes." In *Grundzüge des Umweltrechts*, Eckard Rehbinder & Alexander Schink (eds.), 4–63. Berlin: Erich Schmidt Verlag, 2018.

Hirschnitz-Garbers, Martin, Adrian R. Tan, Albrecht Gradmann & Tanja Srebotnjak. "Key drivers for unsustainable resource use e categories, effects and policy pointers." *Journal of Cleaner Production 132*, 2016: 13–31.

IEA. *World Energy Outlook 2012. Executive Summary.* Paris: OECD/International Energy Agency, 2012.

IIRC. *The International <IR> Framework.* International Integrated Reporting Council (IIRC), 2013.

IMF. *How Large are Global Energy Subsidies? (WP/15/105).* Washington, DC: International Monetary Fund, 2015.

IPCC. *Climate Change 2014: Synthesis Report. Contribution of Working Groups I, II and III to the Fifth Assessment Report of the Intergovernmental Panel on Climate Change [Core Writing Team, R.K. Pachauri & L.A. Meyer (eds.)].* Geneva: IPCC, 2014.

Jonas, Hans. *Das Prinzip Verantwortung.* Frankfurt: Insel (also: Suhrkamp, stw 1085), 1984.

Kaiser, Florian G., Adrian Brügger, Terry Hartig & Heinz Gutscher. "Appreciation of nature and appreciation of environmental protection: How stable are these attitudes and which comes first?" *Revue européenne de psychologie appliquée 64*, 2014a: 269–277.

———, Oliver Arnold & Siegmar Otto. "Attitudes and defaults save lives and protect the environment jointly and compensatorily: Understanding the behavioral efficacy of nudges and other structural interventions." *Behavrioral Sciences* 2014b: 202–212.

Krausmann, Fridolin, Christian Lauk, Willi Haas & Dominik Wiedenhofer. "From resource extraction to outflows of wastes and emissions: The socioeconomic metabolism of the global economy, 1900–2015." *Global Environmental Change 52*, 2018: 131–140.

Ledgard, S. F., M. LIeffering, M. A. Zonderland-Thomassen & M. Boyes. "Life cycle assessment – A tool for evaluating resource and environmental efficiency of agricultural products and systems from pasture to plate." *Proceedings of the New Zealand Society of Animal Production*, 2011: 139–148.

Luppi, Barbara, Francesco Parisi & Shruti Rajagopalan. "The rise and fall of the polluter-pays principle in developing countries." *International Review of Law and Economics 32* 2012: 135–144.

McDonough, William & Michael Braungart. *The Upcycle. Beyond Sustainability – Designing for Abundance.* New York: North Point Press, 2013.

OECD. *Recommendation of the Council on Guiding Principles Concerning International Economic Aspects of Environmental Policies.* Paris: OECD, 1972.

———. *Recommendation of the Council on the Implementation of the Polluter-Pays Principle (OECD/LEGAL/0132).* Paris: OECD, 1974.

———. *The Polluter-Pays Principle (OCDE/GD(92)81).* Paris: OECD, 1992.

Otto, Siegmar & Florian G. Kaiser. "Ecological behavior across the lifespan: Why environmentalism increases as people grow older." *Journal of Environmental Psychology 40*, 2014: 331–338.

——— & Pamela Pensini. "Nature-based environmental education of children: Environmental knowledge and connectedness to nature, together, are related to ecological behaviour." *Global Environmental Change 47*, 2017: 88–94.

Paech, Nico. *Befreiung vom Überfluss. Auf dem Weg in die Postwachstumsökonomie.* Munich: oekom, 2012.

Pauli, Gunter. *Upsizing. The Road to Zero Emissions.* Greenleaf Publishing, 1998.

———. *Neues Wachstum. Wenn grüne Ideen nachhaltig 'blau' werden.* Berlin: Konvergenta, 2010.

Persson, Erik. "What are the core ideas behind the Precautionary Principle?" *Science of the Total Environment 557–558*, 2016: 134–141.

PUMA SE. *PUMA's Environmental Profit and Loss Account for the Year Ended 31 December 2010.* Herzogenaurach: PUMA SE, 2011.

Rockström, Johan, Will Steffen, Kevin Noone, Asa Person, F. Stuart Chapin III, et al. "A safe operating space for humanity." *Nature 461*,24. 09 2009: 472–475.

Santarius, Tilmann. *Green Growth Unravelled. How Rebound Effects Baffle Sustainability Targets When the Economy Keeps Growing.* Berlin: Heinrich Böll Stiftung/Wuppertal Institute for Climate, Environment, and Energy, 2012.

Schaub, Georg & Thomas Turek. *Energy Flows, Material Cycles and Global Development.* Springer International Publishing Switzerland, 2016.

Schmidt-Bleek, Friedrich. Factor10-institute.org. 01 2000. www.factor10-institute.org/files/F10_Manifesto_d.pdf (Accessed 15. 02 2019).

Schmidtchen, Dieter, Christian Koboldt, Jenny Monheim, Birgit E. Will & Georg Haas. *The Internalisation of External Costs in Transport: From the Polluter Pays to the Cheapest Cost Avoider Principle.* Herausgeber: Saarland University Center for the Study of Law and Economics 06. 03 2008. http://ssrn.com/abstract=1069622 (Accessed 12. 06 2019).

Schneidewind, Uwe & Angelika Zahrnt. *The Politics of Sufficiency. Making It Easier to Live the Good Life.* Munich: oekom, 2014.

Soga, Masashi, Kevin J. Gaston, Tomoyo F. Koyanagi, Kiyo Kurisu & Keisuke Hanaki. "Urban residents' perceptions of neighbourhood nature: Does the extinction of experience matter?" *Biological Conservation 203*, 2016: 143–150.

Spiegel Online. „So soll Deutschlands Kohleausstieg ablaufen." 23. 01 2019. www.spiegel.de/wirtschaft/soziales/kohleausstieg-so-soll-der-kohleausstieg-ablaufen-a-1249467.html (Accessed 11. 05 2019).

Steffen, Will, Johan Rockström, Katherine Richardson, Timothy M. Lenton, Carl Folke, et al. "Trajectories of the Earth system in the Anthropocene." *PNAS 14*, 08 2018: 8252–8259.

———, Katherine Richardson, Johan Rockström, Sarah E Cornell, Ingo Fetzer, et al. "Planetary boundaries: Guiding human development on a changing planet." *Science 347*, 13. 02 2015: 736.

Stuchtey, Martin R., Per-Anders Enkvist & Klaus Zumwinkel. *A Good Disruption. Redefining Growth in the Twenty-First Century.* London, Oxford, New York, New Delhi, Sydney: Bloomsbury, 2016.

Tekayak, Deniz. "From 'polluter pays' to 'polluter does not pollute'." *Geoforum 71*, 2016: 62–65.

The Guardian. "UK government spells out plan to shut down coal plants." *The Guardian.* 01 2018.

Theis, Tom & Jonathan Tomkin. *Sustainability: A Comprehensive Foundation.* 2012.N/A: Open Stax CNX.

Tom, Michelle S., Paul S. Fischbeck & Chris T. Hendrickson. "Energy use, blue water footprint, and greenhouse gas emissions for current food consumption patterns and dietary recommendations in the US." *Environ Syst Decis 36 (1)*, 2016: 92–103.

Umweltbundesamt. https://www.umweltbundesamt.de. 2018. www.umweltbunde samt.de/daten/verkehr/fahrleistungen-verkehrsaufwand-modal-split#textpart-1 (accessed 28. 04 2019).

UNCCD. *The Land in Numbers. Livelihoods at a Tipping Point.* Bonn: UNCCD, 2014.

UNCED. *"Agenda 21 (*Rio Declaration)*." Report of the United Nations Conference on Environmenta and Development; Annex I: Rio Declaration on Environment and Development.* New York: UN, 12. 08 1992.

UNEP. *Decoupling Natural Resource Use and Environmental Impacts from Economic Growth. A Report of the Working Group on Decoupling to the International Resource Panel.* Paris: UNEP, 2011.

Valenti, JoAnn M. & Gaugau Tavana. "Report: continuing science education for environmental journalists and science writers." *Science Communication 27 (2)*, 02. 12 2005: 300–310.

von Weizsäcker, Erich, Karlson Hargroves, Michael H. Smith, Cheryl Desha, and Peter Stasinopoulos. *Factor Five: Transforming the Global Economy through 80 % Improvements in Resource Productivity*, London: Earthscan,2009.

von Weizsäcker, Ernst Ulrich, Amory B. Lovins & L. Hunter Lovins. *Faktor Vier. Doppelter Wohlstand - halbierter Naturverbrauch.* Munich: Droemer, 1995.

———, Karlson Hargroves & Michael Smith. *Faktor Fünf. Die Formel für nachhaltiges Wachstum.* Munich: Droemer, 2010.

WBGU. *World in Transition: A Social Contract for Sustainability.* Berlin: German Advisory Council on Global Change (WBGU), 2011.

Worldbank. *The Political Economy of Energy Subsidy Reform.* Washington, DC: Worldbank, 2017.

14

PERSONAL PRINCIPLES

Contents

14.1 Why personal principles matter

Why are personal principles important for sustainability? First of all, simply because of their existence. If a person follows principles it means that she or he is not merely driven by interests or desires.[1] To be sure, personal principles and secondary virtues such as diligence, reliability, or bravery can be empty forms and can be corrupted by ideologies – the Nazis were praising many of these secondary virtues. Furthermore, in times of rising right-wing populism all around the globe, we need to be aware of the ambiguity of principles. Nevertheless, facilitating and practicing personal principles is a decisive countermeasure against the dominance of interests.

One is not surprised if a rationale by which people are driven by interests is applied to business, since the rational choice model has been so influential in economics and wants to explain people's behaviour in terms of their preferences. This model, originally applied in economics, has also been applied in politics, sociology and the interpretation of everyday politics.[2] As McKinnon has argued, the "popularity of rational choice thinking stems in part from its consonance with the 'new common sense' neo-liberal politics has created." The driver for human action is then ultimately seen as self-interest. "No longer is there theoretical scope for human action that is not calculatingly self-serving" (McKinnon 2011, 540).

Such strong attention to preferences as drivers for personal action stands in contrast to most of our great philosophical and religious traditions, certainly to the ancient Greek and Judeo-Christian traditions. Although important currents in these traditions also emphasize an underlying rationality as motivation for moral behaviour – consider, for instance, the reciprocity of the demand to love the stranger in the Hebrew Bible: "Love ye therefore the stranger: for ye were strangers in the land of Egypt." (Deuteronomy 10:19) – the motivation for human action is for the largest part not to be confined to preferences.

For Plato, the measure of personal well-being, of a successful life, that everything will eventually work out for good, was not preference or interest but *justice*. The person who is just will ultimately see that things work out for good for her or him:

> Then this must be our notion of the just man, that even when he is in poverty or sickness, or any other seeming misfortune, all things will in the end work together for good to him in life and death: for the gods have a care of any one whose desire is to become just and to be like God, as far as man can attain the divine likeness, by the pursuit of virtue.
>
> *(Plato,* Republic, *613)*

For Plato, it is clear that the just person will also experience advantages and rewards because of her or his being just. But this is secondary and not the main argument why justice is to be sought:

> And thus, I said, we have fulfilled the conditions of the argument; we have not introduced the rewards and glories of justice ... but justice in her own nature has been shown to be best for the soul in her own nature.
>
> *(Plato,* Republic, *612)*

For Aristotle, in turn, it is the "activity of the soul in accordance with virtue" which is to be pursued in order to reach *eudaimonia* ("happiness"). Aristotle did know that *eudaimonia* would also require external goods, "for it is impossible, or not easy, to do noble acts without the proper equipment" (Aristotle, *Nicomachean Ethics*, 1099a). However, ultimately "human good turns out to be activity of soul in accordance with virtue" (Aristotle, *Nicomachean Ethics*, 1098a).

These two great minds have significantly shaped the Western tradition,[3] but Eastern traditions have placed no less value on the inner path. The Noble Eightfold Path of Buddhism contains eight practices by which the individual can reach redemption.

In the great Tao Teh King, Lao Tzu praises the "holy man" who "attends to the inner": "Racing and hunting will human hearts turn mad, treasures high-prized make human conduct bad. Therefore – the holy man attends to the inner and not to the outer. He abandons the latter and chooses the former" (Lao Tzu 1913, 12). Lao Tzu depicts the saint in a way which is *the antipode of rational choice*:

> 1. Heaven endures and earth is lasting. And why can heaven and earth endure and be lasting? Because they do not live for themselves. On that account can they endure. 2. Therefore – The holy man puts his person behind and his person comes to the front. He surrenders his person and

his person is preserved. Is it not because he seeks not his own? For that reason he can accomplish his own.

(Lao Tzu 1913, 7)

Modern contemporaries also know about the value of personal maturity and growth. I submit it is this holding fast to personal principles, the attitude of surrender and of service which the great wise men and women of our time have in common – Martin Luther King, Mahatma Gandhi, Mother Teresa, or Nelson Mandela. Gandhi reportedly said that our greatness as humans lies not so much in being able to remake the world as in being able to remake ourselves. This may sound a little dated, but "remaking ourselves" is essential, I submit, not only because sustainable consumption patterns will require a change in behaviour. This is only a part of resource consumption. It is also essential because it can help us rediscover dimensions of life which have been buried by our focus on the external, on the product, on consumption.

14.2 Practice praxis and contemplation

Important philosophical and religious traditions have valued contemplation more than action. In modernity, however, the rise of technology symbolizes the modern praise of action – contemplation has fallen behind. Furthermore, it is a particular kind of action which prevails, as Hannah Arendt criticized – fabrication. The focus of action as fabrication is entirely on the result of the action, the product, while the notion of an action which is valuable in itself (praxis) has vanished.

Many of our philosophical and religious traditions hold contemplation in high esteem, certainly ancient Greek philosophy and the Christian tradition. While Plato, Aristotle and the tradition following them saw the contemplative life of the philosopher, the *bios theoretikós* as fulfilment of earthly life, the ultimate form of living, they rather disregarded practical work (*bios praktikós*). With the rise of science and technology in modernity, two important shifts occurred, as Hannah Arendt elaborated (Arendt 1998). On the one side, contemplation (the *vita contemplativa*) was disregarded in light of the preeminent success of action (*vita activa*). The active, producing, visible output generating form of life dominates over the contemplative one. Success and reward are measured in tangible, visible results, which can be counted and priced.

On the other side – modern activity – the modern concept of action misses even the critical element of the Aristotelian concept of action: its purposelessness. What Aristotle called *praxis* were those actions which are *good in themselves*, which realize their end by carrying it out. *Praxis* is the good, meaningful life which is oriented at the moral virtues, practicing reflection and philosophy. It is this concept of *praxis* which has been suppressed by the fabrication and work logic of modernity. Arendt criticizes that in modernity, human activity has lost its freedom. Both work (the activity of *homo faber*) and labour (the activity of *animal laborans*) are subdued by necessity; they do not rise from freedom, as *praxis* does.

Modern activity is determined by necessity, for it either serves the purpose of the product (*homo faber*) which outlasts the activity itself, or, even worse, it creates nothing of permanence but must be renewed perpetually for it is needed to maintain life itself (*animal laborans*).

The maker and fabricator, the *homo faber*, has become the outstanding characteristic of the modern age and his features are omnipresent in modern life:

> his instrumentalization of the world; his confidence in tools and in the productivity of the maker of artificial objects; his trust in the all-comprehensive range of the means-end category, his conviction that every issue can be solved and every human motivation reduced to the principle of utility; his sovereignty, which regards everything given as material and thinks of the whole of nature as of 'an immense fabric from which we can cut out whatever we want to resew it however we like' (Bergson) ... finally, his matter-of-course identification of fabrication with action.
>
> *(Arendt 1998, 305f.)*

If action is only fabrication, the end of action is the product. Once the product is fabricated, the goal is reached. In contrast to the concept of meaning, the goal ceases to exist once it is reached:

> For an end, once it is attained, ceases to be an end and loses its capacity to guide and justify the choice of means, to organize and produce them. It has now become an object among objects, that is, it has been added to the huge arsenal of the given from which *homo faber* selects freely his means to pursue his ends. Meaning, on the contrary, must be permanent and lose nothing of its character, whether it is achieved or, rather, found by man or fails man and is missed by him.
>
> *(154f.)*

The instrumental logic of modernity which turns everything into a means creates a void of ends, a void of meaning. It is in this consequence that Arendt concludes that *homo faber* can do all sorts of great things – but one thing he cannot do is assert himself, ground himself:

> Nothing perhaps indicates clearer the ultimate failure of *homo faber* to assert himself than the rapidity with which the principle of utility, the very quintessence of his world view, was found wanting and was superseded by the principle of 'the greatest happiness of the greatest number' (Jeremy Bentham).
>
> *(307f.)*

Extrapolating from this line of argument, one can say that there is a need to resist this instrumental logic of modernity, which turns everything – and often enough also every being – into a mean but has lost the sense for ends. Everything is measured, quantified and maximized – but for what purpose? This situation is tellingly summarized by a dictum of famous *Pogo* cartoonist Walt Kelly: "Having lost sight of our objectives we redoubled our effort." We need to resist the pervasiveness of economic thinking and regain a sense for that which is an end in itself, which does not need any more justification.

Fromm also criticizes this modern "sprit of instrumentality". He see the greatest problem in the corresponding alienation of the individual from himself

> Not only the economic, but also the personal relations between men have this character of alienation; instead of relations between human beings, they assume the character of relations between things. But perhaps the most important and the most devastating instance of this spirit of instrumentality and alienation is the individual's relationship to his own self. Man does not only sell commodities, he sells himself and feels himself to be a commodity.
>
> *(Fromm 1942, 103)*

The Trappist monk Thomas Merton points to the mismatch between personal alienation and technological achievements:

> What can we gain by sailing to the moon if we are not able to cross the abyss that separates us from ourselves? This is the most important of all voyages of discovery, and without it all the rest are not only useless but disastrous.
>
> *(Merton 1970, 11)*

The principle suggested here invites us to contemplate the essential, the truly important dimensions of life, reflect on that "abyss that separates us from ourselves" and how it could be bridged. It invites us to practice those activities which do not require any justification because they are ends in themselves. As such, this "principle", practicing praxis and contemplation, is not a mere means towards a more sustainable society but – in line with this school of thought – an end in itself, it is life at its best, it is fulfilment, it is happiness (*eudaimonia*).

At the same time, however, practicing contemplation and praxis has multiple beneficial effects for sustainability:

- It regains neglected aspects of human existence which important traditions consider as its true fulfilment.
- It creates a sense of mindfulness and awareness for the other, for the one in need.
- It thereby strengthens mutual understanding social cohesion and tolerance.
- It builds the soil on which the seed of the call for sufficiency can sprout.

In another text, Merton calls for contemplation in his own American society:

> He who attempts to act and do things for others or for the world without deepening his own self-understanding, freedom, integrity and capacity to love, will not have anything to give others. He will communicate to them nothing but the contagion of his own obsessions, his aggressiveness, his ego-centered ambitions, his delusions about ends and means, his doctrinaire prejudices and ideas. There is nothing more tragic in the modern world than the misuse of power and action to which men are driven by their own Faustian misunderstanding and misapprehensions. We have more power at our disposal today than we have ever had, and yet we are more alienated and estranged from the inner ground of meaning and of love than we have ever been. The result of this is evident. We are living through the greatest crisis in the history of man; and this crisis is centered precisely in the country that has made a fetish out of action and has lost (or perhaps never had) the sense of contemplation. Far from being irrelevant, prayer, meditation, and contemplation are of utmost importance in America today.
>
> *(Merton 1971, 164).*[4]

14.3 Be not too certain – and apply policies cautiously

Nobody can predict the outcome of an action or a measure in the long run; our knowledge about the means by which we intend to reach sustainability is always limited. There is ample evidence of tragic and obverse effects of best-intended measures. We should therefore not be too certain about the measures we apply and apply them cautiously.

When I graduated from school, I thought I basically understood the world. When I received my first master's degree (in physics), I no longer thought so. Today, three decades and several academic degrees later, I am much more aware of my ignorance and (hopefully) much more cautious in my statements. This is probably a normal development as one gets older.

It is one of the irritating and frustrating lessons in sustainability that our knowledge is always preliminary and that convictions which were communicated to the public as the scientific solution later appear much less certain, sometimes even wrong. As a society, we need to act under uncertainty because inaction would also have (potentially even worse) consequences. The case of the cane toad in Australia described earlier demonstrates how erroneous this can be – despite best intentions. We constantly need to make decisions "under great uncertainty. This makes the implementation of many seemingly wise and straightforward concepts of sustainability difficult and impossible" (Brown 2008, 143).

Moreover, since sustainability is such a significant and noble goal, people are tempted to implement respective measures with utmost forcefulness, as for instance, the abovementioned example of Allan Savory illustrates (see 1.5.4).

A more recent example of the devastating side effects of a well-intended policy measure is the subsidization of bioenergy. As discussed above, in the 2000s, the use of energy-crops for the production of bio-energy was seen much more positively than today (see 2.2). Even in 2009, when first concerns about this kind of energy production had been raised, a study of the International Energy Agency (IEA), IEA Bioenergy, the international collaboration on bio-energy within the IEA, was almost enthusiastically praising the potential of bioenergy:

> Bioenergy is already making a substantial contribution to meeting global energy demand. This contribution can be expanded very significantly in the future, providing greenhouse gas savings and other environmental benefits … Bioenergy could sustainably contribute between a quarter and a third of global primary energy supply in 2050.
>
> *(IEA 2009)*

To be sure, the study does mention that "[e]xpansion of intensive farming may have an impact on biodiversity through the release of nutrients and chemicals which can lead to changes in species composition in the surrounding ecosystems" (IEA 2009, 24). However, only one decade later, the overall account of this type of energy production is much more critical. Today, we now know about the great threat to biodiversity which the intensive monocultures for energy-crops pose. The cropland expansion for bioenergy can even offset positive effects for climate change mitigation for global vertebrate diversity (Hof et al. 2018). This example therefore illustrates that sometimes the side effects are not (sufficiently) anticipated.

Finally, rebound effects can frustrate the intended effect. Germany was one of the first countries to introduce a circular economy regulation (1991), which established a recycling system for packaging materials. Since Germans have, on average, a high environmental awareness and exhibit a relatively high compliance rate towards regulation, the required separation of garbage into its different elements has become a matter of course for a large proportion of the population. However, almost three decades later, Germany has now become one of the world's top consumers of packaging material; in the European Union, it is number one for packaging waste per capita, presumably because people do not care too much about buying packaged stuff because they think it will be recycled anyhow – and in fact the absolute quantity of recycling per capita is also the highest in Europe (EU 2018, 159). But unfortunately, only a minor fraction of it is actually being materially recycled; the rest is incinerated or exported (Bethge, et al. 2019).

What does this all imply? First, any ecosystem intervention needs to be executed gradually and with utmost caution: the larger the scale and the heavier the impact, the more caution is needed. Second, Karl Popper's appeal to constantly try the falsification of one's own best-loved theories becomes more important the more practical implications scientific accounts have (Popper 1971). Apart from the devastating ecological effect of the examples above, it is even more worrying that such cases undermine the credibility of science and politics in general. If people get the impression that the scientific understanding has fundamentally changed on a certain topic, they might cease to trust the scientific account at all. In the complexity of today's world, with the media looking out for David-against-Goliath stories, a few insights from a study out of context are sufficient to irritate the public. In my view, it is critical (and actually the de facto practice in most cases already) that scientists do not pretend more than they can argue with good reasons, that they communicate openly and admit to mistakes frankly if they do occur.

14.4 Celebrate frugality

Sufficiency calls for reduced consumption, which is not easy to sell in a consumerist society. But frugality and simplicity can actually help us concentrate on and rediscover truly human characteristics. They can help experience the reduction of consumption not as a limitation but as liberation.

Photographers use black-and-white techniques to concentrate on structure, to emphasize contrast, to focus the senses. Concentration, focus, reduction function as stylistic devices not only in photography but also in painting, sculpture, music, literature, arts. Important domains of life can only thrive because people exercise voluntary discomfort. People push themselves hard for physical fitness in workouts and sports. They test and expand their limits, they reach out for the extremes and take on all kinds of adversity to achieve their aims.

Religious traditions have long known the concept of fasting. While it is mostly related to food, abstaining from something can be a good exercise because it trains the will and enhances self-restraint. It can even have liberating power because one can realize the reward which lies in not needing to follow each and every impulse to follow instead what one *actually, truly* wants.

When are you really free? Many would reply: when you do what you want. But who or what determines what you want? Most often our senses, our desires, wishes. Everybody knows that our desires and wishes sometimes tempt us into doing things which we *actually* do not really want. One of my great philosophical teachers, Jörg Splett, always taught us: "You are not free if you do what you want – but if you want what you ought to do!" Of course, the question is then how to determine the "ought to do". For Kant, this would be to freely subordinate yourself to moral law.

The philosopher Harry Frankfurt deliberates on free will and distinguishes between desires and volitions and between first-order and higher-order volitions. The first-order desire often follows the senses and aims at the fulfilment of immediate needs and wants. The second-order volition corresponds to that which a person *truly* wants (Frankfurt 1988, 48). The will is free if first-order desire and second-order volition concur – regardless of the actual choices a person may have:

> A person may act freely when he is not free to act differently. From the fact that X did A freely, in other words, it does not follow that X was free to refrain from doing A. Thus a person may be free to do what it happens that he wants to do, and do it freely, without enjoying freedom in the sense of being in a position to do whatever he might both want and be inherently able to do.
>
> *(Frankfurt 1988, 56)*

With Frankfurt, we could say the will is free if a person has the volition that she or he wants to have (Guckes 2001, 10).

When we fast, we are exercising our second-order volition – to refrain from the impulse to follow our immediate desires. This is a truly uplifting experience, because it is this second-order volition which characterizes us as humans. It is something which does not require any other reason, it is a good thing in and of itself. It is here, in the congruence of the true will and the wish which is realized in concrete actions, where human life is taking place. It is here where a life of abundance becomes conceivable even in light of limited options.

This perspective is light years away from the position of resistance against the call for sufficiency on the grounds that it would confine personal freedom. On the contrary. The call for a frugal lifestyle is an invitation to experience and exercise true human capacities.

Celebrating reduction and frugality can lead the way to an experience of liberation, to a state in which one is "on good terms with oneself", as Nico Paech argued above (see 13.2). It is this "being on good terms with oneself", which describes the second-order volition. We have therefore best reason to celebrate simplicity and frugality.

These ideas would need to qualify the concept of sufficiency and its dissemination: not in the patronizing and moralistic style which tells people what to do – and even worse, condemns everybody who does not follow; but instead one which attracts people to a goal which far transcends immediate satisfaction.

This then matches well the insights of some of the greatest spiritual leaders. Lao Tzu quoted an even older saying:

> Hold fast to that which will endure,
> show thyself simple, preserve thee pure
> And lessen self with desires fewer.
>
> *(Lao Tzu 1913, 19)*

Notes

1 We leave out the special case of a purely hedonistic principle which would always seek the maximization of (one's own) pleasure. The understanding of "principle" applied here leans towards Steven R. Covey's concept, which he explicates in his

Seven Habits of Highly Effective People: the thinking he proposes is "a principle-centered, character-based, 'Inside-Out' approach to personal and interpersonal effectiveness. 'Inside-Out' means to start first with self; even more fundamentally, to start with the most inside part of self – with your paradigms, your character, and your motives" (Covey 1989).

2 Max Weber might be seen as a pathfinder in this respect because of his distinction between an "ethic of ultimate ends" (*Gesinnungsethik*) and an "ethic of responsibility" (*Verantwortungsethik*). Weber argues that only the "ethic of responsibility" would be appropriate for politicians, since

> [n]o ethics in the world can dodge the fact that in numerous instances the attainment of 'good' ends is bound to the fact that one must be willing to pay the price of using morally dubious means or at least dangerous ones – and facing the possibility or even the probability of evil ramifications.
>
> *(M. Weber, Politics as a Vocation 1958, 121); see Norkus 2000)*

3 Alfred North Whitehead even said the "safest general characterization of the European philosophical tradition" would be "that it consists of a series of footnotes to Plato" (Whitehead 1979, 39).

4 I am indebted to Dennis Frank from St. Bonaventure University for providing a copy of the English version of this quote.

Bibliography

Arendt, Hannah. *The Human Condition*. Chicago, IL; London: University of Chicago Press, 1998.

Aristotle. *Nicomachean Ethics*. Herausgeber: Web Atomics The Internet Classics Archive by Daniel C. Stevenson, Übersetzung: W. D. Ross. 1999.http://classics.mit.edu//Aristotle/nicomachaen.html

Bethge, Philipp, Annette Bruhns, Nils Klawitter, & Simone Salden. "Die Müll-Lüge." *Der Spiegel 19*, 2019: 01.

Brown, Casey. "Emergent sustainability: The concept of sustainable development in a complex world." In *Globalization and Environmental Challenges*, Hans Günter Brauch, Navnita Chadha Behera, Béchir Chourou, Pál Dunay John Grin et al. eds., 141–149. Berlin, Heidelberg, New York: Springer, 2008.

Covey, Stephen R. *The Seven Habbits of Highly Effecitve People*. New York: Free Press, 1989.

EU. *Energy, transport and environment indicators, 2018 edition*. Brussels: European Commission, 2018.

Frankfurt, Harry G. "Three concepts of free action." In *The Importance of What We Care About*, Harry G. Frankfurt ed., 47–57. Cambridge, New York, New Rochelle, NY, Melbourne, Sydney: Cambridge University Press, 1988.

Fromm, Erich. *The Fear of Freedom*. London: Kegan Paul, 1942.

Guckes, Barbara. ""Willensfreiheit trotz Ermangelung einer alternative? Harry G. Frankfurts hierarchisches Modell des Wünschens.". In *Freiheit und Selbstbestimmung*, Harry G. Frankfurt. (ed.), 1–17. Berlin: Akademie Verlag, 2001.

Hof, Christian, Alke Voskamp, Matthias F. Biber, Katrin Böhning-Gaese & Eva Katharina Engelhardt et al. "Bioenergy cropland expansion may offset positive effects of climate change mitigation for global vertebrate diversity." *PNAS (Proceedings of the National Academy of Science)*, 26. 12 2018: 13295–13299.

IEA. *Bioenergy – a Sustainable and Reliable Energy Source*. Rotorua, New Zealand: IEA Bio-energy, 2009.

Kelly, Walt. www.azquotes.com/. kein Datum. www.azquotes.com/quote/1269192 (Zugriff am 13. 06 2019).

Lao-Tzu. „Tao Teh Ching. "*The Canon of Reason and Virtue*. Translated by D.T. Suzuki & Paul Carus. 1913.

McKinnon, Andrew M. "Ideology and the market metaphor in rational choice theory of religion: A rhetorical critique of 'religious economies." *Critical Sociology* 39 (4), 2011: 529–543: *39*, (4).

Merton, Thomas. *Wisdom of the Desert*. New York: New Directions Publishing, 1970.

Merton, Thomas. *Contemplation in a World of Action*. Garden City, New York: Doubleday & Company, 1971.

———. *Wisdom of the Desert*. New York: New Directions Publishing, 1970 (1960).

Norkus, Zeonas. ""Max Weber's interpretative sociology and rational choice approach." *Rationality and Society* 12(3), 2000: 259–282: *12*, (3).

Plato. *Republic*. The Internet Classics Archive by Daniel C. Stevenson, Web Atomics, 1994–2000.http://classics.mit.edu//Plato/republic.html

Popper, Karl R. *Logik der Forschung*. Tübingen: J.C.B, Mohr (Paul Siebeck). 1971.

Weber, Max. "Politics as a Vocation." In *Essays in Sociology*, (translated, edited and with an introduction by H. H. Gerth, C. Wright Mills & Max Weber), 77–128. New York: Oxford University Press, 1958.

Whitehead, Alfred. *North. Process and Reality*. New York: The Free Press, 1979.

15

SOCIETY-RELATED PRINCIPLES

Contents

A third group of principles is society-related. Since the early beginnings of religious and philosophical thinking, people have sought principles, virtues and attributes that coordinate human living-together. The insights gained in these long and diverse streams of tradition have shaped modern history and they remain the foundation of any quest for sustainability.

The classic "cardinal virtues" prudence (phronesis, prudentia), courage (andreia), temperance (sophrosyne/temperantia), justice (dikaiosyne/iustitia), the Christian theological virtues of faith, hope and charity, or the goals of the French revolution liberty, equality, fraternity (Liberté, Égalité, Fraternité) were the sources of the modern constitutional state and the foundation of the Universal Declaration of Human Rights. The latter are based on "the recognition of the inherent dignity and of the equal and inalienable rights of all members of the human family" as "the foundation of freedom, justice and peace in the world" (UN 1948). The 2030 Agenda is dedicated to its own goal of peace, justice and strong institutions (SDG 16).

In addition, we will suggest a few principles which partly overlap in their goals with the above. Living together in harmony with fellow human beings near and far, today and tomorrow, and in harmony with nature – this is what sustainability is all about. The following principles are meant as suggestions to foster the societal, communal values needed for an overall sustainable development. As self-evident as they might seem, I briefly mention a few of them here. It would be a great step forward if they became common practice among all agents, starting from small-scale personal situations up to the very large geopolitical contexts.

15.1 Grant the least privileged the greatest support

Social cohesion is at stake if a society does not care about the least privileged. The former German Federal President Gustav Heinemann is one of the many people who have reportedly said: "One recognizes the value of a society in how it deals with its weakest members."[1] The challenge is, however, that there are not only significant differences in the intra-national distribution of wealth and income but even greater ones in an international perspective. The consumerist resource-heavy lifestyles of the global North raise important ethical questions because they can certainly not be universalized. Rawls' difference principle, according to which "economic and social inequalities are to be judged in terms of the long-run expectations of the least advantaged social group" (Rawls 1999, 39), can help to argue for a reconsideration of a fair global distribution of opportunities, goods, and wealth.

As discussed above, Wilkinson and Pickett have shown that too much inequality weakens community life (Wilkinson & Pickett 2011, 230). By correlating income inequality and wellbeing indicators, they state that inequality has "pernicious effects on societies", it would erode trust, increase anxiety and illness, and encourage excessive consumption.

Donella Meadows discusses the winner-takes-all principle, which occurs if the "winners of a competition are systematically rewarded with the means to win again, a reinforcing feedback loop is created by which, if it is allowed to proceed uninhibited, the winners eventually take all, while the losers are eliminated" (Meadows 2008, 130). This is pretty much our situation, both regarding individual level of wealth but also within countries. The way out which Meadows proposes is:

> Diversification, which allows those who are losing the competition to get out of that game and start another one; strict limitation on the fraction of the pie any one winner may win (antitrust laws); policies that level the playing field, removing some of the advantages of the strongest players or increasing the advantage of the weakest; policies that devise rewards for success that do not bias the next round of competition.
>
> *(ibid., 130)*

15.2 Seek mutual understanding, trust and multiple wins

Situations of conflicting interests require negotiations among the respective parties. Negotiations and dialogue, however, require mutual understanding and trust. It is this rather basic human capability which is often neglected but which can bring forth great results. Künkel sees it even at heart of a new form of "collective leadership": "At the core of collective leadership is the human capacity to dialogue and transform differences into progress. It enables the transcendence of self-centered views, a prerequisite for successfully addressing the challenges of sustainability" (Künkel 2019, 19).

Sincere open dialogue does not only help to cognitively understand the other position, it will also help build up trust. As trust is a "major requirement of well-functioning social systems", it is a "current crucial challenge for sustainability". Trust

> is vital for sustainability since it is ultimately the only basis by which humans can interact to ensure human basic needs can be met. The corollary is that trust provides the basis for cooperative behaviours that can be maintained in the long run.
>
> *(Krabbe 2015, 69)*

Dealing with community currency systems, Krabbe has shown that community exchange, in particular insofar as it involves face-to-face relationships as a basis for interpersonal trust, provides a basis for generalized trust, and "is proposed as having significant potential to increase trust and hence sustainability" (Krabbe 2015, 69).

Mutual understanding and trust will thus not only facilitate understanding on a cognitive level – by building up trust, this will also lead to understanding of where the other is coming from, it will help to understand the other's needs and demands. This, in turn, will then also allow us to see things from another perspective, it will help to "walk in the moccasins of the other", to find ways to meet the other's demands and needs. Knowing the needs of the other enables *and demands* the seeking of mutual benefit, of win-win situations lest the newly built-up trust should vanish.[2]

Mutual understanding is an essential ingredient of achieving common goals, from basic personal human interaction in personal relationships to inter- and transdisciplinary research and the highest political and diplomatic contacts. Support for the strengthening of dialogue capabilities comes from quite diverse schools of thought.

In scientific discourse, dialogue is, of course, a vital necessity. Mauser et al. strive for a new research model in global sustainability. They call for a

> sectoral integration of knowledge ... between actors from the state, knowledge institutions, market and civil society sectors so as to achieve a mutual understanding of the kinds of research questions that need to be addressed and the ways of doing so.
>
> *(Mauser et al. 2013, 426)*

This would enhance mutual understanding and mutual responsibility (ibid., 427). Spangenberg's above mentioned "basic law of interdisciplinarity", which demands to respect well-corroborated insights of other disciplines, does certainly require dialogue.

Similar messages come from quite a different angle. Marshall B. Rosenberg, a trained psychologist, developed the concept of non-violent communication, which he used in peace programs in conflict zones to deal with ethnic, cultural, or political conflicts in many regions around the globe. The idea that all humans share the same basic needs is fundamental for understanding his approach. By differentiating observation from evaluation and by discussing universal human needs without implying judgement, criticism, or punishment, non-violent communication creates the basis for mutual understanding and dialogue (CNVC 2019).

Yet another source, also a practitioner, is the management consultant Steven R. Covey. Covey argues that seeking win-win, seeking mutual benefit between negotiation partners, is one of the habits which make people effective and successful (Covey 1989).

Finally, according to the Report of the Kingdom of Bhutan on Happiness, it is mindfulness, as the "cultivation of non-judgmental, non-reactive, metacognitive awareness of present-moment experience", which is an important "happiness skill". Other such skills are "loving-kindness, compassion-meditation and the conscious practice of gratitude, empathy, and patience" (NDP Steering Committee and Secretariat 2013, 35; see Göpel 2016, 137). Businesses have meanwhile discovered the importance of mindfulness for cooperation and teamwork, which is remarkable especially in very competitive environments. Several large IT companies have introduced mindfulness programs, in which employees are taught how to reflect on their own life, and their own and other people's needs (Gelles 2016).

15.3 Strengthen social cohesion and collaboration

Working towards a common goal requires some understanding of what "common" means. The very idea of politics depends upon some kind of understanding of *polis*, of community, some sense of commonwealth. Aristotle even defined the human individual as a political being (*zoon politikón*). Any political discourse, surely the one on sustainability, therefore requires the safeguarding and protection of social cohesion. Social cohesion is not only a precondition for political action – it is also related to a number of other appreciated social phenomena. The Bertelsmann Foundation investigated social cohesion in Germany and found, for instance, that it is positively correlated to happiness. The researchers found that regions "with a strong sense of community are home to people who are happier and more satisfied" (Bertelsmann Foundation 2017, 22). Other interesting results are:

a) Social cohesion is weaker in areas with high unemployment, which is particularly true for high youth unemployment.
b) A strong focus on achievement in society correlates negatively with social cohesion.
c) Being open to new developments correlates positively with social cohesion.
d) Humanistic values, such as honesty, and a willingness to assume responsibility, correlate positively with higher social cohesion.
e) The number of foreigners or migrants living in a region or federal state does not influence social cohesion.

f) Social cohesion is threatened by strong inequalities; practical solutions for strengthening the degree of cohesion should involve measures which reduce social inequality and prevent poverty. (21f.)

This matches well with the results of Wilkinson and Pickett, who found "that the quality of social relations deteriorates in less equal societies" (Wilkinson & Pickett 2011, 51). The authors quote the political scientist Robert Putnam, who stated that community and equality are mutually enforcing (54).

The call for social cohesion also comes from anti-utilitarians who criticize the prevailing postulate of utility in economic theory. In contrast to utility, they emphasize the critical role of the social, of social cohesion, social ties compared to self-interest (Romano 2015). Strengthening social bonds and social cohesion would therefore not only contribute to mutual understanding and win–win situations, it would also coincide with the de-growth critique of the utilitarian welfare maximization of economic theory.

Dittmer argues that local currencies have the potential to build up and strengthen local communities, which points to the great potential of social cohesion in local contexts (Dittmer 2015). There are promising initiatives especially on local levels which are mutually reinforcing with social cohesion:

a) Free-riding phenomena are less severe in smaller communities with communication among members (see 5.1).
b) Local value creation contributes to resource preservation and lesser degrees of specialization.
c) Local community currency systems facilitate a level of trust.

The latest PISA study has shed some interesting light on collaboration – with regards to both gender and income differences. The 2015 PISA study was the first one to investigate collaborative skills, it was "the world's first international assessment of collaborative problem-solving skills, defined as the capacity of students to solve problems by pooling their knowledge, skills and efforts with others" (OECD 2017, 5).[3] Referring to their own studies with school children as well as to the respective literature, the authors assert "that it is students of lower socio-economic status who more commonly exhibit behaviour consistent with co-operation and consideration of others" (113)

The belief that a higher level of education would also increase the sense for otherness and social values is proven wrong: the authors refer to a study in the USA, which detected that

> university students who were the first in their family to attend university were more likely to be other-focused (as opposed to self-oriented) than university students whose parents had also attended university ... Intriguingly, brain scans show that those of higher socio-economic status actually display reduced neural responses of empathy.
>
> *(OECD 2017, 113)*

On top of this, students from higher socio-economic status not only show lower empathy compared to students of lower status, they also consistently *thought* they would be *more compassionate* than their poorer peers: "Interestingly", as Varnum et al. show, higher socio-economic status "was positively correlated with self-reported trait empathy, suggesting that those higher in status may not realize that they are actually lower in empathy" (Varnum et al. 2015, 122). The PISA study states: "It appears that those of higher socio-economic status might overstate the degree to which they display certain positive attributes, with the same outcome as if they displayed higher levels of social desirability" (OECD 2017, 113).

However, for many readers it will get even more awkward: It is not only the rich who underperform in social values, it is men, too. The study is pretty clear in its account: "Girls perform significantly better than boys in collaborative problem solving in every country and economy that participated in the assessment" (OECD 2017, 90). The recommendation offered by authors is that "education systems should look into fostering boys' appreci-ation of others" (166).

As a side-note, this fact alone calls for a much greater role for women in executive positions, since this better performance of girls in collaborative prob-lem solving presumably does not stem from any privileged position they would have in their educational system or society. In most countries, women are significantly more empathetic and more conscientious than men (OECD 2017, 93).

Finally, Western countries with their competitive environments and strong individualistic values are apparently less successful in teaching student collabor-ation. While the share of students in the OECD who can *at best* manage very simple tasks – i.e. "straightforward collaborative problems, if any at all" – is 28%, this percentage is significantly lower at not even 16% in Estonia, Hong Kong (China), Japan, Korea, Macao (China) and Singapore – apart from Estonia all countries from (South) East Asia (OECD 2017, 17).

15.4 Engage the stakeholders

Diversity is an important feature of complex systems that maintains stability and innovation at the same time, as will be discussed below (see 16.2). This principle also holds for social systems. Diversity can contribute to both innovation and stability: "Diversity is key because systems depend on it to innovate" (Stroh 2015, 79). Talking about multi-stakeholder[4] collaboration, Künkel points to diversity as a crucial requirement for the resilience of systems. According to her, systems even become more sustainable over time the greater their diversity:

> Many authors (Berry, 1999; Elgin, 2001; Capra, 2003) have argued that in order for that which emerges in dialogue and deliberation to be considered collectively meaningful, diversity must be seen as an asset and endeavours

must belong to the collective. The importance of dialogue in quality communication has long since been adopted in the corporate world.

(Künkel 2019, 35)

Sala, Ciuffo and Nijkamp (2015) list broad participation as one of the principles in their systematic framework for sustainability assessments because this could "strengthen legitimacy and relevance, engaging early on with users of the assessment, reflecting the views of the public while providing active leadership" (Sala, Ciuffo & Nijkamp 2015).

Stakeholder participation has, in fact, long been practiced by public agencies; it has become "an increasingly accepted component of natural resources and environmental planning processes" in many parts of the world (NRC 2004, 73). Integrating stakeholders in the planning process of public projects has several advantages if done sincerely and effectively (and not just for obtaining public blessing for what had long been decided) because it broadens the horizon of the risk radar. Ordinary people might have a more realistic and down-to-earth view on risks than experts; they can indicate (if representatively chosen) their appetite for levels of acceptable risk, something which experts cannot determine *ex ante*; therefore, better decisions are possible; local knowledge is integrated, conflicts between stakeholders can be reduced and new dialogue among them facilitated, etc. (see, e.g. NOAA 2015; NRC 2004).

An African Development Bank Handbook emphasizes that participation is a *mindset* or *attitude*. A mindset of participation would mean:

a) *Focusing on people* – recognizing that people are at the center of development;
b) *Being humble* – realizing that local knowledge is as valid as "expert" knowledge;
c) *Learning to listen* – accepting that stakeholders have wisdom and a right to be heard;
d) *Sharing control* – sharing influence and control with project stakeholders (This can be frightening for development experts that are accustomed to "being in control");
e) *Empowering others* – focusing on building the capacity of marginalized stakeholders to find their own solutions to development problems, enabling beneficiaries to become active owners rather than passive recipients of development and;
f) *Valuing process* – understanding development as a "process", not just a "product". (ADB 2001, Chapter 2, p. 3; original emphases)

This also points to the fact that diversity is not appreciated by everybody and the call for it will cause resistance, for those in power might not be interested in change. In addition, there is also a risk contained because by nature one cannot predict the kind of novelty that emerges. You cannot control the outcome of an innovation process.

15.5 Foster education – share knowledge and collaborate

"Foster education" is actually a measure rather than a principle, but is of such fundamental and general importance for the future development of humankind that I nevertheless include it here. Understanding, addressing and overcoming the barriers to sustainability as much as the principles for sustainable action require sound education. This pivotal role of education for sustainable development (ESD) was already recognized in the Agenda 21, it led to the United Nations Decade of Education for Sustainable Development (2005–14); the 2030 Agenda dedicated SDG 4 to "quality education" and target 4.7 particularly to ESD,[5] and the Incheon Declaration and Implementation plan (adopted in 2015) also recognizes the important role of education as a main driver of development (UNESCO 2015).

Despite all these general agreements and initiatives, progress is challenging, and partly non-existent: almost 69 million additional teachers were needed to reach the 2030 Education Goals (UNESCO Institute for Statistics 2016). In 2018, three years after the adoption of SDG 4 and the promise to provide universal primary and secondary education, UNESCO admits that "there has been no progress in reducing the global number of out-of-school children, adolescents and youth" (UNESCO Institute for Statistics 2018).

It was possible to reduce gender differences in out-of-school rates, but strong regional differences remain: "The world is moving towards gender parity in out-of-school rates, although inequalities persist at regional and country levels." Africa, in particular sub-Saharan countries, is still worst off, with out-of-school rates of approximately 20% at primary school age and more than 50% out-of-school youth of upper secondary school age (ibid.). In many LICs, the educational situation is worsened by displacement. LICs host "10% of the global population but 20% of the global refugee population, often in their most educationally deprived areas" (UNESCO 2019, xvii). On the other hand, well-educated people are more likely to migrate – triggering a "brain drain" which challenges LICs because they constantly lose a large share of their best-educated people: "Domestically, those with tertiary education are twice as likely to migrate as those with primary education; internationally, they are five times as likely" (ibid.).

The strong regional differences, in particular the challenging situation in sub-Saharan Africa endangers global efforts for sustainability for a number of reasons – apart from the fact that education is already a dedicated sustainable development goal;

- Sub-Saharan Africa is likely to see significant population growth in the coming decades, actually the highest of all global regions. In the next thirty years, the population of sub-Saharan Africa alone will increase from approximately 1 billion today to 2–2.5 billion, and between 2.7–5.5 billion by 2100 (UN, World Population Prospects 2017). Since many of the respective countries rank low on human development, such a population growth will cause additional strain on resources and pollution – and also affect economic development.

- Education is key for reducing fertility rates – both directly and indirectly. Directly, because there is a positive correlation between the mean years at school and the age of first marriage (Gapminder 2019), and indirectly, because better education facilitates higher income, and higher income is related to lower fertility rates. Education has thus a substantial return – but this return is unfortunately measured in decades or generations, not in quarters or legislation periods. A 2018 OECD Report on education indicates the internal rate of return on tertiary education is approximately 15% as an average of all OECD countries (OECD 2018, 104). Better-educated adults "pay higher income taxes and social contributions and require fewer social transfers" (OECD 2018, 104), leading to a substantial net financial return for the public.
- Public spending on education is already relatively high in countries like Kenya, Niger, or Ghana, relatively in the same range as OECD countries but due to lower GDP correspondingly lower in absolute terms (World Bank 2019 (Data from 2015)).

A substantial improvement of the educational situation in Sub-Saharan Africa should therefore be seen as a common mandate for humanity. Improving the educational situation will help improve the economic situation and by way of demographic transition (i.e. reduction of fertility rate) also diminish population growth and thereby lessen the stress level on the environment mid-term.

Notes

1 Quoted by the Federal President Frank-Walter Steinmeier (Bundespräsidialamt 2017).
2 Wilkinson and Pickett point to the relation of trust and inequality, "people who trust others are more likely to donate time and money to helping other people. 'Trusters' also tend to believe in a common culture, that America is held together by shared values, that everybody should be treated with respect and tolerance. They are also supportive of the legal order" (Wilkinson & Pickett 2011, 56).
3 A more comprehensive definition reads that this was defined as "the capacity of an individual to effectively engage in a process whereby two or more agents attempt to solve a problem by sharing the understanding and effort required to come to a solution and pooling their knowledge, skills and efforts to reach that solution" (OECD, PISA 2015 Results (Volume V): Collaborative Problem Solving, PISA, (OECD 2017, 47).
4 Stakeholders are "people and organizations that affect and are affected by the issue" (Stroh 2015, 79).
5 SDG 4.7 calls to "ensure that all learners acquire the knowledge and skills needed to promote sustainable development, including, among others, through education for sustainable development and sustainable lifestyles, human rights, gender equality, promotion of a culture of peace and non-violence, global citizenship and appreciation of cultural diversity and of culture's contribution to sustainable development" (UN 2015, §4.7).

Bibliography

ADB. *Handbook on Stakeholder Consultation and Participation in ADB Operations*. African Development Bank, 2001. https://www.afdb.org/fileadmin/uploads/afdb/Docu ments/Policy-Documents/Handbook%20on%20Stakeholder%20Consultaion.pdf (accessed 13 September 2019).

Bertelsmann Foundation. *Sozialer Zusammenhalt in Deutschland 2017*. Gütersloh: Bertelsmann Stiftung, 2017.

Bundespräsidialamt. *150-jähriges Jubiläum Bethel*. 17. 04 2017. www.bundespraesident.de/SharedDocs/Reden/DE/Frank-Walter-Steinmeier/Reden/2017/04/170417-Bethel.html (accessed 13. 06 2019).

CNVC. *Cnvc.org*. 2019. www.cnvc.org/ (accessed 13. 06 2019).

Covey, Stephen R. *The Seven Habbits of Highly Effective People*. New York: Free Press, 1989.

Dittmer, Kristofer. "Community currencies." In *Degrowth. A Vocabulary for A New Era*, Giacomo D'Alisa, Federico Demaria & Giorgos Kallis (eds.), 149–151 Abingdon, Oxon and New York: Routledge, 2015.

FAZ. "Bombendrohung gegen CDU-Politiker Voss." 15. 03 2019. www.faz.net/aktuell/feuille ton/bombendrohung-gegen-cdu-politiker-voss-16092259.html (accessed 13. 06 2019).

Gapminder. *Gapminder.org*. Stockholm: Stockholm, 2019.

Gelles, David. *Mindful Work. How Meditation Is Changing Business from inside Out*. Boston, MA, New York: Eaman Dolan, 2016.

Göpel, Maja. *The Great Mindshift. How a New Economic Paradigm and Sustainability Transformations Go Hand in Hand*. Springer Nature, 2016.

Krabbe, Robbin. "Building trust: Exploring the role of community exchange and reputation." *IJCCR – International Journal of Community Currency Research 19 (Special Issue, Section D)*, 2015: 61–71.

Künkel, Petra. *Stewarding Sustainability Transformation. An Emerging Theory and Practice of SDG Implementation*. Springer Nature, 2019.

Mauser, Wolfram, Gernot Klepper, Martin Rice, Bettina Susanne Schmalzbauer, Heide Hackmann et al. "Transdisciplinary global change research: the co-creation of knowledge for sustainability," In *Current Opinion in Environmental Sustainability 5*, 2013: 420–431.

Meadows, Donella H. *Thinking in Systems. A Primer, Ed. By Diana Wright*. White River Junction, VT: Chelsea Green Publishing, 2008.

Müller, Jan-Werner. *Was Ist Populismus?* Berlin: Suhrkamp, 2016.

NDP Steering Committee and Secretariat. *Happiness: Towards a New Development Paradigm. Report of the Kingdom of Bhutan*. Kingdom of Bhutan: Report of the Kingdom of Bhutan, 2013.

NOAA. *Introduction to Stakeholder Participation*. Washington, DC: Office for Coastal Management, 2015.

NRC. *Analytical Methods and Approaches for Water Resources Project Planning*. Washington, DC: National Research Council: The National Academies Press, 2004.

OECD. *PISA 2015 Results (Volume V): Collaborative Problem Solving, PISA*. Paris: OECD Publishing, 2017.

———. *Education at a Glance 2018: OECD Indicators*. Paris: OECD Publishing, 2018.

Pörksen, Bernhard. *Die Große Gereiztheit. Wege Aus Der Kollektiven Erregung*. Munich: Carl Hanser, 2018.

Rawls, John. *A Theory of Justice*. (Revised Edition), Cambridge, MA: Belknap Press of Harvard University Press, 1999.

Romano, Onofrio. "Antiutilitarianism." In *Degrowth. A Vocabulary for A New Era*, von Giacomo D'Alisa, Federico Demaria & Giorgos Kallis (eds.), 21–24. Abingdon, Oxon, New York: Routledge, 2015.

Sala, Serenella, Biagio Ciuffo & Peter Nijkamp. "A systemic framework for sustainability assessment." *Ecological Economics 119*, 2015: 314–325.

Spangenberg, Joachim. "Sustainability science: A review, an analysis and some empirical lessons." In *Environmental Conservation 38 (3)*: 275–287.

Stroh, David Peter. *Systems Thinking for Social Change*. White River Junction, VT: Chelsea Green Publishing, 2015.

UN. *Universal declaration of human rights*. 1948. www.un.org/en/universal-declaration-human-rights/index.html (Accessed 28. 03 2019).

———. *Transforming Our World: the 2030 Agenda for Sustainable Development (A/RES/70/1)*. New York: United Nations, 2015.

———. *World Population Prospects*. New York: United Nations, 2017.

UNESCO. *Education 2030 Incheon Declaration and Framework for Action for the Implementation of Sustainable Development Goal 4*. Paris: UNESCO, 2015.

———. *Global Education Monitoring Report 2019: Migration, Displacement, and Education: Building Bridges, Not Walls*. Paris: UNESCO, 2019.

UNESCO Institute for Statistics. *UIS Factsheet No. 39*. Montreal: UNESCO, 2016.

———. *Fact Sheet No. 48 (UIS/FS/2018/ED/48)*. Montreal: UNESCO Institute for Statistics, 2018.

Varnum, Michael E.W., Chris Blais, Ryan S. Hampton & Gene A. Brewer. "Social class affects neural empathic responses." *Culture and Brain 3 (2)*, 2015: 122–130.

Wilkinson, Richard & Kate Pickett. *The Spirit Level. Why Greater Equality Makes Societies Stronger*. New York, Berlin, London, Sydney: Bloomsbury Press, 2011.

World Bank. https://data.worldbank.org/. Washington, DC, 2019 (Data from 2015).

16

SYSTEMS-RELATED PRINCIPLES

Contents

A final group of principles take a systems perspective. As several examples have shown, the complexity of natural, social, political, or economic systems has often been underestimated, human intervention has often led to unintended results, new trends have arisen and surprising developments have irritated the "experts". Nobody anticipated the fall of the Iron Curtain, hardly anybody anticipated the subprime crisis, and many people can still not believe how populists could have reached presidential office and attempt to subordinate the common good under their vested or selfish interests. It is therefore critical to understand that issues of sustainability involve complex systems which require very careful consideration and action. Thus, systems thinking needs to be applied. There are only a few aspects directly relating to sustainability which need to be considered here. Furthermore, since diversity is a precondition for the complexity of living systems, fostering diversity is suggested as a second systems-related principle. Finally, increasing transparency on the publicly relevant is, in my view, one of the most important action principles for sustainability because only by means of transparency can a fair negotiation of conflicting interests be achieved (see 16.4).

16.1 Apply systems thinking

Among the typical features of a complex system are the interdependencies of their parts, feedback loops, non-linear behaviour, emergence, self-organization and adaptability. It is important to understand the behaviour of systems because it is almost certain that a complex system will produce unintended effects if these features are not considered. Complex systems behave remarkably differently from what one could expect. It is essential to understand the levers of the system's dynamics, identify buffering as well as critical elements, and know how to increase their resilience to perturbation and capacity for recovery. This understanding becomes all the more important as the interaction with systems becomes more impactful – in both time and space. Out of the extensive material available, we can again only give a selective snapshot of aspects which seem to be particularly important these days.

Systems thinking has been characteristic for the work of the Club of Rome since its inception, and several of the reports to the Club of Rome take a strong systems view (Club of Rome 2019). An unprecedented milestone was the first report in 1972 to the Club of Rome, *The Limits to Growth*. This book describes results of a computer-generated world model in which the authors ran several different scenarios. They concluded that the system of human civilization might well collapse if the then-current trends continued: "The basic behavior mode of the world system is exponential growth of population and capital, followed by collapse" (Meadows, Meadows, Randers & Behrens 1972, 142).

Botkin et al. complemented the systems view of Meadows et al. 1972 with a *social* perspective in their book, *No Limits to Learning. Bridging the Human Gap* (Botkin, Elmandjra & Malitza 1979). The "human gap" which they see in need of closing "is the distance between growing complexity and our capacity to cope with it" (6). According to the authors

> global problems, currently the chief manifestations of complexity, are first and foremost human problems … It is a profound irony that we should be confronted with so many problems at the same time in history when humanity is at a peak of its knowledge and power.
>
> *(ibid., 7)*

Learning, they continue, would be the process of preparing to deal with new situations. This is exactly, I submit, what is needed for a proper dealing with complex systems.

The authors suggest "Features of Integrative Thinking", which resonate well with what we are discussing in this book:

- evaluation of the long-term future consequences of present decisions;
- consideration of second-order consequences (what are called either side effects or surprise effects);
- ability to make plans and strategies for the future, to monitor and modify plans (called "rolling planning"), and to conduct evaluations to detect early-warning signs of possible problems;
- skill in "systemic" thinking, which is the capacity to see the whole as well as its parts, and to see multiple rather than single causes and effects;
- capacity to detect interrelationships and to assess their importance, which is often greater than that of the elements they interlink. (Botkin, Elmandjra & Malitza 1979, 98)

Donella H. Meadows wrote a commendable book, *Thinking in Systems*, which was only posthumously published (Meadows 2008). While listing leverage points for influencing complex systems in the desired direction, she points to the importance of delays – the "lengths of time relative to the rates of system changes" (151). It can be critical if the feedback process of the system exhibits delay compared to the change in the stocks that the feedback loop is trying to control: "Overlong delays in a system with a threshold, a danger point, a range past which irreversible damage can occur, cause overshoot and collapse" (152). The concept of planetary boundaries mentioned earlier (see 1.5) indicated zones for a number of variables in which irreversible damage could happen. According to the current state of knowledge, we are already in the zone of higher risk for such irreversible developments. But can we be sure about the delay parameters for these indicators? In many instances, our anthropogenic changes of natural ecosystems are much more rapid than the system's response time. Ignoring this latency can easily bring us into the realm of irreversibility.

In one of the more recent Club of Rome reports, *The Seneca Effect. Why Growth is Slow but Collapse is Rapid*, the Italian chemist Ugo Bardi illustrates the irreversible overshoot characteristic of complex systems, which occur for instance in fishing grounds (Bardi 2017), as we have discussed above. Once the exploitation of a species has reached overshoot, it will not recover (see Figure 3.3).

It is this kind of systems thinking which needs to be fostered. Three foundational principles of systems thinking are in a way, core principles of the Club of Rome: think holistically, long term and global.

16.1.1 Think holistically

Complexity is an important barrier to sustainability (see 2.2). Our highly interdependent world has become so complex that anticipating the long-term effect of actions is difficult or even impossible. The advice to think holistically therefore sounds impracticable. People are already overwhelmed by the complexity of our world and the variety of options. Complexity needs rather to be reduced than expanded by considering additional system components. Consumers, for instance, who want to purchase consciously, can rely on labels which they consider to be reliable. This sufficiently reduces complexity to be practicable. However, the more impactful a decision is, the more severe its consequences are, the more important it is to consider at least first-order effects in adjacent fields.

The awareness of the importance of a more holistic understanding of issues has luckily increased in the last few decades and is manifested in quite different contexts.

16.1.2 Think long-term and decelerate

Short-termism was another barrier discussed in Part 1, and it often goes hand-in-hand with acceleration. Sustainability is about long-term thinking. Many sustainability issues would be resolved if we had full transparency about the implications of actions in the long run and in a global perspective, and acted accordingly. Keynes already knew, however, that long-term explanations cannot replace the need for action in the short term: "*In the long run* we are all dead" (Keynes 1923, 80; italics in original). This points to the problem of "positive time preference" discussed above (see 5.1). People prefer consumption today versus the same consumption tomorrow. But because this seems to be a principle inherent in any economic activity, we need institutional support to compensate for this tendency. We need (a) to adjust our incentive structures for long-term thinking, (b) we need to introduce fraction parameters for highly regenerative (socio-technical) systems, and (c) we need to foster and cultivate deceleration in our personal life.

a) *Adjust incentive structures for long-term thinking*

Long-term thinking, particularly including the demands of future generations, needs to be supported by institutional measures (e.g. incentive structures). There are multiple suggestions as to how this can be provided on a state level, ranging from the integration of a sustainability principle as a state goal in the constitution, to a panel of wise women and men as solicitors of the future, and longer legislation periods. In the business world, there are also indications that more long-term incentives might be effective, although under the current market framework they will presumably only work through regulatory measures – at least for listed companies. Executive compensation is increasingly tied to mid- or long-term success (although "success" is often still defined in a relatively elementary manner).

b) *Introduce friction parameters in highly regenerative systems (e.g. a financial transaction tax)*

Systems with feedback loops can produce harmful effects and even system breakdown. The example of a loudspeaker amplified by a too-closely placed microphone, resulting in a piercing screech, demonstrates this. Many natural systems can adapt their behaviour accordingly, for instance in predator-prey contexts, the diminishing of one party automatically reduces the other as well. Technical systems often introduce some friction or damping to avoid extreme system reactions. In some cases, however, this has not yet been done and harmful feedback loops occur. The financial markets, for instance, with their automated trading system (e.g. automatic purchase or sales orders) have repeatedly seen stock market crashes (e.g. the 2010 "Flash Crash") which were caused or at least increased in severity by self-energizing feedback loops. A friction parameter for this system would be a financial transaction tax (FTT). A financial transaction tax could, as proponents argue, reduce speculation and the overall volatility of the market – but it would still allow for proper hedging. There has been a lot of discussion around its potential effects and its practicability – especially if it is not implemented globally. The debate as to whether it can really reduce speculation and volatility is intense (IDS

2010). However, it has long been mooted in the international political arena and has received sound backing from scientists and economists. In 2011, a thousand leading economists, the Nobel laureates Joseph Stiglitz and Paul Krugman among them, urged Bill Gates and the G20 leaders to introduce a FTT (Makefinance-work 2011; see Krugman 2010). The feasibility and effects of an FTT need be studied by economists – but from a systems-theoretical point of view we know that systems with positive feedback loops can easily get out of control if this feed-back is not damped.

c) *Cultivate deceleration in personal life*

On a personal level, it is also possible to decelerate. As Stephen R. Covey stresses, we are so used to working on efficiency and doing things right – but that does not replace doing the right thing (Covey 1994). Who would regret on their deathbed, as Covey asks, not having spent more time in the office? How can you make free time to reflect on what really matters if you are largely driven by external forces, if your day is predetermined by workload, deliverables, emails and project proposals – but also by workouts at the gym, soccer training, personal network cultivation, or choir rehearsals?It can be a great act of liberation to resist the calls of hurry and hustle, of the constant struggle for ever-greater achievements, of the hamster wheel of accumulating and consuming. Deceleration in daily life, slow food, slow travel are becom-ing more and more attractive in times of an ever-growing need for speed (Schneidewind & Zahrnt 2014, 55), and monasteries have become popular venues for managers taking downtime (Grün & Zeitz 2010).

16.1.3 Think global – promote local

Thinking globally has long been a prime advice of environmentalists. But what does this mean in concrete situations? In politics, what is the effect of agriculture or trade policies on natural habitats in other regions of the globe? In business, is global sourcing always the best option? Could near-shore, regional, or local alternatives to global sour-cing maybe exhibit a better risk profile than global suppliers? How can local sourcing, local swapping platforms, and complementary, local currencies support the transition to a more sustainable global economy? Local currencies have the potential to regional-ize production and consumption cycles (Dittmer 2015).

"Promote local" means to consider the local context, the neighbourhood and adjacent systems. This is one of the conclusions for stewarding sustainability transformations which Künkel develops:

> *Taking care of adjacent systems … is a natural consequence of strengthening overall and individual resilience …* The search for mutual consistency, a negotiated balance between individual and collective interests, or resilience will never end. The future is constantly under negotiation and construction. But what

counts is a heightened awareness and increasing knowledge of which patterns of behavior and interaction among humans and between humans and nature may strengthen overall systems as well as individual resilience.

(Künkel 2019, 73; original emphasis)

16.2 Foster diversity

Nature exhibits a breathtaking diversity – diversity of species, of forms, of colours, of adaptation strategies and many more. Diversity is a fascinating feature of complex systems. As such, it has a bi-directional causal relationship to other features: it is both a precondition and a result of novelty and innovation. By definition, it is a result of processes of innovation because something which is new, a new feature, a new idea, an innovation is, of course, something which has not existed before. But diversity is also a precondition for innovation and novelty, because a monotonous environment which only depends on few parameters follows simple laws and does not allow for any change and novelty. This can be inferred from the theory of complex systems. On the one side of the spectrum, in systems with linear relationships, everything is predictable and regular; one can calculate the system states back and forth as one wishes, no novelty occurs. On the other side of the spectrum, in a chaotic system, there is no order anymore; future system states are totally erratic and unpredictable. There is a fine zone in between these extremes, however, in which both regularity and novelty can occur – this is, as it is called, the edge of chaos, "the state of an organization from which both stability and novelty arises" (Create Advantage Inc. 2019).

It is here, at the edge of chaos, where systems optimize – as can be shown with computer models of "cellular automatons": "Eventually, systems optimize at the border between order and chaos. The state at the border is critical in the sense of phase transitions" (Ito & Gunji 1992, 138) "Life seems to be at the border between order and chaos" (135).

It can be shown, in a variety of different fields, that complex systems exhibit the most fascinating characteristics at the edge of chaos. Rai finds edge-of-chaos phenomena in population dynamics (Rai 2004). Jørgensen investigates the growth rates of zooplankton in ecological models and finds "that systems at the edge of chaos have the highest level of (thermodynamic) information, which supports the hypothesis that systems at the edge of the chaos can coordinate the most complex behaviour" (Jorgensen 1995, 13). Hung and Lai investigate innovation processes of printers and find also that innovation occurs on the edge of chaos: positive feedback loops induce chaotic behaviour in the innovation process and boost innovation. To illustrate their main message, the authors quote Nietzsche in saying "You must have chaos within you to give birth to a dancing star" (Hung & Lai 2016, 31).

This creative zone between chaos and order is used by creative thinkers and artists. Design thinking, which has seen remarkable attention in management practice, capitalizes on this edge of chaos in a variety of ways – which likely explains its success: For instance, teams need to produce results under high time pressure, following the motto: "If everything is under control, you're just not going fast enough." Furthermore, the teams should be as diverse as possible. Rules of the game are, for instance, build on the idea of others (which requires listening, of course), defer judgement (which requires respect for different views and concept), produce tangible results and think with the end in mind (Brown 2008; HPI Academy 2019).

I submit we can derive three conclusions from this

a) The diversity of our *natural ecosystems* must be vigorously protected and maintained. Biodiversity loss seriously threatens this diversity, which might actually be more dangerous than climate change (Steffen, Rockström, Richardson, Lenton & Folke 2018). The weight of vertebrate land animals in the wild now accounts for only 1% of all vertebrates on earth, 32% is the weight of all humans and 67% is the weight of livestock animals (Population Matters 2019). Between 1970 and 2014 alone, the wild vertebrate animal population has halved, while the human population has doubled (ibid.). Humans and their livestock are simply displacing wild life.

b) Regarding *socio-natural systems*, i.e. agriculture, we need to ensure a sustainable degree of diversity. The very notion of mono-"cultures" is, from this point of view, an oxymoron: culture can never be mono. We are just beginning to realize the huge ecosystem damage imposed by industrial farming: "Of the 40,000 vertebrate species on the earth, 40 were selected as useful by different human cultures and domesticated. Of these, only 14 species account for over 90% of today's global livestock production" (EU-BDP 2001, 1). The FAO is urging for more livestock diversity, since increasing numbers of livestock breeds are becoming extinct: "Livestock diversity facilitates the adaptation of production systems to future challenges and is a source of resilience in the face of greater climatic variability" (FAO 2015, 14).

c) Regarding *social, cultural, economic but also technological systems*, we need to implement proper measures of diversity management. The corporate world is beginning to understand the importance of it. Diversity in the workforce, for instance, is not only a demand of corporate responsibility but a driver for innovation: "Diversity is key because systems depend on it to innovate" (Stroh 2015, 79). McKinsey investigated the relationship between diversity and financial performance and found a "statistically significant relationship between a more diverse leadership team and better financial performance" (McKinsey & Company 2015, 1). Furthermore, as mentioned above, innovation techniques necessitate diversity. Boston Consulting Group has studied the relationship of diversity and innovation in eight countries and found that "companies that reported above-average diversity on their management teams also reported innovation revenue that was 19 percentage points higher than that of companies with below-average leadership diversity" (BCG 2018). Proper stakeholder management has become indispensable not only for corporations but also in public planning processes, in mediation and conflict resolution and everywhere where people affected by a measure need to be heard. Stroh defines key stakeholders as those "people and organizations that affect and are affected by the issue. They include anyone that can make a contribution to the effort, or anyone that can possibly derail it if not on board" (Stroh 2015, 79). Margaret Robertson, in her 2017 book about principles and practice of sustainability, speaks about diversity as "an essential asset" within communities of all scales: "The more we are open to alternative voices, the stronger our social fabric and the richer our choice of potential futures." She also emphasizes the importance of diversity for the resilience of systems: all living systems which are resilient feature diversity (Robertson 2017, 342).

There is in inherent antagonism between the need for diversity (e.g. as precondition for innovation) and the logic of an economy of scale, which is particularly visible in industrialized mass production. Products become more and more streamlined and standardized. Only by means of state intervention (monopoly commissions) can some diversity be protected, but the dominance of just a few players or systems (e.g. operating systems in IT) weakens the overall system's resilience.

16.3 Increase transparency of the publicly relevant

This principle might raise concern among readers. Do we not already have too much transparency – given that tech companies catch our individual timelines, algorithms know about our preferences even before we are aware of them ourselves, or contractual negotiations are impeded by selective leakage of information and trustworthy proceedings seem to have become impossible? These examples illustrate that there are domains and aspects of life in which a further increase of transparency is to be prevented – or at least tightly bound to strict rules and their application. Nevertheless, I believe that transparency should be increased in publicly relevant cases to facilitate sustainability, at least in three different contexts:

- when public institutions fail to deal with issues of public interests (e.g. whistleblowing);
- when market mechanisms fail insofar as the price does not account for the true costs of a product or service, or if there is no level playing field which is theoretically needed for the market to function;
- when there is concealed violation of law – on any level (from local to international) and whatever the crime (e.g. corruption, organized crime, money laundering, tax avoidance).

These three contexts relate to several of the barriers discussed in Part 1, like market failures (see 5.1), trade-offs (see 3.4), conflicts of interests (see 4.5). For these barriers, transparency improvements will be absolutely key. There will be no improvement in mitigating these barriers without greater transparency.

Lack of transparency often masks injustices. I propose that this is actually an underlying assumption in Rawls' fairness principle. Fairness can only be a distribution on which rational actors, who all *share the same information*, would be able to agree. This is presumably the main reason why lobbyists normally do not like too much public attention and transparency, because lobbyism is often a marginal case between legitimately bringing forward arguments and securing their own party's interest – which might as such not be transparent to a broader public (although it might still be legal).

Trade-offs and conflicting interests can only be fairly resolved if all parties involved get the chance to express their concerns and interests and then jointly discuss and negotiate a priority list for addressing the respective goals and how to tackle the conflicts which arise. In other words, a fair public discourse about goals and means is inevitable. This, however, requires transparency – transparency about the stakeholders, their interests, their agendas, their relationships, transparency about winners and losers of different policy options, and transparency about the governance process, which follows agreement on the way forward.

Increased transparency supports openness and fairness and makes fraud and tortious interference more difficult. It provides the chance for public discourse on politicians' relationships with industry. It helps fight corruption and facilitates fair and equitable negotiations.[1]

Lack of transparency must, of course, not always be due to illegal behaviour. Sometimes it is "just" that nobody has an interest in emphasizing certain conditions. As a 2019 study on behalf of the Green Party in the European Parliament revealed, there are substantial differences between the nominal and the effective tax rates for corporations within the European Union. Most significant is Luxembourg, which has a nominal tax rate of 29% but an effective tax rate of 2.2% (Janský 2019, 3). As the study admits, these data need to be considered with a grain of salt, since they use unconsolidated data, which are "imperfect". However, they still represent "the best available company-level data for the EU" (ibid.). Why does it need a study like this to reveal to the public that in one EU country (actually an EU founding member) the effective tax rates for corporations is 2.2% while the "official", nominal tax rate is almost fifteen times higher?

Finally, there is a lack of transparency in the social and ecological impact of our production and consumption patterns. Readers might ask themselves how much carbon is contained in their shirt, how much dye was used, how many hours of child labour and how many people worked under conditions which violated the basic standards of ILO? You don't know? Have you never wondered why you don't know? "It's difficult to get the data" – is a frequent reply heard from government agencies as well as politicians. However, having worked in the IT industry for more than a decade, I see this as a rather spurious argument. How is the price of a shirt calculated? This is, of course, a complex process in which many factors need to be considered but in the long run (disregarding sales activities) the minimum price needs to be at least the sum of all compensations for all members of the value chain. Since this works in monetary metrics, why should it not work for extra-monetary metrics? Of course, there is the issue of the missing price of public goods and common-pool resources. But this situation is changing. The preferences of consumers, corporate customers and investors, the regulators' enforcement of environmental policies, carbon taxes, cap-and-trade mechanism and many others introduce prices into the system. As discussed above, a growing number of companies report on their emissions and sustainability performance. The case of PUMA mentioned earlier (see 13.3) illustrates that this does explicitly include the supply chain. One can expect that up-to-date IT infrastructure will greatly improve the transparency on the external cost of consumption along the supply chain (see Berg, Hack & Blome 2014), which will allow consumers to make better-informed decisions.

The Economy for the Common Good of Christian Felber suggests publishing the common-good performance of a company, calculated according to a well-defined scheme, on a label for every product or service the company offers. This would provide consumers with full transparency of the degree to which that respective company contributes to the common good (Felber 2018, 41f).

A UN Commission which was mandated to give advice on reforms of the international monetary and financial system concluded: "The lack of transparency is often a symptom of deeper market failures that produces incentives to limit information, and these deeper market failures may have other manifestations" (UN 2009, 56). Prior to this, Joseph Stiglitz, leading the commission, had already called for more transparency in international economic institutions:

> Short of a fundamental change in their governance, the most important way to ensure that the international economic institutions are more responsive to the poor, to the environment, to the broader political and social concerns that I have emphasized is to increase openness and transparency ... Transparency is even more important in public institutions like the IMF, the World Bank, and the WTO, because their leaders are not elected directly.
>
> *(Stiglitz 2002, 227)*

The importance of "increasing transparency" can be seen by the fact that this mission is a key aspect of the work of several NGOs:

- Transparency International (Transparency International 2019a) is fighting corruption globally and, among other publications, produces the Corruptions Perceptions Index (Transparency International 2019b).

- The Extractive Industries Transparency Initiative is a multi-stakeholder organization which targets corruption in their extractive industries, fights nontransparent information processes, monetary flows and dependencies, and facilitates more equitable processes in business and administration (EITI 2019).
- The Environmental Justice Foundation is campaigning against illegal, unreported and unregulated fishing (IUU fishing) (EJF 2018).
- The Global Fishing Watch is tracking the activity of about 60,000 commercial fishing vessels in near real time to fight IUU fishing: Public sharing of VMS (vessel monitoring system) data improves surveillance by encouraging vessels to comply with fisheries regulations; transparency breeds self-correcting behaviour. It is a strong deterrent to illegal operators. By going public with VMS, unauthorized vessels and those that don't have a history of compliance can be easily spotted and prioritized for inspection (Global Fishing Watch 2019).
- The Global Forest Watch, a branch of the World Resources Institute, is a crowd-funded dynamic online forest monitoring and alert system that empowers people everywhere to better manage forests (WRI 2019; Wohlgemuth 2014).

There are also synergies to other principles. There is a positive correlation between the decentralization of processes and structures and increased transparency. As transparency and openness facilitate democratic processes, so secrecy impedes them:

> Secrecy also undermines democracy. There can be democratic accountability only if those to whom these public institutions are supposed to be accountable are well informed about what they are doing – including what choices they confronted and how those decisions were made.
>
> *(Stiglitz 2002, 229)*

Hösle calls for more transparency in expert opinions. Expert judgements should not only include declarations of financial dependencies but also reveal their premises (Hösle 1994, 83).

In conclusion, we can quote Sala, Ciuffo and Nijkamp, who investigated a systemic framework for sustainability assessments. They "advocate transparency as the decisive means to acknowledge the richness and complexity of the sustainability concept" (Sala, Ciuffo & Nijkamp 2015, 316).

16.4 Maintain or increase option diversity

The final principle to be mentioned here is again one which can claim universal validity: always act in a way that the diversity of options is at least maintained, if not increased. Nobody knows the future. Nobody knows exactly what future generations will need. It is therefore important to ensure that future generations will have choices about their way forward. Decisions made today must not pre-determine future behaviour by reducing future options for action. The Austrian-American Heinz von Foerster, one of the founding fathers of cybernetics, called this an "ethical imperative: Act always so as to increase the number of choices" (Foerster 2003, 227). This is, in a way, a fundamental principle in all contexts, not only with regard to systems. It is simply not smart to run into dead ends if one cannot be sure to get out again.

Closely related to option diversity are lock-in effects and path dependencies. A path dependency occurs, for instance, by building a new power plant. Depending on the type of plant, this has implications for the carbon emissions involved. If it is a coal-powered plant, one can calculate the cumulative GHGs this plant will emit in its lifetime, at least during its amortization period. Nuclear energy is one of the most extreme examples of path dependencies: While one or two generations benefit from the respective energy supply, it will be thousands of generations to come who, without their prior consent, will have to deal with the risk of radioactive waste leaking from disposal sites. This is, in my view, a strong argument against nuclear power.

Particularly dangerous path dependencies will occur once humanity exceeds planetary boundaries. By definition, the planetary boundaries indicate thresholds beyond which the likelihood of irreversible change and runaway effects increase because developments might become self-reinforcing, taking us further away from the relative dynamic stability of the present. Of course, we can never be entirely sure about the long-term consequences of our actions. But so great is the evidence from such a diversity of fields (climatology, oceanography, ecology, botany,etc.) that we had better take seriously the advice to stay within the planetary boundaries. The precautionary principle (see 13.6) calls for refraining from an action or a technology if one cannot be sure how to properly handle the consequences.

Maintaining option diversity also carries a preference for decentralized, small and low-impact technologies. Decentralized systems are much more resilient than centralized ones, which was one of the reasons why the Internet was originally conceived – as a communication system which could survive a nuclear attack – although its current form no longer resembles that initial plan (Barabási 2002, 144).

Notes

1 Lack of transparency also hinders a better governance of our global public institutions. As Stiglitz illustrates, backed by his personal experience within the IMF, it is the lack of transparency which precludes a public discourse because certain issues "officially" do not even exist:

> The IMF is not just pursuing the objectives set out in its original mandate, of enhancing global stability and ensuring that there are funds for countries facing a threat of recession to pursue expansionary policies. It is also pursuing the

interests of the financial community. This means the IMF has objectives that are often in conflict with each other.

The tension is all the greater because this conflict can't be brought out into the open: if the new role of the IMF were publicly acknowledged, support for that institution might weaken, and those who have succeeded in changing the mandate almost surely knew this. Thus the new mandate had to be clothed in ways that *seemed* at least superficially consistent with the old. Simplistic free market ideology provided the curtain behind which the real business of the "new" mandate could be transacted … I should be clear: the IMF never *officially* changed its mandate, nor did it ever formally set out to put the interests of the financial community over the stability of the global economy or the welfare of the poor countries they were supposed to be helping.

(Stiglitz 2002, 206f.; original emphasis)

Bibliography

Barabási, Albert-László. *Linked. The New Science of Networks.* Cambridge, MA: Perseus Publishing, 2002.

Bardi, Ugo. *The Seneca Effect. Why Growth is Slow but Collapse is Rapid.* Springer, 2017.

BCG. *"How diverse leadership teams boost innovation."* 23. 01 2018. www.bcg.com/publica tions/2018/how-diverse-leadership-teams-boost-innovation.aspx (Accessed 15. 07 2019).

Berg, Christian. *Vernetzung als Syndrom. Risiken und Chancen von Vernetzungsprozessen für eine nachhaltige Entwicklunge.* Frankfurt: Campus, 2005.

———, Stefan Hack, & Constantin Blome. "How IT can enable sustainability throughout supply chains." In *Beyond Sustainability*, von Christian Scholz, and Joachim Zentes, (eds.), 184–202. Baden-Baden: Nomos, 2014.

Botkin, James W., Mahdi Elmandjra, & Mircea Malitza. *No Limits to Learning. Bridging the Human Gap.* Oxford, New York, Toronto, Sydney, Paris, Frankfurt: Pergamon Press, 1979.

Brown, Tim. "Design thinking." *Harvard Business Review 86(6)*, 2008: 84–92.

Club of Rome. *"Clubofrome.org/."* 2019. www.clubofrome.org/ (accessed 15. 07 2019).

Covey, Stephen R. *First Things First.* New York: Simon & Schuster, 1994.

Create Advantage Inc. *Managingresearchlibrary.org.* Herausgeber: Create Advantage Inc, 2019. https://managingresearchlibrary.org/glossary/edge-chaos (accessed 02. 05 2019).

Dittmer, Kristofer. "Community currencies." In *Degrowth. A Vocabulary for a New Era*, von Giacomo D'Alisa, Federico Demaria, & Giorgos Kallis (eds.). 149–151. Abingdon, Oxon and New York: Routledge, 2015.

EITI. *EITI.org.* 2019. https://eiti.org/ (Accessed 02. 05 2019).

EJF. *Out of the Shadows. Improving Transparency in Global Fisheries to Stop Illegal, Unreported and Unregulated Fishing.* London: Environmental Justice Foundation, 2018.

EU-BDP. *Biodiversity in Development.* Brussels: European Commission, 2001.

FAO. *Animal and Genertic Resources for Food and Agriculture in brief.* Rome: FAO – Food and Agriculture Organization of the United Nations, 2015.

Felber, Christian. *Gemeinwohlökonomie.* München: Piper, 2018.

Foerster, Heinz von. *Understanding Understanding.* New York: Springer, 2003.

Global Fishing Watch. "Global Fishing Watch." 2019. https://globalfishingwatch.org/ vms-transparency/ (accessed 02. 05 2019).

Grün, Anselm, & Jochen Zeitz. *Gott, Geld und Gewissen.* Münsterschwarzach: Vier-Türme-Verlag, 2010.

Hösle, Vittorio. *Philosophie der ökologischen Krise.* Munich: Beck, 1994.

HPI Academy. *Hpi-academy.de*. Herausgeber: HPI Academy, 2019. https://hpi-academy. de/design-thinking/bibliothek.html (accessed 13. 06 2019).

Hung, Shih-Chang, & Jiun-Yan Lai. "When innovations meet chaos: Analyzing the technology development of printers in 1976–2012." *Journal of Engineering and Technology Management 42*, October–December. 2016: 31–45.

IDS. *IDS IN FOCUS POLICY BRIEFING 14.2*. Brighton: Institute of Development Studies, 2010.

Ito, Keisuke, & Yukio-Pegio Gunji. "Self-organization toward criticality in the Game of Life." *BioSystems 26*, 1992: 135–138.

Janský, Petr. *Effective Tax Rates of Multinational Enterprises in the EU*. Brussels, 22. 01 2019. A REPORT COMMISSIONED BY THE GREENS/EFA GROUP IN THE EUROPEAN PARLIAMENT and https://www.greens-efa.eu/files/doc/docs/ 356b0cd66f625b24e7407b50432bf54d.pdf (accessed 13 September 2019).

Jorgensen, Sven Erik. "The growth rate of Zooplankton at the edge of chaos: Ecological models." *Jorunal of theoretical Biology 175 (1)*, 1995: 13–21.

Keynes, John M. *A Tract on Monetary Reform*. London, New York: Macmillan, 1923.

Krugman, Paul. "*Taxing the speculators.*" 27. 11 2009. www.nytimes.com/2009/11/27/opin ion/27krugman.html (accessed 28. 04 2019).

Künkel, Petra. *Stewarding Sustainability Transformation. An Emerging Theory and Practice of SDG Implementation*. Springer Nature, 2019.

Makefinancework. *Makefinancework*. 2011. www.makefinancework.org/financial-transac tion-tax/1000-economists-for-a-financial-transaction-tax/ (accessed 29. 04 2019).

McKinsey & Company. *Diversity Matters*. McKinsey&Company, 2015.

Meadows, Donella H. *Thinking in Systems. A Primer, ed. by Diana Wright*. White River Junction, VT: Chelsea Green Publishing, 2008.

———, Dennis L. Meadows, Joergen Randers, & William W. Behrens, III. *The Limits to Growth. A Report for the Club of Rome's Project on the Predicament of Mankind*. London: Earth Island Ltd., 1972.

Population Matters. "*Population matters.*" 2019. https://populationmatters.org/the-facts/bio diversity. (accessed 02. 05 2019).

Rai, Vikas. "Chaos in natural populations: edge or wedge?" *Ecological Complexity 1*, 2004: 127–138.

Robertson, Margaret. *Sustainability Principles and Practices*. London, New York: Routledge, 2017.

Sala, Serenella, Biagio Ciuffo, & Peter Nijkamp. "A systemic framework for sustainability assessment." *Ecological Economics 119*, 2015: 314–325.

Schneidewind, Uwe, & Angelika Zahrnt. *The Politics of Sufficiency. Making it easier to live the good life*. Munich: oekom, 2014.

Steffen, Will, Johan Rockström, Katherine Richardson, Timothy M. Lenton, & Carl Folke "Trajectories of the earth system in the anthropocene." *PNAS 115 (33)*, 14. 08 2018: 8252–8259.

Stiglitz, Joseph. *Globalization and its Discontents*. New York, London: W.W. Norton, 2002.

Stroh, David Peter. *Systems Thinking for Social Change*. White River Junction, VT: Chelsea Green Publishing, 2015.

Transparency International. 2019a. www.transparency.org/ (accessed 02. 05 2019).

———. *CORRUPTION PERCEPTIONS INDEX 2018*. 2019b. www.transparency.org/ cpi2018 (Accessed 15. 07 2019).

UN. *Report of the Commission of Experts of the President of the United Nations General Assembly on Reforms of the International Monetary and Financial System.* New York: United Nations, 2009.

Wohlgemuth, Erik. "Not just good on paper: How businesses and NGOs can protect rainforests." 14. 01 2014. www.theguardian.com/sustainable-business/business-power-ngo-protect-rainforests-paper (Accessed 15. 07 2019).

WRI. "Global forest watch." 2019. www.wri.org/our-work/project/global-forest-watch (accessed 02. 05 2019).

17

CONCLUSION

Sustainable action principles trigger phase transition

Contents

17.1 Summary: overcoming the barriers

Sustainability is a hugely complex challenge. Despite several decades of global discourse, despite numerous global conferences and several international agreements and treaties, progress is anything but sufficient; humanity's path is unsustainable and the indications calling for urgent action increase by the day. The lack of progress in sustainability causes different responses – from cynicism to ever-fiercer and more urgent calls for action, to lethargy, desperation, or ignorance.

Ironically, this sobering view on sustainability comes at a time when the globally agreed 2030 Agenda has just set out hugely ambitious and noble goals – 17 SDGs and 169 targets, each with its own justification. However, nobody knows whether it will ever be possible to reach all these goals together, given the severe trade-offs between them.

Furthermore, we can hardly say with any certainty whether a measure will really be proven as sustainable in the long run or not. This should make us cautious and humble. It is much easier to say what is *not* sustainable; it is much easier to specify the *barriers* to sustainability than to propose solutions or policy *measures* which will be sustainable in the long run.

The starting point of this book is the conviction that sustainability is more needed than ever – but a more comprehensive view of its barriers is necessary. By drawing attention to single issues, neglecting both the complexity of the challenge and the underlying root causes (called barriers here), success will be limited, sometimes even impeded. While the public was still debating whether the Paris Agreement and the related NDCs would be sufficient, populism appeared on the scene as a potentially greater threat to the climate crises than unambitious NDCs. Given that there is no single solution to complex issues, that there is no single actor capable of redesigning the global system, it is high time for a comprehensive view of the barriers to sustainability, one which considers as many disciplines and backgrounds as possible.

Moreover, it is important to understand the differences and complexities of the various barriers. Even in the simple case of the phase transition of water from liquid to gaseous state, boiling depends on two parameters: temperature and pressure. If all energy is focussed on just one parameter, if all effort invested in increasing the temperature, the boiling (i.e. phase transition) can still be prevented if somebody else changes the surrounding pressure at the same time. Even this trivial example illustrates that a one-dimensional focus can prevent success. However, relatively small changes in temperature *and* pressure can induce the phase transition rather quickly.

How much more must we as a society be aware of the complexity of the issues! We should refrain from the single-issue focus suggested by public discourse which neglects the complexity of the barriers behind the scenes. The rise of populism and the threat it poses to sustainability should make us aware of the need for a comprehensive view of the barriers that need to be overcome.

In order to facilitate the "phase transition" towards sustainability, a comprehensive look at the barriers is therefore critical. Some of the barriers relate to physical reality. Complexity, for instance, is an inherent feature of natural (eco-)systems, which makes long-term predictions difficult or impossible and often implies unpleasant surprises. Other barriers, like cognitive or moral limitations, relate to the human condition, still others to social reality (e.g. inequalities or conflicting interests).

This first group of barriers was described here as *intrinsic* barriers because they are intrinsic to the concept of sustainability as such. They will always be present when sustainability issues are tackled.

Other barriers are *extrinsic* to the concept of sustainability; they are, for instance, related to our institutional set-up. Institutional change is something which takes time and is resistant to change (since one function of institutions is to provide a framework which provides long-term security of expectations for agents). The good news is that in principle institutions can be formed and shaped, however gradual and troublesome this might be in detail. Neither the market order nor the role of global public institutions will be easy to change, but gradual reforms are conceivable and have already taken place. A better market framework is possible – several economies have proven that a social welfare state is possible. This idea needs to be expanded to include the ecological dimension and to include the markets globally. Better global governance is conceivable – we have made some progress in closing enforcement gaps and, given political will, much more will be possible here. Rising global awareness on issues of climate change, biodiversity, pollution, or migration will increase the pressure on policy makers to react.

Each barrier has specific challenges, in principle or practice, but none of the barriers needs to be simply accepted. For each barrier, solution perspectives have been discussed to give some indication how each can be addressed. Some cross-references to other barriers or to action principles have been given, but a more systematic and complete view of the interrelations of both barriers and principles would be a valuable task for future research.

While the first part of the book took a systems perspective, the second part focusses on the actors' views. As the barriers to sustainability cannot be addressed in isolation, and due to the absence of a central coordinating mechanism, it is the multitude of actors on multiple levels and in diverse sectors which are the critical levers for systemic change. Focussing on principles provides the actors with operational guidance, although the principles suggested cannot claim to be complete or flawless. The categories according to which the principles were discussed mirrored the categories of the barriers, i.e. there are nature-related principles, personal principles, society-related and system-related principles. These principles do not intend to provide any new ethics. Rather, by pointing to their origins in different currents of human cultural tradition (with a certain bias, to be sure, by the author's own heritage), it was evidenced that many of these principles are already quite present in cultural traditions, but the respective areas of application need to be expanded. Furthermore, it is important to note that the principles do not just address individuals but also corporate entities like corporations, or governments.

Nature-related principles include, for instance, the demand to decarbonize as well as the polluter-pays and the precautionary principles. Special attention was paid to the personal principles. Not only does all change start with committed individuals, more importantly, any peaceful transition towards a more sustainable society will have to provide a convincing answer to the question of how personal fulfilment can be possible despite decreasing resource consumption.

The barriers of the first part and the action principles of the second part are the two pillars of the book. They both point to *measures* (e.g. carbon tax) but from different angles (see Figure 1.5). For instance, internalizing external cost is suggested as a solution perspective for the problem of externalities, for which a carbon tax would be one possible realization. The same measure is supported from the point of view of the actor: the polluter-pays principle also calls for a measure of internalizing externalities. The measures in practical terms were not focussed on in the current book for three reasons: first, the measures are context dependent. What works effectively in one region might not be appropriate in another. Second, the detailed realization of a solution perspective or an action principle in a concrete measure requires substantial technical background which I could not provide. Third, even if that background were given (or additional authors consulted), the readability of the book would suffer from that extra scope. However, I intended to look at both barriers and action principles with as much clarity and brevity as possible.

17.2 The goal: future of terra and humanity – Futeranity

The following section makes some tentative suggestions which are much less substantiated than the rest of this book. However, they might nevertheless stimulate the discussion on how we can make sustainability operational and effective. First, three challenges to the pursuit of the SDGs are described. Then a new concept for the utopian vision of sustainability will be introduced: Futeranity. Finally, it is shown how this new concept can help, in combination with the action principles discussed above, to address these challenges to the SDG process.

17.2.1 Three challenges to the SDG process

Notwithstanding what was just said about barriers and action principles, the current endeavours towards sustainability and the pursuit of the SDGs faces three kinds of challenge: a substantive challenge, an addressee challenge, and a conceptual challenge.

1. *Substantive challenge*

   The 2030 Agenda explicitly states that the 17 SDGs are "integrated and indivisible" (UN 2015). One could rephrase this by saying that sustainability will only be achieved to the extent that its goals will be jointly reached. Conversely, sustainability is *not* fully reached if only a subset of SDGs is met. However, whether or not it is ever possible to jointly reach all 17 SDGs and their 169 targets *is up in the air.* Some authors are rather sceptical about this.

   For practical reasons, no actor can truly address all 17 SDGs at once, no actor can address sustainability in all its dimensions. This means, however, that each actor will have to target at a *subset* of what sustainability means. In case of the SDGs, each actor will have to concentrate on a selection of SDGs. In doing so – and assuming only the best intentions – the actor can well claim to be contributing to the Sustainable Development Goals – how could it be otherwise?[1] However, best-intended actions often produce unintended results. In complex systems, it is a huge challenge to reach the intended effect (and not, in the worst case, its opposite). It is *hardly possible, however, to pursue a subset of the SDGs without also affecting other SDGs* in one way or another. Quite likely the actions of one actor pursuing a subset of the SDGs will impede the achievement of other actors to reach other subsets – maybe slight improvements in a few areas are countered by setbacks in several others. Nobody can guarantee that the net effect of a policy measure on the SDGs in total is positive.

   The much-needed phase transition towards a sustainable society might never occur if not all critical parameters are considered – just as water will not boil by heating it up if somebody else increases the surrounding pressure.

2. *Addressee challenge*

   As important as goals (such as the SDGs) are on a systemic (or at least: nation state) level, they do not provide guidance for actors as to how these goals can be reached by concrete actions here and now.

It was the nation states that agreed on the 2030 Agenda and most of the targets are measured at state level. The 2030 Agenda calls for the involvement of all stakeholders; but for individuals, corporations, or NGOs, it is not easy to operationalize the 17 SDGs and their 169 targets. The goals give little advice about how to follow them in everyday actions. In many cases, it is very difficult (or even impossible) to anticipate the long-term effects of policy measures and the extent to which they actually foster progress towards any of the 169 targets − or frustrate the success of others. This is, due to the complexity of the issues, hardly possible even for the best-equipped research teams, but the vast majority of actors do not have access to this knowledge. This not only leaves many willing actors on their own in their wish to contribute to the SDG process, it also exacerbates the risk that the effects of different actors counteract or impede each other.

Moreover, I submit that even the best organized government with the best monitoring system cannot *ensure* target achievement because (in a liberal society) it cannot control which actor contributes to what extent. What if uncoordinated actions make things even worse?

3. *Conceptual challenge*

The current sustainability discourse also faces a conceptual problem which arises from the definition of sustainability. The *utopian vision of sustainability, the ultimate goal* − which I assume to be something like a state in which all humans can thrive in harmony with nature − is described *in the same conceptual terms as the measures to get there. We target sustainability and the means to get there are supposed to be sustainable!* This problem lies at the heart of the Brundtland definition (and presumably others) because a development which is supposed to start here and now − simply because of its inherent imperative − is labelled sustainable, while sustainability is also the ultimate goal we are seeking.

Is this not merely a battle of words? To some extent it is. However, the fact that we cannot conceptually distinguish means and end is a challenge − for four reasons:

1. Firstly, if the best-intended actors strive for sustainability, they will follow measures which should enhance that goal, which should be sustainable. However, they can never be sure that these measures actually *are* sustainable in the sense that they support the SDGs. Now nobody can prohibit people from using the age-old adjective "sustainable", which was in use long before the Brundtland Report and the discussion of the last fifty years. Therefore, the different usages of the concept − the traditional semantic one and the Brundtland concept − are ambiguous and inconsistent.

2. Secondly, and partly due to this ambiguity, there are many actors, corporations in particular, who are playing fast and loose with the word "sustainable". Everything can be called sustainable − subtly playing with the positive connotations of the concept of sustainable development. How can

it be that there are hundreds of products with labels proclaiming their sustainability? How can it be that there are stock indexes of corporations which list "sustainable companies", even though they just follow a best-in-class approach, which basically means the industry might be dirty, but as long as one company proves to be better than its peers, it will be listed? How can it be that several companies with controversial practices and/or from controversial industries (e.g. oil and gas) decorate themselves with listings in such "sustainability indexes"?

3. Thirdly, if the best-intentioned actors realize that some of their actions which were supposed to facilitate sustainability have actually *not* achieved that – sometimes even impeded it – they might become frustrated about the concept and turn away from it. This is even worsened by the fact that individual actions carry the heavy burden of the ultimacy of sustainability.

4. Fourthly, populists, climate sceptics and cynics might make fun of the concept of sustainable development as such, because it is hard to believe that the overall goal is ever achievable if the concrete measures are flawed and difficult to enforce anyhow. If well-intended measures turn out to be faulty, this can facilitate scepticism in the validity of the scientific account and can ridicule the proponents of sustainability as ideological know-it-alls.

Some of the difficulties with the current endeavours for sustainability (e.g. in the SDG process) are illustrated in Figure 17.1.

17.2.2 The utopian ideal of sustainability is Futeranity

In the introduction I suggested that sustainability be viewed as a utopian ideal. This suggestion has now been validated. The noble ideal of humanity living in harmony with each other and with nature is more needed than ever because humanity is on non-sustainable paths in many different respects. The reasons for this slow progress towards sustainability are manifold – and it was the goal of the first part of the book to provide a comprehensive overview of the most important barriers and suggest solution perspectives how to overcome them.

As a common goal, as a utopian ideal, sustainability is needed. However, can the three challenges mentioned in the section above be resolved – the substantive challenge, the addressee challenge and the conceptual challenge?

In my view, the substantive challenge and the addressee challenge can best be addressed by thoughtfully considering both parts of the book: by taking a comprehensive view of the barriers on a systemic level and guiding the concrete action of agents from an actors' point of view. It is particularly the latter which is promising in light of the absence of anybody coordinating the multitude of activities on different levels. However, does a principle for sustainable action deliver *sustainable results*? If the principle is rightly chosen, this should be the case. However, would this action deliver *sustainability* – in the meaning of the Brundtland concept? Most likely not – at least not on its own, because

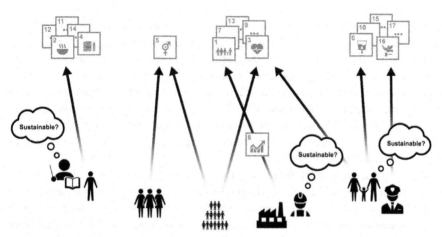

FIGURE 17.1 Challenges to current SDG process 1. Substantive challenge: Although SDGs are conceived as "integrated and indivisible", in practice actors often follow subsets of SDGs since no single actor can target all SDGs at once. This increases the risk of trade-offs between the different goals which individual actors pursue, the overall goal gets lost. 2. Addressee challenge: What is the guidance for different actors on different levels to contribute to target achievement? Nation states committed to SDGs – but how can they get their multiple actors going, ensure their synergistic effect and minimize trade-offs? 3. Conceptual challenge: Goal and measures are labelled with the same term. Sustainability is the great vision for humanity's future. At the same time, the measures to reach the Sustainable Development Goals ought to be "sustainable" here and now (which can never be proven as true). This charges each single measure with the ultimacy of the great goal – which is likely to fail.

Source: Own illustration.

sustainability as the *utopian vision for human development* can certainly not be reached by any single measure, principle, or policy.

Obviously some tension arises because we use the same concept for the ultimate goal and the means – the conceptual problem. We are using the same concept for a utopian vision as well as for concrete measures, which confuses the discourse – even more so given that the measures are often proven to be non-sustainable and the same concept is also used in much weaker everyday contexts.

Using the same concept for means and end does not sound very helpful, particularly since we use an existing word to describe a new phenomenon. It is no wonder that people feel the concept is "exhausted", because its constant use for everyday activities dissipates its energy and frustrates its users. A state in which all people can prosper and live their lives in harmony with each other and with the whole of creation definitely has a religious connotation (e.g. the prophetic tradition of ancient Israel). This religious connotation is charged to every action which is performed for the sake of sustainability. Every single act which is said to be sustainable carries this huge soteriological weight.

Why is that a problem?

- It is a problem because the people of "good will" become exhausted and overstrained – realizing that they can never reach it, they will try harder and harder but can never truly claim to live a sustainable life, thereby creating a constant bad conscience. Constant failure in reaching noble goals can lead to frustration or cynicism. I am often distressed to see how people who have fought for their ideals for a long time eventually give up and become cynical, or addicted, or opt to adapt. It is also counterproductive because we can never motivate the number of people needed for a sustainability transition by such an outlook.
- If we can only offer failure and bad conscience, the masses will not follow. Rather, they will realize the do-gooders have failed, make fun of them and lean towards populist views – which exploit people's anxiety and their resistance against the imposition of standards they do not understand and they do not want to adhere to. In fear of being taken in by the "arrogant elite" and their "totalitarian-like" claims, they would rather flee into evident totalitarian promises – which they understand better and which seem to address "real problems".

Maybe it would help to rationalize and substantiate the public sustainability discourse if we can clearly differentiate between the *ultimate goal* and *concrete principles and measures* to reach it. It should be possible to reach agreement about the ultimate goal and there will be impassioned discussions about concrete measures and policies. But if we distinguish the goal conceptually from the means, we can release the discussion of the means and measures from the burden of carrying the entire future of humankind in each and every act. We can – and should – first agree on the goal as such, and then engage in the discussion about the right policies which will bring us closer to it.

Since it is hardly possible to avoid the usage of the English adjective "sustainable", which will be continuously used to describe enduring, lasting effects, we might consider using another word for the sublime goal for humanity. By creating a new term for the goal, we acknowledge that the goal is not something which is already there, in our daily work, our everyday life. Rather it is a noble goal which we might never reach but which is worth every effort in pursuit of it.

The challenges are unprecedented – so maybe we also need a new definition for our goal and an original name to go with it amidst these challenges. What would be the goal?

Following on from an earlier proposal of mine,[2] I would suggest that the goal should be the future of the earth, terra, and the future of humanity: Future of terra and humanity, which one can abbreviate as "Futeranity".

The following aspects are important to me:

- The "future of terra" stands for ecological integrity and a life within the planetary boundaries. This acknowledges the earth as a proxy for the physical preconditions of any human and non-human life and it resists simple anthropocentrism.

- The future of humanity not only refers to the physical existence of human-kind – but includes the humane dimension. Our goal cannot just be the physical survival of the human race. That would be by far too unambitious. I am personally not too concerned about the physical survival of the human species – because I trust that a small group of individuals would manage to survive even after the worst crises have occurred. But I am horrified by the idea how human relationships would deteriorate, how human values would erode and how everything which we call culture would vanish long before the physical existence of the human race is at stake. The truly human char-acteristics – human rights, human culture, liberty, democracy – are much more fragile than the survival of the human race. That is why we need to protect them first.

I am, of course, aware that the reader might object to such a suggestion as impractical. What sense does it make to challenge a concept which is so well established and has reached so much, has convinced so many people? To be sure, I do not really believe that this proposal has the chance of adoption. Fur-thermore, a mere change of words does not resolve the issues at its heart, and it is these which need to be addressed. However, it would already be great pro-gress if the questions raised here can facilitate the discussion and help clarify what we can and what we should not expect and how we become effective in our pursuit of sustainability.

17.2.3 Sustainable action principles facilitate Futeranity

With the help of the action principles as well as the concept of Futeranity, we can now respond to the three challenges of the SDG process:

1. The substantive challenge is addressed because individual actors do not need to target the SDGs – that responsibility lies with dedicated actors, primarily governments. Rather, *individual actors* follow sustainable action principles. Trade-offs will still be there but once suitable action principles are estab-lished, these will be minimized.
2. The addressee challenge is addressed because the action principles provide concrete guidance for actors on all levels and for different contexts. While the contribution of single actors to the overall SDG achievement can hardly be measured anyhow, the pursuit of action principles is more tangible from an actor's point of view.
3. The conceptual challenge is addressed by introducing a new concept for the goal of humanity which is "integrated and indivisible": Futeranity. The term "sustainable" can still be used to describe more or less effective prin-ciples or measures towards that great goal, but the day-to-day actions are no longer charged with ultimacy.

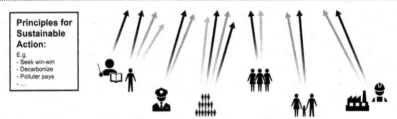

FIGURE 17.2 Action Principles for Sustainability facilitate Futeranity: 1. The substantive challenge is addressed because individual actors do not need to target the SDGs – that responsibility lies with dedicated actors, primarily governments. Rather, individual actors follow sustainable action principles. Trade-offs will still be there but once suitable action principles are established, these will be minimized. 2.The addressee challenge is addressed because the action principles provide concrete guidance for actors on all levels and for different contexts. While the contribution of single actors to the overall achievement of SDGs can hardly be measured anyhow, the pursuit of action principles is more tangible from an actor's point of view. 3. The conceptual challenge is addressed by introducing a new concept for the goal of humanity which is "integrated and indivisible": Futeranity. The term "sustainable" can still be used to describe more or less effective principles or measures towards that great goal, but the day-to-day actions are no longer charged with ultimacy.

Source: Own illustration.

Figure 17.2 summarizes the main ideas.

If the reader is sceptical about my conceptual suggestion, she or he might drop it and stay with the idea of sustainability as a utopian ideal – as a guiding frame for concrete action, which we should wholeheartedly strive for while being aware that we will never fully achieve it, or at least that we cannot be certain about it. Those who follow the idea that conceptual clarity in distinguishing means and end would be beneficial might however speak of Futeranity – or suggest another phrase.

The concept of Futeranity would certainly not impede the pursuit of the SDGs. On the contrary, by elaborating sustainable action principles, actors will be supported on their route towards sustainability. The action principles will certainly not resolve all the issues described in this book. What has been said about barriers, about the principal difficulties of addressing them remains true; trade-offs will remain a challenge as will ignorance about the long-term effects and the lack of control mechanisms. But mature action principles (again, those proposed are only first tentative suggestions and will need refinement) could facilitate the transition towards sustainability if a sufficient number of actors followed.

Futeranity would not cease to be relevant beyond 2030. Even if all SDGs were met – something I do hope for but frankly speaking doubt – it would remain a guiding vision for humanity: to live in harmony with each other and with nature.

17.2.4 The critical role of the actors for a transition towards sustainability

Addressing the sustainability barriers requires politicians at once visionary and wise, policy makers and business women who manage to combine a global and long-term view with charisma and commitment to centre their people behind the idea that global challenges require global collaboration, a long-term perspective and the open and frank discussion on questions of equity and justice. Unilateralism will not succeed in a multilateral world. There are no isolated islands in our densely networked global markets and societies. The few cases in which states have tried to isolate themselves from the rest of the world illustrate dramatically how this fails and what the power of collaboration and trade can accomplish. This has been true in the past and it will be much more so in the future because the global challenges will strike each and every country. Not all countries will be affected in the same way, of course. But those places which will experience only relatively mild changes due to the global environmental crises will experience unprecedented migration pressure. No wall can be high enough to resist that pressure. And even if it could, this would only be possible by sacrificing the core values of humanity – this would be, in one way or another, be a lose-lose game.

These policy makers depend, on the other hand, on pressure from below, the pressure from the street. They depend on the voters. They depend upon public opinion, which is, in turn, shaped by public discourse, by protesters and activists, by intellectuals and charismatic leaders. Policy makers also depend on the power of the markets and business executives, who, in turn, depend upon investors and consumers. Here we have come full circle. It is the agents which trigger change, agents of multiple kinds, of multiple function and on multiple levels. However powerless each of us might feel, change will always start with the individual. Committed and engaged individuals can encourage others to trigger change in their domains, which will, in turn, encourage others, etc. Like an avalanche, change can spread, remove barriers along the way and affect those who have never considered participation.

There are indications that a global civil society is forming (see Kaldor 2003; Spini 2014).[3] Such a global civil society needs to take shape and develop a common concern about our common future and be on guard to unmask nationalism and unilateralism as immoral and irresponsible dead ends.

No complex system, however, can be "controlled" or "mastered" by an external force. Even a world police – if it were ever to exist – could not really control the course of global action unless it strongly reduces complexity and heavily interferes with the relatively independent subsystems.

However, the problem in the current context, in which an effective regulatory global framework is missing and nobody is coordinating the activities towards the 2030 Agenda, is that no agent can focus on 17 SDGs with 169 targets. It is here where the sustainable action principles become essential. They reduce complexity by providing concrete guidance for agents on all levels, by suggesting more sustainable alternatives for action here and now. If a sufficiently large number of actors follow such principles for sustainable action, the pressure from the bottom will increase and systemic change will occur. The transition towards a sustainable society can happen.

17.3 Outlook: change is coming

Future research needs to complement and refine the suggested lists of barriers as well as action principles, to further illuminate the dependencies among them, and elaborate and contextualize the solution perspectives and substantiate concrete measures. Finally, it will need to look at the question of how we can motivate people to jump on this bandwagon. The readers of this book have great dedication to the topic – otherwise they would not have spent so much time on reading through to this last chapter. But how do we motivate others? That might be the most critical question of all – which we have not even touched here.

What is the way forward? For me, as a privileged individual in the global North, it is clear that we will need to see a transformation of our Western lifestyles. The "American way of life", which is actually pretty much the way of life of the entire global North and which a former US politician called "nonnegotiable", must be negotiated. And it *will* be negotiated, either by insight or by the normative power of the factual. Drawing on the concept of liberty to justify excessive and harmful, even deadly consumption will not work in the long run. The more obvious the damage this consumption creates, the more evident it will be that liberty cannot be invoked to justify the whims of those living in luxury at the cost of harming innocent people and ruining our common livelihoods. Tacitly accepting the harm caused to others and the destruction of nature by invoking rights to liberty needs to be unmasked as an act of injustice.

Let me sketch the case very simply – at the risk of overdoing it. Neglecting the complexity of the issues, focussing too strongly on just one topic, on one approach, on one group of actors or one mechanism, coupled with the forceful and sometimes messianic expectation that this propagated measure will be "truly sustainable" – such an approach cannot but fail. Yet if we vigorously and consistently address the barriers to sustainability, if we follow sustainable action principles and collaborate with a great variety of actors in civil society, media, governments and business, change is much more likely to occur.

Such change is possible. There are promising examples of rapid changes. Nobody expected the Berlin Wall to fall within weeks in 1989. This peaceful revolution is maybe the most impressive example of a very fundamental system change. But there are many more promising examples. Nobody expected that smoking would become so unpopular in many countries of the world within a few years. Nobody in Germany had expected in early 2011 that a government which had just prolonged the runtime of nuclear power would announce its final phase-out within a decade in response to the Fukushima disaster. Nobody expected that the misconduct of one Hollywood celebrity would trigger an avalanche of developments which has had a lasting impact on the way we see and treat each other. And nobody could have expected that a Swedish teenage girl would change the global discourse on global warming. Change is possible and change is coming – maybe faster than we think.

Notes

1 We neglect here the case that some actors (e.g. corporations) might pretend to pursue sustainability but simply select those SDGs which they have most interest in anyhow.
2 I pick up an idea which I originally published in a small Festschrift for Michael F. Jischa in 2002 (Berg 2002).
3 Of course, one cannot, as Kaldor stresses, easily "transpose the concept of civil society into the concept of global civil society, since … the key to understanding what is new about contemporary meaning is precisely their global character" (Kaldor 2003, 7).

Bibliography

Berg, Christian. "Nachhaltigkeit oder Futerumanum? Zur Kritik eines Begriffs zehn Jahre nach 'Rio'." In *Folgenabschätzungen. Resonanzen zum 65. Geburtstag von Michael F. Jischa*, Christian Berg, Ildiko Tulbure & Ralph Charbonnier (eds.), 69–80. Clausthal-Zellerfeld: Papierflieger (Schriftenreihe Forum Clausthal; H 15), 2002.
Kaldor, Mary. *Global Civil Society: An Answer to War*. Cambridge: Polity Press, 2003.
Spini, Debora. *Global Civil Society*. Oxford Bibliographies. 19. 09 2014.
UN. *Transforming Our World: The 2030 Agenda for Sustainable Development (A/RES/70/1)*. New York: United Nations, 2015.

INDEX